12/11

# THE GLORIOUS GOLDEN RATIO

# THE GLORIOUS GOLDEN RATIO

ALFRED S. POSAMENTIER
AND INGMAR LEHMANN

59 John Glenn Drive
Amherst, New York 14228–2119

Published 2012 by Prometheus Books



Inquiries should be addressed to
Prometheus Books
59 John Glenn Drive
Amherst, New York 14228–2119
VOICE: 716–691–0133
FAX: 716–691–0137
WWW.PROMETHEUSBOOKS.COM

16 15 14 13 12 5 4 3 2 1

Library of Congress Cataloging-in-Publication Data

Posamentier, Alfred S.
The glorious golden ratio / by Alfred S. Posamentier and Ingmar Lehmann.
p. cm.
ISBN 978–1–61614–423–4 (cloth : alk. paper)
ISBN 978–1–61614–424–1 (ebook)
1. Golden Section. 2. Lehmann, Ingmar II. Title.

QA466.P67 2011
516.2'04--dc22

2011004920

Printed in the United States of America on acid-free paper

To Barbara for her support, patience, and inspiration.

To my children and grandchildren, whose future is unbounded:
David, Lauren, Lisa, Danny, Max, Sam, and Jack.

And in memory of my beloved parents, Alice and Ernest,
who never lost faith in me.

Alfred S. Posamentier

To my wife and life partner, Sabine, without whose support and patience my work on this book would not have been possible.

And to my children and grandchildren:
Maren, Claudia, Simon, and Miriam.

Ingmar Lehmann

# Contents

# Acknowledgments

The authors wish to extend sincere thanks for proofreading and useful suggestions to Dr. Michael Engber, professor emeritus of mathematics at the City College of the City University of New York; Dr. Manfred Kronfeller, professor of mathematics at Vienna University of Technology, Austria; Dr. Bernd Thaller, professor of mathematics at Karl Franzens University–Graz, Austria; and Dr. Sigrid Thaller, professor of mathematics at Karl Franzens University–Graz, Austria. We are very grateful to Heino Hellwig of the Humboldt University–Berlin for contributing the chapter covering biology and for suggestions offered beyond that chapter. With gratitude we thank Dr. Ana Lucia Braz Dias, professor of mathematics at Central Michigan University, for contributing the chapter on fractals. And to Peter Poole our thanks for some wise suggestions throughout the book. We wish to thank Linda Greenspan Regan for her editorial assistance, Peggy Deemer for her extraordinary expertise in technical editing, and Jade Zora Ballard for putting the book into final shape.

# Introduction

Few mathematical concepts, if any, have an impact on as many aspects of our visual and intellectual lives as the golden ratio. In the simplest form, the golden ratio refers to the division of a given line segment into a unique ratio that gives us an aesthetically pleasing proportion. This proportion is formed in the following way: The longer segment ($L$) is to the shorter segment ($S$) as the entire original segment ($L+S$) is to the longer segment. Symbolically, this is written as $\frac{L}{S}=\frac{L+S}{L}$.

Let us consider a rectangle whose length is $L$ and whose width is $S$, and whose dimensions are in the golden ratio. We call this a golden rectangle, which derives its name from the apparent beauty of its shape: a view supported through numerous psychological studies in a variety of cultures. The shape of the golden rectangle can be found in many architectural masterpieces as well as in famous classical works of art.

When the golden ratio is viewed in terms of its numerical value, it seems to infiltrate just about every aspect of mathematics. We have selected those manifestations of the golden ratio that allow the reader to appreciate the beauty and power of mathematics. In some cases, our endeavors will open new vistas for the reader; in other cases, they will enrich the reader's understanding and appreciation for areas of mathematics that may not have been considered from this unusual vantage point. For example, the golden ratio is a value, frequently referred to by the Greek letter $\phi$ (phi), which has the unique characteristic in that it differs from its reciprocal by 1, that is, $\phi - \frac{1}{\phi} = 1$. This unusual characteristic leads to a plethora of fascinating properties and genuinely connects $\phi$ to such familiar topics as the Fibonacci numbers and the Pythagorean theorem.

In the field of geometry, the applications of the golden ratio are practically boundless, as are their beauty. To fully appreciate their visual aspects, we will take you through a journey of geometric experiences that will include some rather unusual ways of constructing the golden ratio, as well as exploring the many surprising geometric figures into which the golden ratio is embedded. All this requires of the reader is to be merely fortified with nothing more than some elementary high school geometry.

Join us now as we embark on our journey through the many wonderful appearances of the golden ratio, beginning with a history of these sightings dating from before 2560 BCE all the way to the present day. We hope that throughout this mathematical excursion, you will get to appreciate the quotation by the famous German mathematician and scientist Johannes Kepler (1571–1630), who said, "Geometry harbors two great treasures: One is the Pythagorean theorem, and the other is the golden ratio. The first we can compare with a heap of gold, and the second we simply call a priceless jewel."[1] This "priceless jewel" will enrich, entertain, and fascinate us, and perhaps open new doors to unanticipated vistas.

# Chapter 1

# Defining and Constructing the Golden Ratio

As with any new concept, we must first begin by defining the key elements. To define the golden ratio, we first must understand that the ratio of two numbers, or magnitudes, is merely the relationship obtained by dividing these two quantities. When we have a ratio of 1:3, or $\frac{1}{3}$, we can conclude that one number is one-third the other. Ratios are frequently used to make comparisons of quantities. One ratio stands out among the rest, and that is the ratio of the lengths of the two parts of a line segment which allows us to make the following equality of two ratios (the equality of two ratios is called a proportion): that the longer segment ($L$) is to the shorter segment ($S$) as the entire original segment ($L+S$) is to the longer segment ($L$). Symbolically, this is written as $\frac{L}{S} = \frac{L+S}{L}$. Geometrically, this may be seen in figure 1-1:

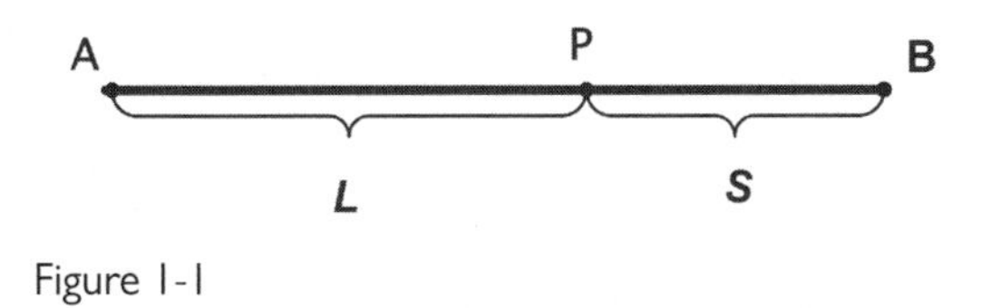


Figure 1-1

This is called the *golden ratio* or the *golden section*—in the latter case we are referring to the "sectioning" or partitioning of a line segment. The terms *golden ratio* and *golden section* were first introduced during the nineteenth century. We believe that the Franciscan friar and mathematician Fra Luca Pacioli (ca. 1445–1514 or 1517) was the first to use the term *De Divina Proportione* (*The Divine Proportion*), as the title of a book in 1509, while the German mathematician and astronomer Johannes Kepler (1571–1630) was the first to use the term *sectio divina* (divine section). Moreover, the German mathematician Martin Ohm (1792–1872) is credited for having used the term *Goldener Schnitt* (golden section). In English, this term, *golden section*, was used by James Sully in 1875.[1]

You may be wondering what makes this ratio so outstanding that it deserves the title "golden." This designation, which it richly deserves, will be made clear throughout this book. Let's begin by seeking to find its numerical value, which will bring us to its first unique characteristic.

To determine the numerical value of the golden ratio $\frac{L}{S}$ we will change this equation $\frac{L}{S}=\frac{L+S}{L}$, or $\frac{L}{S}=\frac{L}{L}+\frac{S}{L}$, to its equivalent, when $x=\frac{L}{S}$, to get[2]: $x=1+\frac{1}{x}$.

We can now solve this equation for $x$ using the quadratic formula, which you may recall from high school. (The quadratic formula for solving for $x$ in the general quadratic equation $ax^2+bx+c=0$ is $x=\frac{-b\pm\sqrt{b^2-4ac}}{2a}$. See the appendix for a derivation of this formula.) We then obtain the numerical value of the golden ratio:

$$\frac{L}{S}=x=\frac{1+\sqrt{5}}{2},$$

which is commonly denoted by the Greek letter, phi[3]: $\phi$.

$$\phi=\frac{L}{S}=\frac{1+\sqrt{5}}{2}\approx\frac{1+2.236067977499789696409173668731276235440}{2}$$

$$\approx\frac{3.236067977499789696409173668731276235440}{2}$$

$$\approx 1.61803.$$

Notice what happens when we take the reciprocal of $\frac{L}{S}$, namely $\frac{S}{L}=\frac{1}{\phi}$:

$$\frac{1}{\phi}=\frac{S}{L}=\frac{2}{1+\sqrt{5}},$$

which when we multiply by 1 in the form of $\frac{1-\sqrt{5}}{1-\sqrt{5}}$, we get

$$\frac{2}{1+\sqrt{5}}\cdot\frac{1-\sqrt{5}}{1-\sqrt{5}}=\frac{2\cdot(1-\sqrt{5})}{1-5}=\frac{2\cdot(1-\sqrt{5})}{-4}=\frac{1-\sqrt{5}}{-2}=\frac{\sqrt{5}-1}{2}=\frac{\sqrt{5}+1}{2}-1=\phi-1$$
$$\approx 0.61803.$$

But at this point you should notice a very unusual relationship. The value of $\phi$ and $\frac{1}{\phi}$ differ by 1. That is, $\phi-\frac{1}{\phi}=1$. From the normal relationship of reciprocals, the product of $\phi$ and $\frac{1}{\phi}$ is also equal to 1, that is, $\phi\cdot\frac{1}{\phi}=1$. Therefore, we have two numbers, $\phi$ and $\frac{1}{\phi}$, whose difference and product is 1—these are the only two numbers for which this is true! By the way, you might have noticed that

$$\phi+\frac{1}{\phi}=\sqrt{5}, \text{ since } \frac{\sqrt{5}+1}{2}+\frac{\sqrt{5}-1}{2}=\sqrt{5}.$$

We will often refer to the equations $x^2-x-1=0$ and $x^2+x-1=0$ during the course of this book because they hold a central place in the study of the golden ratio. For those who would like some reinforcement, we can see that the value $\phi$ satisfies the equation $x^2-x-1=0$, as is evident here:

$$\phi^2-\phi-1=\left(\frac{\sqrt{5}+1}{2}\right)^2-\frac{\sqrt{5}+1}{2}-1=\frac{5+2\sqrt{5}+1}{4}-\frac{2\left(\sqrt{5}+1\right)}{4}-\frac{4}{4}$$

$$=\frac{5+2\sqrt{5}+1-2\sqrt{5}-2-4}{4}=0.$$

The other solution of this equation is

$$\frac{1-\sqrt{5}}{2}=-\frac{\sqrt{5}-1}{2}=-\frac{1}{\phi},$$

while $-\phi$ satisfies the equation $x^2+x-1=0$, as you can see here:

$$(-\phi)^2+(-\phi)-1=\phi^2-\phi-1=\left(\frac{\sqrt{5}+1}{2}\right)^2-\frac{\sqrt{5}+1}{2}-1=0.$$

The other solution to this equation is $\frac{1}{\phi}$.

Having now defined the golden ratio numerically, we shall *construct* it geometrically. There are several ways to construct the golden section of a line segment. You may notice that we appear to be using the terms *golden ratio* and *golden section* interchangeably. To avoid confusion, we will use the term *golden ratio* to refer to the numerical value of $\phi$ and the term *golden section* to refer to the geometric division of a segment into the ratio $\phi$.

## GOLDEN SECTION CONSTRUCTION 1

Our first method, which is the most popular, is to begin with a unit square *ABCD*, with midpoint *M* of side *AB*, and then draw a circular arc with radius *MC*, cutting the extension of side *AB* at point *E*. We now can claim that the line segment *AE* is partitioned into the golden section at point *B*. This, of course, has to be substantiated.

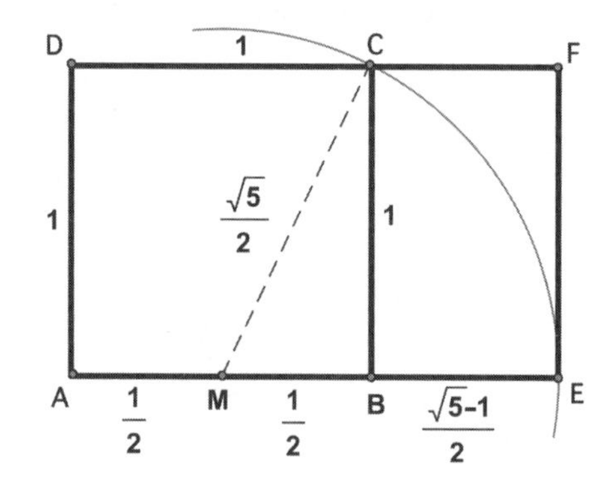


Figure 1-2

To verify this claim, we would have to apply the definition of the golden section: $\frac{AB}{BE}=\frac{AE}{AB}$, and see if it, in fact, holds true. Substituting the values obtained by applying the Pythagorean theorem to $\Delta MBC$ as shown in figure 1-2, we get the following:

$$MC^2 = MB^2 + BC^2 = \left(\frac{1}{2}\right)^2 + 1^2 = \frac{1}{4} + 1 = \frac{5}{4}; \text{ therefore, } MC = \frac{\sqrt{5}}{2}.$$

It follows that

$$BE = ME - MB = MC - MB = \frac{\sqrt{5}}{2} - \frac{1}{2} = \frac{\sqrt{5}-1}{2}, \text{ and}$$

$$AE = AB + BE = 1 + \frac{\sqrt{5}-1}{2} = \frac{2}{2} + \frac{\sqrt{5}-1}{2} = \frac{\sqrt{5}+1}{2}.$$

We then can find the value of $\frac{AB}{BE}=\frac{AE}{AB}$, that is,

$$\frac{1}{\frac{\sqrt{5}-1}{2}} = \frac{\frac{\sqrt{5}+1}{2}}{1},$$

which turns out to be a true proportion, since the cross products are equal. That is,

$$\left(\frac{\sqrt{5}-1}{2}\right)\cdot\left(\frac{\sqrt{5}+1}{2}\right) = 1\cdot 1 = 1.$$

We can also see from figure 1-2 that point $B$ can be said to divide the line segment $AE$ into an inner golden section, since

$$\frac{AB}{AE} = \frac{1}{1+\frac{\sqrt{5}-1}{2}} = \frac{1}{\frac{\sqrt{5}+1}{2}} = \frac{\sqrt{5}-1}{2} = \frac{1}{\phi}.$$

Meanwhile, point *E* can be said to divide the line segment *AB* into an outer golden section, since

$$\frac{AE}{AB}=\frac{1+\frac{\sqrt{5}-1}{2}}{1}=\frac{\sqrt{5}+1}{2}=\phi.$$

You ought to take notice of the shape of the rectangle *AEFD* in figure 1-2. The ratio of the length to the width is the golden ratio:

$$\frac{AE}{EF}=\frac{\frac{\sqrt{5}+1}{2}}{1}=\frac{\sqrt{5}+1}{2}=\phi.$$

This appealing shape is called the *golden rectangle*, which will be discussed in detail in chapter 4.

## GOLDEN SECTION CONSTRUCTION 2

Another method for constructing the golden section begins with the construction of a right triangle with one leg of unit length and the other twice as long, as is shown in figure 1-3.[4] Here we will partition the line segment *AB* into the golden ratio. The partitioning may not be obvious yet, so we urge readers to have patience until we reach the conclusion.

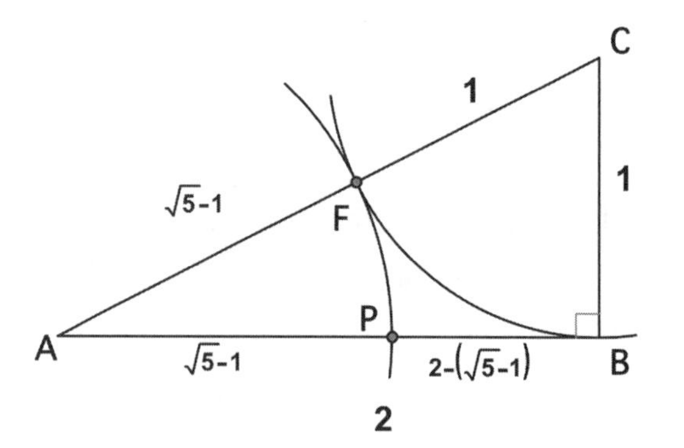


Figure 1-3

With $AB = 2$ and $BC = 1$, we apply the Pythagorean theorem to $\Delta ABC$. We then find that $AC = \sqrt{2^2 + 1^2} = \sqrt{5}$. With the center at point $C$, we draw a circular arc with radius 1, cutting line segment $AC$ at point $F$. Then we draw a circular arc with the center at point $A$ and the radius $AF$, cutting $AB$ at point $P$.

Because $AF = \sqrt{5} - 1$, we get $AP = \sqrt{5} - 1$. Therefore, $BP = 2 - (\sqrt{5} - 1) = 3 - \sqrt{5}$.

To determine the ratio $\frac{AP}{BP}$, we will set up the ratio $\frac{\sqrt{5}-1}{3-\sqrt{5}}$, and then to make some sense of it, we will rationalize the denominator by multiplying the ratio by 1 in the form of $\frac{3+\sqrt{5}}{3+\sqrt{5}}$.

We then find that

$$\frac{\sqrt{5}-1}{3-\sqrt{5}} \cdot \frac{3+\sqrt{5}}{3+\sqrt{5}} = \frac{3\sqrt{5}+5-3-\sqrt{5}}{3^2-(\sqrt{5})^2} = \frac{2\sqrt{5}+2}{9-5} = \frac{2(\sqrt{5}+1)}{4} = \frac{\sqrt{5}+1}{2} = \phi \approx 1.61803,$$

which is the golden ratio! Therefore, we find that point $P$ cuts the line segment $AB$ into the golden ratio.

## GOLDEN SECTION CONSTRUCTION 3

We have yet another way of constructing the golden section. Consider the three adjacent unit squares shown in figure 1-4. We construct the angle bisector of $\angle BHE$. There is a convenient geometric relationship that will be very helpful to us here; that is, that the angle bisector in a triangle divides the side to which it is drawn proportionally to the two sides of the angles being bisected.[5] In figure 1-4 we then derive the following relationship: $\frac{BH}{EH} = \frac{BC}{CE}$. Applying the Pythagorean theorem to $\Delta HFE$, we get $HE = \sqrt{5}$. We can now evaluate the earlier proportion by substituting the values shown in figure 1-4:

$\frac{1}{\sqrt{5}} = \frac{x}{2-x}$, from which we get $x = \frac{2}{\sqrt{5}+1}$, which is the reciprocal of $\frac{\sqrt{5}+1}{2} = \phi$.

Therefore, $x = \frac{1}{\phi} \approx 0.61803$.

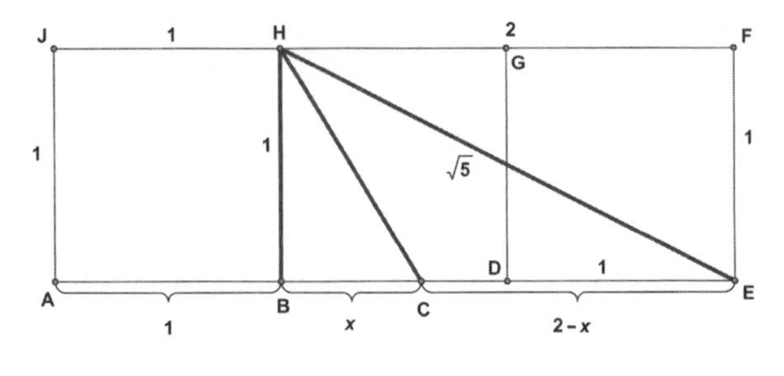


Figure 1-4

Thus, we can then conclude that point $B$ divides the line segment $AC$ into the golden section, since

$\frac{AB}{BC} = \frac{1}{x} = \frac{\sqrt{5}+1}{2} \approx$ 1.61803, the recognized value of the golden ratio.

## GOLDEN SECTION CONSTRUCTION 4

Analogous to the previous construction is one that begins with two congruent squares as shown in figure 1-5. A circle is drawn with its center at the midpoint, $M$, of the common side of the squares, and a radius half the length of the side of the square. The point of intersection, $C$, of the circle and the diagonal of the rectangle determines the golden section, $AC$, with respect to a side of the square, $AD$.

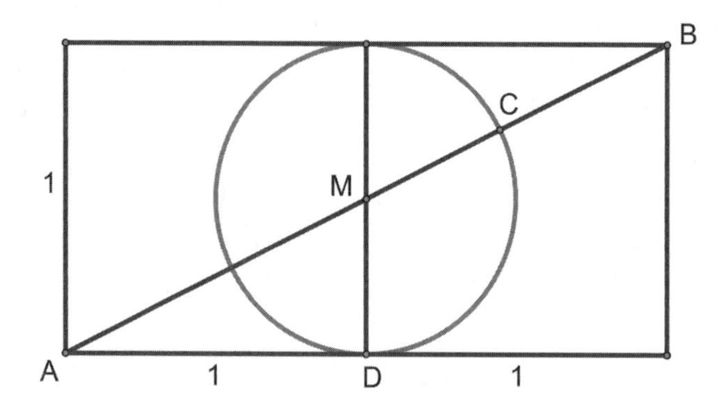


Figure 1-5

With $AD = 1$ and $DM = \frac{1}{2}$, we get $AM = \frac{\sqrt{5}}{2}$ by applying the Pythagorean theorem to triangle $AMD$. (See fig. 1-6.) Since $CM$ is also a radius of the circle, $CM = DM = \frac{1}{2}$. We can then conclude that

$$AC = AM + CM = \frac{\sqrt{5}}{2} + \frac{1}{2} = \frac{\sqrt{5}+1}{2} = \phi.$$

Furthermore,

$$BC = AB - AC = \sqrt{5} - \frac{\sqrt{5}+1}{2} = \frac{\sqrt{5}-1}{2} = \frac{1}{\phi}.$$

We have thus constructed the golden section and its reciprocal.

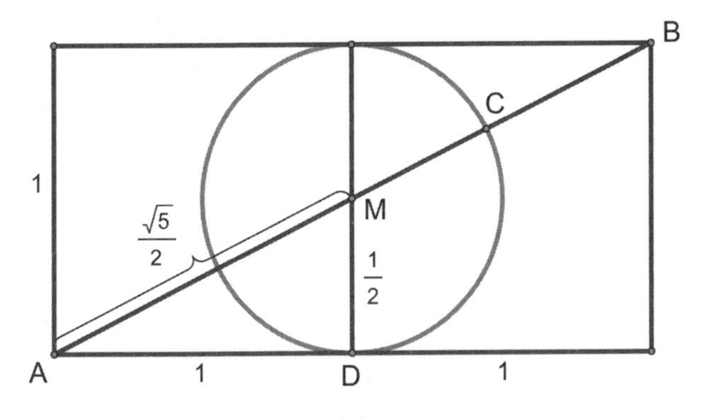


Figure 1-6

## GOLDEN SECTION CONSTRUCTION 5

In this rather simple construction we will show that the semicircle on the side (extended) of a square, whose radius is the length of the segment from the midpoint of the side of the square to an opposite vertex, creates a line segment where the vertex of the square determines the golden ratio. In figure 1-7, we have square $ABCD$ and a semicircle on line $AB$ with center at the midpoint $M$ of $AB$ and radius $CM$. We encountered a similar situation with Construction 1, where we concluded that $\frac{AB}{BE} = \phi$ and $\frac{AE}{AB} = \phi$.

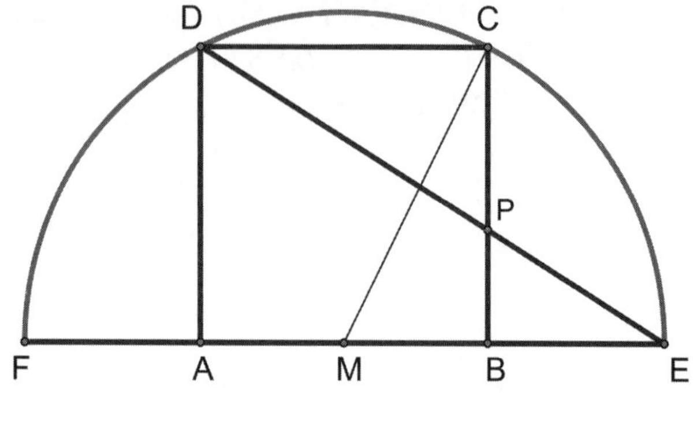


Figure 1-7

However, here we have an extra added attraction: *DE* and *BC* partition each other into the golden section at point *P*. This is easily justified in that triangles *DPC* and *EBP* are similar and their corresponding sides, *DC* and *BE*, are in the golden ratio. Hence, all the corresponding sides are in the golden ratio, which here is $\frac{CP}{PB} = \frac{DP}{PE} = \phi$.

## GOLDEN SECTION CONSTRUCTION 6

Some of the constructions of the golden section are rather creative.[6] Consider the inscribed equilateral triangle *ABC* with line segment *PT* bisecting the two sides of the equilateral triangle at points *Q* and *S* as shown in figure 1-8.

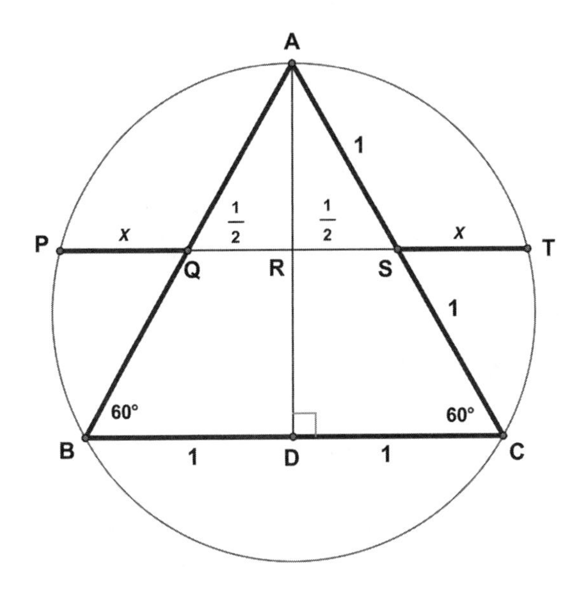


Figure 1-8

We will let the side length of the equilateral triangle equal 2, which then provides us with the segment lengths as shown in figure 1-8. The proportionality there gives us $\frac{RS}{CD}=\frac{AS}{AC}$, which then by substituting appropriate values yields $\frac{RS}{1}=\frac{1}{2}$, and so $RS=\frac{1}{2}$.

A useful geometric theorem will enable us to find the length of the segments $PQ = ST = x$ due to the symmetry of the figure. The theorem states that the products of the segments of two intersecting chords of a circle are equal. From that theorem, we find

$$PS \cdot ST = AS \cdot SC$$

$$(x+1) \cdot x = 1 \cdot 1$$

$$x^2 + x - 1 = 0$$

$$x = \frac{\sqrt{5}-1}{2}.$$

Therefore, the segment $QT$ is partitioned into the golden section at point $S$, since

$$\frac{QS}{ST}=\frac{1}{x}=\frac{2}{\sqrt{5}-1}=\frac{\sqrt{5}+1}{2}\approx 1.61803,$$

which we recognize as the value of the golden ratio. We can generalize this construction by saying that the midline of an equilateral triangle extended to the circumcircle is partitioned into the golden section by the sides of the equilateral triangle.

## GOLDEN SECTION CONSTRUCTION 7

This is a rather easy construction of the golden ratio in that it simply requires constructing an isosceles triangle inside a square as shown in figure 1-9. The vertex $E$ of $\Delta ABE$ lies on side $DC$ of square $ABCD$, and altitude $EM$ intersects the inscribed circle of $\Delta ABE$ at point $H$. The golden ratio appears in two ways here. First, when the side of the square is 2, then the radius of the inscribed circle $r=\frac{1}{\phi}$, and second when the point $H$ partitions $EM$ into the golden ratio as $\frac{EM}{HM}=\phi$.

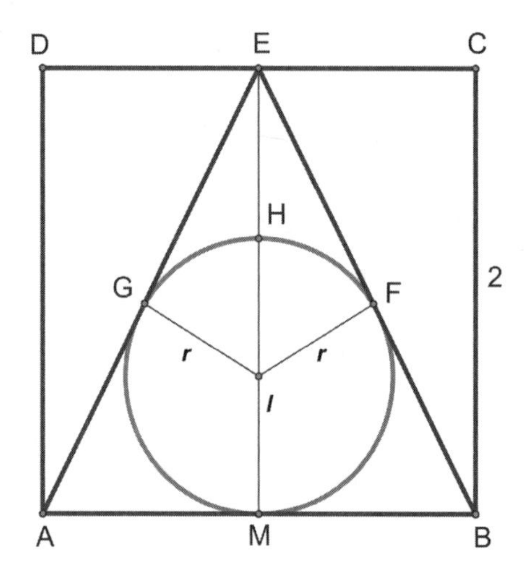


Figure 1-9

To justify this construction, we will let the side of the square have length 2. This gives us $BM = 1$ and $EM = 2$. Then, with the Pythagorean theorem applied to triangle $MEB$, we derive $AE = BE = \sqrt{5}$, whereupon we recognize that $GE = \sqrt{5} - 1$ (fig. 1-10).[7]

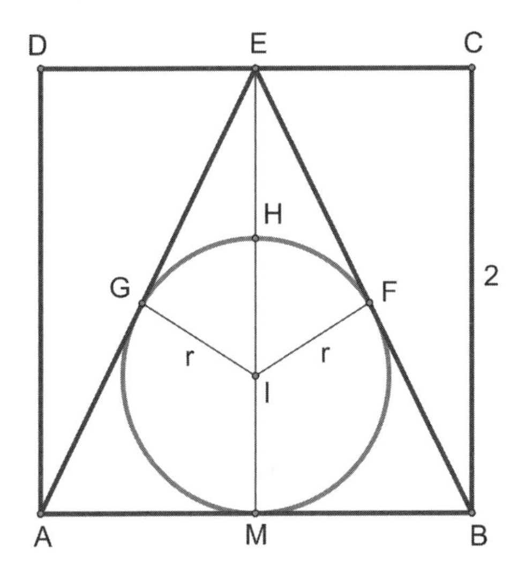


Figure 1-10

For the second appearance, again we apply the Pythagorean theorem, this time to $\Delta EGI$, giving us $EI^2 = GI^2 + GE^2$. Put another way, $(2-r)^2 = r^2 + (\sqrt{5}-1)^2$; therefore, $4 - 4r + r^2 = r^2 + 5 - 2\sqrt{5} + 1$. This determines the length of the radius of the inscribed circle

$$r = \frac{\sqrt{5}-1}{2} = \frac{1}{\phi}.$$

Now, with some simple substitution, we have $EM = 2$ and $HM = 2r$, yielding the ratio $\frac{EM}{HM} = \frac{2}{2r} = \frac{1}{r} = \phi$.

## GOLDEN SECTION CONSTRUCTION 8

A somewhat more contrived construction also yields the golden section of a line segment. To do this, we will construct a unit square with one vertex placed at the center of a circle whose radius is the length of the diagonal of the square. On one side of the square we will construct an equilateral triangle. This is shown in figure 1-11.

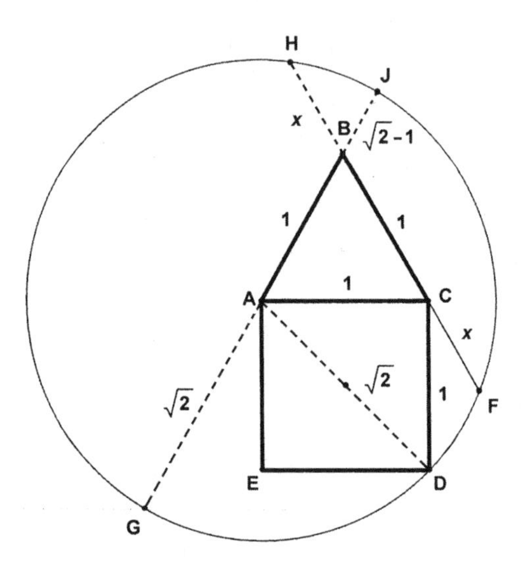


Figure 1-11

Again applying the Pythagorean theorem to triangle *ACD*, we get the radius of the circle as $\sqrt{2}$, which gives us the lengths of *AD*, *AG*, and *AJ*. Because of symmetry, we have $BH = CF = x$. Again applying the theorem involving intersecting chords of a circle (as in Construction 6), we get the following:

$$GB \cdot BJ = HB \cdot BF$$

$$(\sqrt{2}+1)(\sqrt{2}-1) = x(x+1)$$

$$x = \frac{\sqrt{5}-1}{2}.$$

Once again we find the segment *BF* is partitioned into the golden section at point *C*, since

$$\frac{BC}{CF}=\frac{1}{x}=\frac{2}{\sqrt{5}-1}=\frac{\sqrt{5}+1}{2}\approx 1.61803,$$ which we recognize as the value of the golden ratio.

## GOLDEN SECTION CONSTRUCTION 9

We can derive the equation $x^2+x-1=0$, the so-called *golden equation*, in a number of other ways, one of which involves constructing a circle with a chord *AB*, which is extended to a point *P* so that when a tangent from *P* is drawn to the circle, its length equals that of *AB*. We can see this in figure 1-12, where $PT=AB=1$.

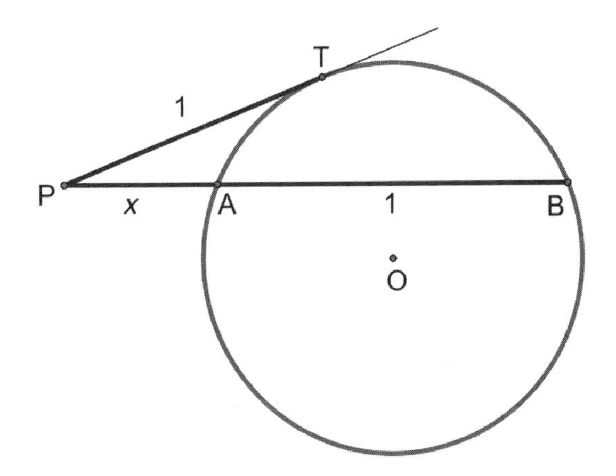

Figure 1-12

Here we will apply a geometric theorem which states that when, from an external point, *P*, a tangent (*PT*) and a secant (*PB*) are drawn to a circle, the tangent segment is the mean proportional between the entire secant and the external segment, that is, $\frac{PB}{PT}=\frac{PT}{PA}$. This yields $PT^2=PB\cdot PA$, or $PT^2=(PA+AB)\cdot PA$. If we let $PA=x$, then $1^2=(x+1)x$, or $x^2+x-1=0$, and, as before, we can conclude that point

*A* determines the golden section of line segment *PB*, since the solution to this equation is the golden ratio.

The next method we present is a bit convoluted. Yet, it begins with the famous 3-4-5 right triangle—probably one of the earliest to be recognized as a true right triangle, going back to the so-called *rope-stretchers* of ancient Egypt.[8]

## GOLDEN SECTION CONSTRUCTION 10

In figure 1-13 we have the 3-4-5 right triangle *ABC*. The bisector of $\angle ABC$ intersects side *AC* at point *G*. With *G* as its center, a circle of radius *GC* is drawn and can be shown to be tangent to both *BC* and *AB*.

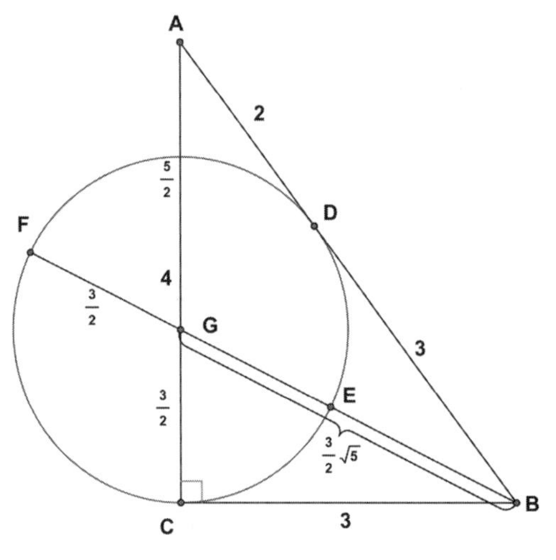


Figure 1-13

As we noted earlier, the bisector of an angle of a triangle divides the side to which it is drawn proportionally to the angle's two sides. Therefore, $\frac{AG}{GC}=\frac{AB}{BC}=\frac{5}{3}$ or $AG=\frac{5}{3}GC$.

With $AG+GC=4$, we get $\frac{5}{3}GC+GC=\frac{8}{3}GC=4$, or $GC=\frac{3}{2}$. So we can determine that $AG=\frac{5}{2}$.

$GC=GD=GE=GF$ are radii of the circle, so we then have $FG=\frac{3}{2}$, and $GE=\frac{3}{2}$. Applying the Pythagorean theorem to $\Delta GBC$, we get $GB^2=BC^2+GC^2=9+\frac{9}{4}=\frac{45}{4}$. Therefore, $GB=\frac{3}{2}\sqrt{5}$.

We are now ready to show that the point $E$ partitions the line segment $BF$ into the golden ratio:

$$\frac{BF}{FE}=\frac{GF+GB}{GF+GE}=\frac{\frac{3}{2}+\frac{3}{2}\sqrt{5}}{\frac{3}{2}+\frac{3}{2}}=\frac{\sqrt{5}+1}{2}\approx 1.61803,$$

which by now is easily recognizable as the golden ratio.

A similar construction with a 3-4-5 right triangle was discovered by Gabries Bosia while pondering the knight's moves in chess.[9]

## GOLDEN SECTION CONSTRUCTION 11

In figure 1-14, we see three concentric circles with radii of lengths 1, 2, and 4 units, respectively. $PR$ is tangent to the inner circle at $T$ and cuts the other circles at points $P$, $Q$, and $R$.

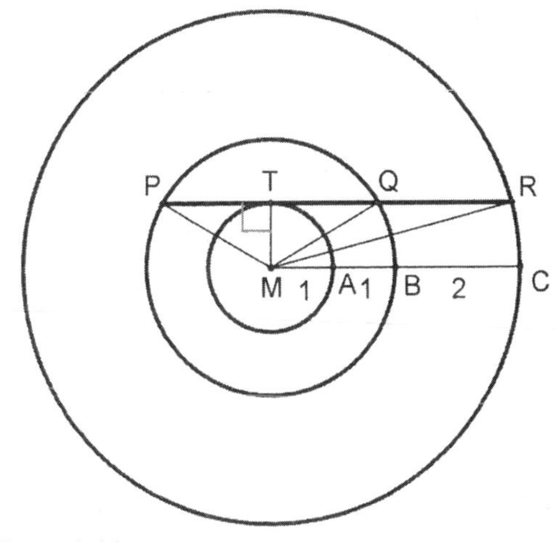


Figure 1-14

With $AM=AB=1$ and $BC=2$, we apply the Pythagorean theorem to $\Delta MQT$ and $\Delta MRT$, and get $QT=\sqrt{2^2-1^2}=\sqrt{3}$, and $RT=\sqrt{4^2-1^2}=\sqrt{15}$.

As we have $PR=RT+PT=RT+QT=\sqrt{15}+\sqrt{3}=\sqrt{3}(\sqrt{5}+1)$, and $PQ=PT+QT=2\sqrt{3}$, we derive

$$\frac{PR}{PQ}=\frac{\sqrt{3}\left(\sqrt{5}+1\right)}{2\sqrt{3}}=\frac{\sqrt{5}+1}{2}\approx 1.61803,$$

which is again recognizable as the golden ratio.

## GOLDEN SECTION CONSTRUCTION 12

We have yet another way of constructing the golden section, this time with three circles. Consider the three adjacent congruent circles with radius $r=1$, as shown in figure 1-15.

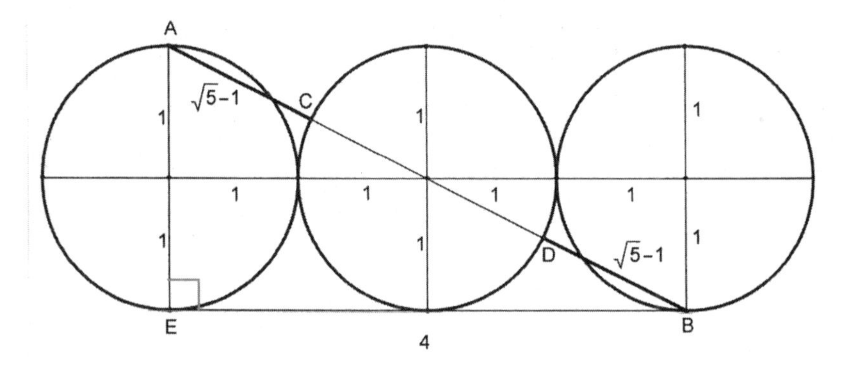


Figure 1-15

In figure 1-15, we have $AE=2$ and $BE=4$. We apply the Pythagorean theorem to $\Delta ABE$ to get $AB=\sqrt{2^2+4^2}=\sqrt{20}=2\sqrt{5}$. Because of the symmetry, $AC=BD$ and $CD=2$, we then have $AB=AC+CD+BD=2AC+BD=2AC+2$. Therefore, $2AC+2=2\sqrt{5}$. It then follows that $AC=\sqrt{5}-1$ and $AD=AB-BD=AB-AC=2\sqrt{5}-(\sqrt{5}-1)=\sqrt{5}+1$.

The ratio $\frac{AD}{CD}=\frac{\sqrt{5}+1}{2}\approx 1.61803$ again denotes the golden ratio.

You may notice that each time we have been using a unit measure as our basis. We could have used a variable, such as $x$, and we would have gotten the same result; however, using 1 rather than $x$ is just a bit simpler.

## GOLDEN SECTION CONSTRUCTION 13

When we place the three equal unit circles tangent to each other and tangent to the semicircle, as shown in figure 1-16, we have the makings for another construction of the golden section.

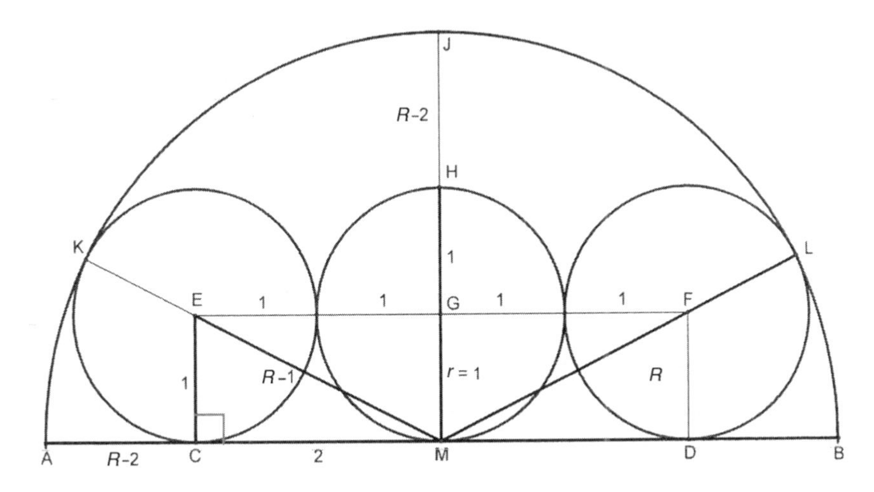


Figure 1-16

First, we note that $AM = BM = JM = KM = LM = R$, and $GH = GM = CE = DF\,(=r) = 1$ (and also $CM = DM = EG = FG = 2$) and $EM = R - r = R - 1$. When we apply the Pythagorean theorem to $\Delta CEM$ in figure 1-16, we get $EM^2 = CM^2 + CE^2$, or $(R-1)^2 = 2^2 + 1^2$.

When we solve this equation for $R$, we get

$$R^2 - 2R + 1 = 5$$
$$R^2 - 2R - 4 = 0$$
$$R = 1 \pm \sqrt{5}.$$

Since a radius cannot be negative, we only use the positive root of $R$; therefore, $R = 1 + \sqrt{5}$.

We then take the ratio $\frac{R}{r} = \sqrt{5} + 1$. Yet, half this ratio will give us the golden ratio:

$$\frac{1}{2}\left(\frac{R}{r}\right)=\frac{\sqrt{5}+1}{2}.$$

Therefore, $\dfrac{LM}{HM}=\dfrac{R}{2r}=\dfrac{R}{2}=\dfrac{\sqrt{5}+1}{2}\approx 1.61803.$

Additionally, the ratios $\frac{HM}{HJ}$ and $\frac{CM}{AC}$ also produce the golden ratio, since with $R-2r=R-2=1+\sqrt{5}-2=\sqrt{5}-1$, which then gives us

$$\frac{HM}{HJ}=\frac{CM}{AC}=\frac{2r}{R-2r}=\frac{2}{\sqrt{5}-1}=\frac{\sqrt{5}+1}{2}.$$

## GOLDEN SECTION CONSTRUCTION 14

Another construction of the golden section was popularized by Hans Walser,[10] who placed the three circles on a coordinate grid as shown in figure 1-17. This construction can be further expanded as we show here. A circle with radius length 1 is enclosed by two circles of radius length 3.

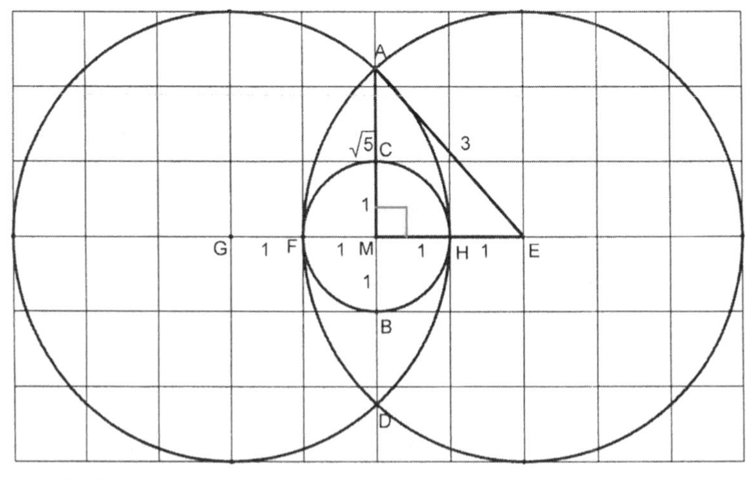


Figure 1-17

With $AE=EF=GH=3$ and $BC=2$, we can find the length of $AM$ by applying the Pythagorean theorem to $\Delta AEM$, whereupon $AM=\sqrt{3^2-2^2}=\sqrt{5}$. Since $AB=AM+BM=\sqrt{5}+1$, then we can establish

$$\frac{AB}{BC} = \frac{\sqrt{5}+1}{2} = \phi \approx 1.61803,$$

which is again recognizable as the golden ratio.

Also, the ratio $\frac{BC}{AC}$ demonstrates the golden section:

$$\frac{BC}{AC} = \frac{BC}{AM - CM} = \frac{2}{\sqrt{5}-1} = \frac{\sqrt{5}+1}{2}.$$

We now present the classic construction of the golden section based on the work of Euclid, which is a pleasant variation of the first construction we offered. Perhaps one of the greatest contributions to our knowledge of mathematics is *Elements* by Euclid, a work divided into thirteen books that covers plane geometry, arithmetic, number theory, irrational numbers, and solid geometry. It is, in fact, a compilation of the knowledge of mathematics that existed up to his time, approximately 300 BCE. We have no records of the dates of Euclid's birth and death, and little is known about his life, though we do know that he lived during the reign of Ptolemy I (305–285 BCE) and taught mathematics in Alexandria, now Egypt. We conjecture that he attended Plato's Academy in Athens, studying mathematics from Plato's students, and later traveled to Alexandria. At the time, Alexandria was the home to a great library created by Ptolemy, known as the Museum. It is believed that Euclid wrote his *Elements* there since that city was also the center of the papyrus industry and book trade. To date, *Elements*, after over one thousand editions, presents synthetic proofs for his propositions and thereby set a standard of logical thinking that impressed many of the greatest minds of our civilization. Notable among them is Abraham Lincoln, who carried a copy of *Elements* with him as a young lawyer and would study the presented propositions on a regular basis to benefit from its logical presentations.

## GOLDEN SECTION CONSTRUCTION 15

So now we come to Euclid's construction of the golden section. In figure 1-18, a right triangle, $\Delta ABC$, is constructed with legs of length 1 and $\frac{1}{2}$. An arc is drawn with center $C$ and radius of length $BC$, and $AC$ is extended to point $D$. A second arc is drawn with center $A$ and tangent to the first arc, naturally passing through point $D$. Using the Pythagorean theorem, we can see that $BC=\frac{\sqrt{5}}{2}$; we will let the length of $AD$ be $x$.

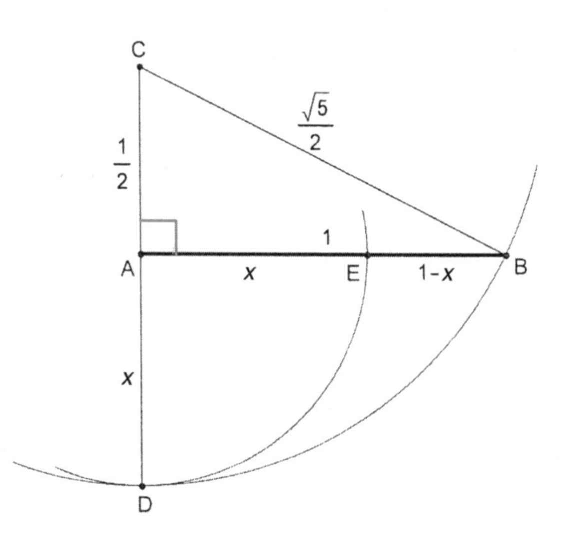


Figure 1-18

$$x=AE=AD=CD-AC=BC-AC=\frac{\sqrt{5}}{2}-\frac{1}{2}=\frac{\sqrt{5}-1}{2}, \text{ and}$$

$$BE=AB-AE=1-x=1-\frac{\sqrt{5}-1}{2}=\frac{3-\sqrt{5}}{2}.$$

This sets up the ratio

$$\frac{AE}{BE} = \frac{\frac{\sqrt{5}-1}{2}}{\frac{3-\sqrt{5}}{2}} = \frac{\sqrt{5}-1}{3-\sqrt{5}} = \frac{\sqrt{5}-1}{3-\sqrt{5}} \cdot \frac{3+\sqrt{5}}{3+\sqrt{5}}$$

$$= \frac{3\sqrt{5}+5-3-\sqrt{5}}{9-5} = \frac{2\sqrt{5}+2}{4} = \frac{\sqrt{5}+1}{2} = \phi \approx 1.61803,$$

which is again recognizable as the golden ratio.

## GOLDEN SECTION CONSTRUCTION 16

The last in our collection of constructions of the golden section is one that may look a bit overwhelming but actually is very simple, as it uses only a compass! All we need is to draw five circles.[11]

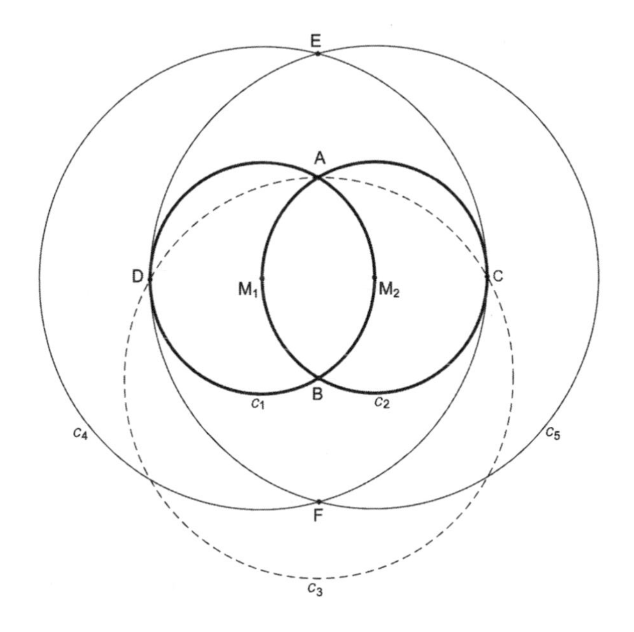

Figure 1-19

In figure 1-19, we begin by constructing circle $c_1$ with center $M_1$ and radius $r_1 = r$. Then, with a randomly selected point $M_2$ on circle $c_1$, we construct a circle, $c_2$, with center $M_2$ and radius $r_2 = r$; naturally $M_1M_2 = r$. We indicate the points of intersection of the two circles, $c_1$ and $c_2$, as $A$ and $B$. Constructing circle $c_3$ with center $B$ and radius $AB = r_3$ will intersect circles $c_1$ and $c_2$ at points $C$ and $D$. (Note that the points $D$, $M_1$, $M_2$, and $C$ are collinear.) We now construct circle $c_4$ with center at $M_1$ and radius $M_1C = r_4 = 2r$. Finally, circle $c_5$ with center $M_2$ and radius $M_2D = r_5 = r_4 = 2r$ is constructed so that it intersects circle $c_4$ at points $E$ and $F$.

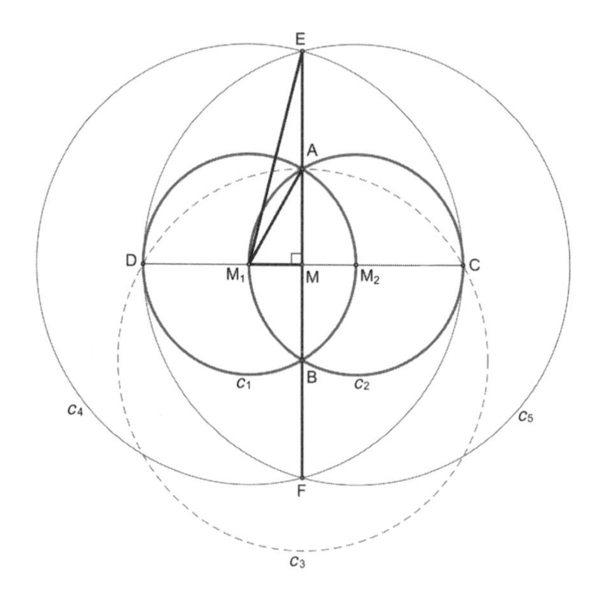


Figure 1-20

From figure 1-20, as a result of obvious symmetry, $AE = BF$, $AF = BE$, $AM = BM$, $EM = FM$, and $CM = DM$, $MM_1 = MM_2$. We can then get $\frac{AB}{AE} = \frac{BE}{AB} = \phi$ (or analogously, $\frac{AB}{BF} = \frac{AF}{AB} = \phi$).

This can be justified rather simply by inserting a few line segments. The radius of the first circle is $r_1 = r = AM_1$, and the radius of the fourth circle is $r_4 = 2r = CM_1 = EM_1$. We can apply the Pythagorean theorem to $\Delta AMM_1$ to get $AM_1^2 = AM^2 + MM_1^2$, or

$$r^2 = AM^2 + \left(\frac{r}{2}\right)^2,$$

which then determines $AM = \frac{r}{2}\sqrt{3}$. Then, applying the Pythagorean theorem to $\Delta EMM_1$, we get $EM_1^{\ 2}(=CM_1^{\ 2}) = EM^2 + MM_1^{\ 2}$, or

$$(2r)^2 = EM^2 + \left(\frac{r}{2}\right)^2,$$

whereupon $EM = \frac{r}{2}\sqrt{15}$.

We now seek to show that the ratio we asserted above is in fact the golden ratio.

$$\frac{AB}{AE} = \frac{AM + BM}{EM - AM} = \frac{2AM}{EM - AM} = \frac{2 \cdot \frac{r}{2}\sqrt{3}}{\frac{r}{2}\sqrt{15} - \frac{r}{2}\sqrt{3}}$$

$$= \frac{2\sqrt{3}}{\sqrt{3}(\sqrt{5}-1)} = \frac{2}{\sqrt{5}-1} \cdot \frac{\sqrt{5}+1}{\sqrt{5}+1} = \frac{\sqrt{5}+1}{2} = \phi.$$

Now the second ratio that we must check is

$$\frac{BE}{AB} = \frac{EM + BM}{AM + BM} = \frac{EM + AM}{2AM}$$

$$= \frac{\frac{r}{2}\sqrt{15} + \frac{r}{2}\sqrt{3}}{2 \cdot \frac{r}{2}\sqrt{3}} = \frac{\sqrt{3}(\sqrt{5}+1)}{2\sqrt{3}} = \frac{\sqrt{5}+1}{2} = \phi.$$

In both cases we have shown that the golden ratio is in fact determined by the five circles we constructed.

We should not want to give the impression that we have covered all possible constructions of the golden section. There currently exist about forty such constructions of the golden section—with new methods being developed continually. As we mentioned, there exist a host of curious geometric configurations where the golden section can be found, but we shall leave these hiding places for later in the book. Notice, however, that our goal for construction of the golden section is to somehow get a length equal to $\sqrt{5}$. For now, we simply want to introduce the numerical value of the golden ratio and its sightings algebraically and geometrically, as it can be seen partitioning a line segment.

Chapter 2

# The Golden Ratio in History

One can never say with certainty where the golden ratio first appeared in the civilized world. To the best of our knowledge, the earliest use of the golden ratio occurred in the ancient Egyptians' construction of the Great Pyramid at Giza, the only one of the "seven wonders of the ancient world" that still exists today. We might someday yet discover older examples where this ratio may be seen. What is still unknown to this day is whether the architect of this structural wonder, Hemiunu (ca. 2570 BCE), consciously chose the dimensions that yield the golden ratio, as he strove to achieve beauty in this structure, or it arose simply by chance. This and other questions about the structure of the pyramid have prompted the writing of numerous books, and yet the issue is still without a definitive conclusion.

This colossal structure, built about 2560 BCE, is the oldest and largest of three pyramids in the Giza Necropolis near modern-day Cairo, Egypt.

The Great Pyramid of Giza. Photo courtesy of Wolfgang Randt.

For years archaeologists have studied this famous pyramid inside and out. Yet for our purposes, we shall focus on its outer dimensions. We will use the cubit as the unit measure, since that was what was used in the time of the construction. (A *cubit* is the first recorded unit of length used in ancient times. It is the measure from the elbow to the tip of the middle finger, and was assumed to be a length of 52.25 cm.) The diagram of the pyramid (fig. 2-1) shows its height to be 280 cubits, half its base length as 220 cubits, and its slant height as 356 cubits.[1]

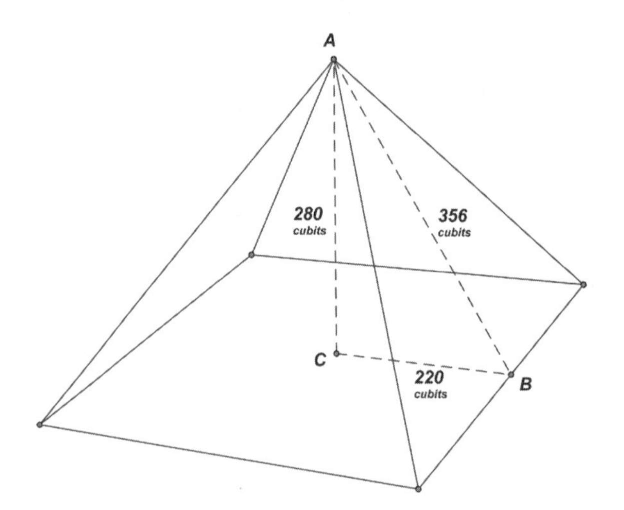


Figure 2-1

The ratio of the slant height to half the base length is $\frac{AB}{BC}=\frac{h_\Delta}{a/2}=\frac{356}{220}=\frac{89}{55}$ $\approx 1.61818$, which is approximately equal to the golden ratio of 1.61803.

Furthermore, as if that isn't enough to convince you that this marvelous pyramid is based on the golden ratio, consider the ratio of the height of the pyramid to half the base length, namely $\frac{280}{220}=\frac{14}{11}=1.27272\ldots$, which is very close to the square root of $\phi$, or approximately 1.2720196.... Were we to divide each of the dimensions of triangle *ABC* (fig. 2-1) by 220, we would get a triangle with the dimensions shown in figure 2-2:

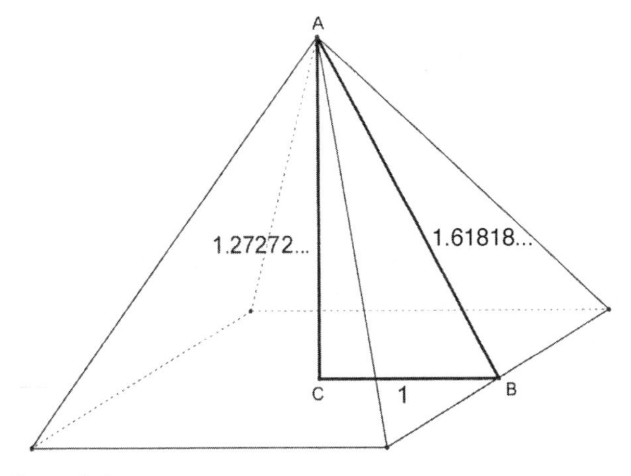


Figure 2-2

These values approximate the golden ratio in various forms, as we can see in figure 2-3a, where we have in terms of $\phi$ the dimensions of a similar right triangle.

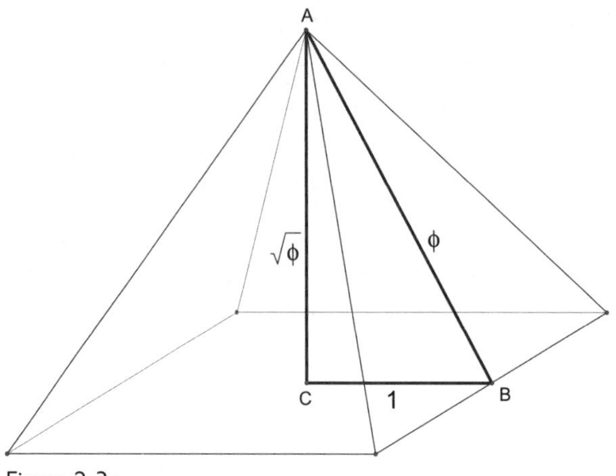


Figure 2-3a

According to the Greek historian Herodotus (ca. 484–424 BCE), the Khufu (Cheops) Pyramid at Giza was constructed in such a way that the square of the height of the pyramid is equal to the area of one of the lateral sides.[2] Once we analyze this curious relationship, we will get some rather surprising results.

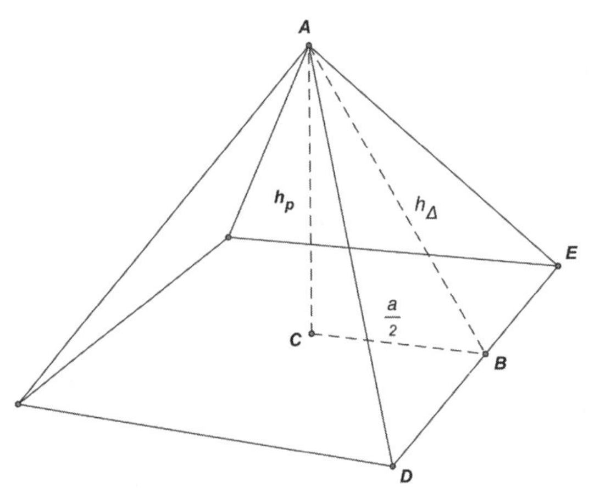


Figure 2-3b

We begin by applying the Pythagorean theorem to triangle $ABC$ in figure 2-3b to arrive at the following: $h_\Delta^2 = \frac{a^2}{4} + h_P^2$.

The area of one of the lateral triangles is $A = \frac{a}{2} \cdot h_\Delta$.

Using the curious relationship Herodotus ascribed to the Giza Pyramid, we get: $h_P^2 = h_\Delta^2 - \frac{a^2}{4} = A = \frac{a}{2} \cdot h_\Delta$.

If we divide both sides of the equation $\frac{a}{2} \cdot h_\Delta = h_\Delta^2 - \frac{a^2}{4}$ by $\frac{a}{2} \cdot h_\Delta$, we get

$1 = \frac{h_\Delta}{\frac{a}{2}} - \frac{\frac{a}{2}}{h_\Delta}$. By letting $x = \frac{h_\Delta}{\frac{a}{2}}$, and then taking the reciprocal, $\frac{\frac{a}{2}}{h_\Delta} = \frac{1}{x}$,

and then substituting in the equation above, we get a simplified equation: $1 = x - \frac{1}{x}$, which just happens to generate the (golden) equation $x^2 - x - 1 = 0$, whose solution is the golden ratio $x_1 = \phi$ and $x_2 = -\frac{1}{\phi}$. By now you realize that $x_2$ is negative and holds no real meaning for us geometrically; so we won't consider it here.

Using today's measurement capabilities, this great pyramid has the following dimensions (see fig. 2-1):

| Cheops pyramid | Length of the side of the base: $a$ | Height of lateral triangle: $h_\Delta$ | Pyramid height $h_P$ | $\frac{h_\Delta}{\frac{a}{2}}$ | $\frac{C}{2h_P}$ |
|---|---|---|---|---|---|
| measurements | 230.56 m | 186.54 m | 146.65 m | 1.61813471 ($\approx\phi$) | 3.144357313 ($\approx\pi$) |

Lo and behold, the ratio of the height of the lateral triangle to half its base is

$$\frac{h_\Delta}{\frac{a}{2}} = 1.61813471$$

Was this done intentionally by the design of a genius architect? No one knows. We can only point out what has been found by measurement and historical clues.

In the history of the golden ratio, the next significant sighting would be that of the Pythagoreans, who, among other applications, used it in their music investigations. The first recorded direct reference to this famous ratio is found in Euclid's *Elements*, which as we mentioned earlier, was a compilation of everything that was known about mathematics at the time of its writing, which was about 300 BCE. This monumental work consisted of thirteen books, in which there are two references made to the golden ratio: In book 2, proposition 11, he constructs a straight line (segment), which is cut so that a rectangle is formed by the whole segment and one of its parts (segment), that is equal to the square on the remaining segment. This can be demonstrated pictorially, as shown in figure 2-4. There we begin with the line segment *ACB*, where point *C* cuts the segment so that *CB* is used to form rectangle *ABHF*, where $CB = HB$, and square *ACGD* has the same area as the rectangle *ABHF*. This equality of areas can be expressed as $AC^2 = AB \cdot CB$, which then can be converted to $\frac{AB}{AC} = \frac{AC}{CB}$, and which then is referred to in a second citation in *Elements*.[3]

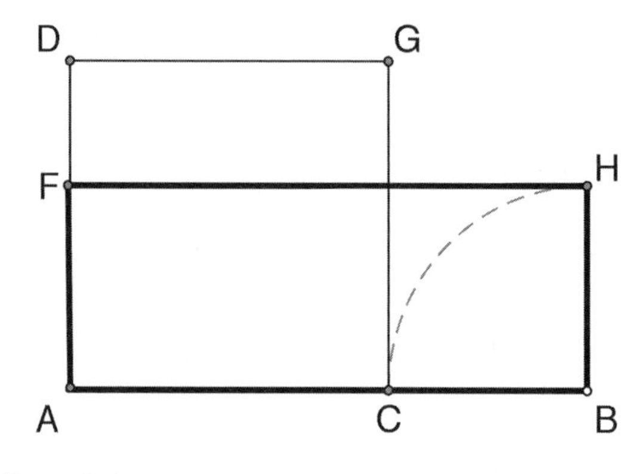


Figure 2-4

Here Euclid refers to a given straight line that is cut—or sectioned—into a mean and extreme ratio,[4] namely for the line segment *AB*, containing point *C*, thus we get $\frac{AB}{AC} = \frac{AC}{CB}$. This is precisely our definition of the golden section.

As we survey the history of the golden ratio, we find its next prominent display in the works of the great Greek sculptor Phidias (ca. 490–ca. 430 BCE). His design for the construction of the Parthenon in Athens, Greece (fig. 2-5), as well as the sculptures he made to adorn this structure, such as the famous statue of Zeus, are said to be reflective of this beautiful ratio. As a matter of fact, the Greek letter $\phi$ is used by many mathematicians today (as well as in this book) to represent the golden section, as it is the first letter of Phidias's name when written in Greek as *Φειδίας*.[5] As you can see in figure 2-5, the

Parthenon in Athens, Greece, fits nicely into a golden rectangle—that is, a rectangle where the quotient of the sides is the golden ratio (see chapter 4). Furthermore, in figure 2-5a you will notice a number of additional golden ratios. Yet even today, no one can say with certainty that Phidias had the golden ratio in mind when he designed the structure.

Figure 2-5. Parthenon in Athens, Greece.

And some more:

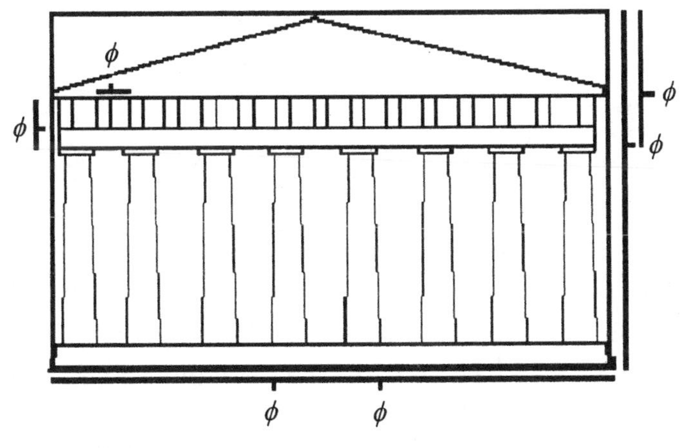


Figure 2-5a

As we continue to trace the history of the golden ratio, we find the next significant sighting in a three-volume book titled *De Divina Propor-*

*tione* (*The Divine Proportion*), written in 1509 by the Franciscan friar and mathematician Fra Luca Pacioli (ca. 1445–1514 or 1517). The book contains drawings by the Italian painter, sculptor, architect, and also mathematician[6] Leonardo da Vinci (1452–1519) of the five Platonic solids. DaVinci also drew the Vitruvian Man (fig. 2-6), in about 1487. This is a picture of a man's body, which clearly exhibits a very close approximation to the golden ratio.

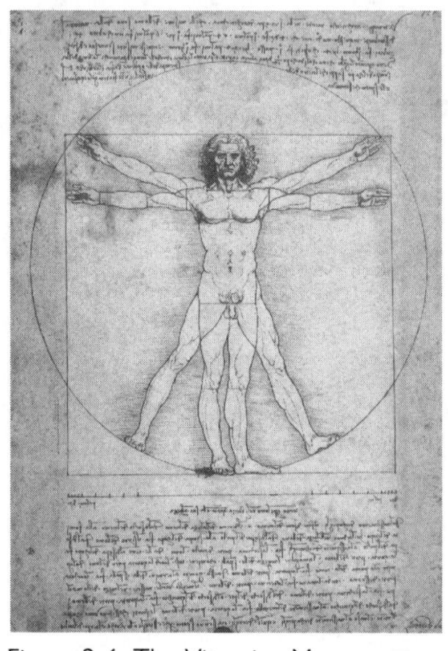

Figure 2-6. The Vitruvian Man.
© Wood River Gallery.

Da Vinci provided notes based on the work of Marcus Vitruvius Pollio (ca. 84–ca. 27 BCE), an ancient Roman writer, architect, and engineer. The drawing, which is in the possession of the Gallerie dell' Accademia in Venice, Italy, is often considered one of the early breakthroughs of pictorially depicting a perfectly proportioned human body. Apparently da Vinci derived these geometric proportions from Vitruvius's treatise *De Architectura*, book 3.

The drawing shows a male figure in two superimposed positions with his arms and legs apart and inscribed in a circle and square, which are tangent at only one point. The golden ratio is exhibited in that the distance from the navel to the top of his head divided by the distance from the soles of the man's feet to his navel (which appears to be at the center of the circle, as shown in figure 2-7), which is about 0.656, approximates the golden ratio (which we know is 0.618...).

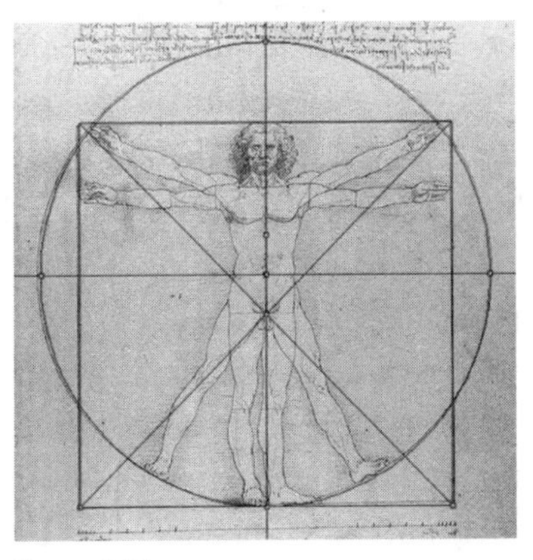

Figure 2-7

Had the square's upper vertices been somewhat closer to the circle, then the golden ratio would have been attained. This can be seen in figure 2-8, where the radius of the circle is selected to be 1, and the side of the square is 1.618, approximately equal to $\phi$, the golden ratio.

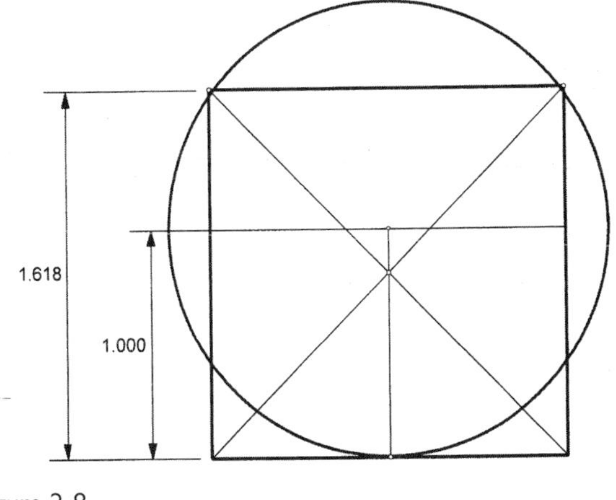


Figure 2-8

To what extent architects and artists consciously used the golden section in their work will remain a mystery because there are no documents that clearly establish its use. Our desire to find examples of the golden ratio might also play into the many sightings. Still, there are countless examples of where it is believed that the golden ratio appears in art and architecture. Many such examples appear throughout the Internet.[7] There are also those who are skeptical about the authenticity of these sightings, such as Dan Pedoe,[8] George Markowsky,[9] Marguerite Neveux,[10] and Roger Herz-Fischler.[11] On the other hand, there are structures (for example, by Le Corbusier),[12] sculptures (such as by Étienne Béothy),[13] or paintings and graphics (for example, by Jo Niemeyer)[14] that document the use of the golden section in their design.[15]

In chapter 3 we will explore the extraordinary relationship between the golden ratio and the Fibonacci sequence, which was popularized by the French mathematician Edouard Lucas (1842–1891) and named after the Italian mathematician Fibonacci (or Leonardo of Pisa, ca. 1175–after 1240). Not only is Lucas credited with discovering characteristics of the numbers based on the regeneration of rabbits problem found in chapter 12 of Fibonacci's book *Liber Abaci* (1202), but he is also responsible for establishing an analogous sequence of numbers that carries his name. Among the relationships he discovered is that which relates the *Lucas numbers* and the *Fibonacci numbers* to the golden ratio. These relationships and much more will be explored in chapter 3.

# Chapter 3

# The Numerical Value of the Golden Ratio and Its Properties

In the previous chapters, we have established that the golden ratio between $a$ and $b$ is $\frac{a+b}{b}=\frac{b}{a}$, where $a$ and $b$ are positive real numbers. As with all ratios, this one has a very specific numerical value. To get the numerical value of this ratio, we first must set up the equation that we get from this ratio by equating the product of the means and extremes, namely $b^2=a(a+b)=a^2+ab$. This equation can be written as $b^2-ab-a^2=0$ and can be solved for either $a$ or $b$; say, we solve for $b$. Using the formula for solving quadratic equations,[1] we find that

$$b=\frac{a(1+\sqrt{5})}{2}=a\frac{\sqrt{5}+1}{2}.$$

Since a length ($a$, $b$) cannot be negative, we ignored the negative root

$$\frac{a(1-\sqrt{5})}{2}=-a\frac{\sqrt{5}-1}{2}.$$

Therefore, by dividing both sides of this equation by $a$, we get

$$\frac{b}{a}=\frac{\sqrt{5}+1}{2},$$

which is then the value of the golden ratio, $\phi$. Numerically, this is approximately[2] equal to:

$$\phi = \frac{\sqrt{5}+1}{2}$$

$\approx$1.6180339887498948482045868343656381177203091798057628621
3544862270526046281890244970720720418939113748475408807538
6891752126633862223536931793180060766726354433389086595939
5829056383226613199282902678806752087668925017116962070322
2104321626954862629631361443814975870122034080588795445474
9246185695364864449241044320771344947049565846788509874339
4422125448770664780915884607499887124007652170575179788341
6625624940758906970400028121042762177111777805315317141011
7046665991466979873176135600670874807101317952368942752194
8435305678300228785699782977834784587822891109762500302696
1561700250464338243776486102838312683303724292675263116533
9247316711121158818638513316203840052221657912866752946549
0681131715993432359734949850904094762132229810172610705961
1645629909816290555208524790352406020172799747175342777592
7786256194320827505131218156285512224809394712341451702237
3580577278616008688382952304592647878017889921990270776903
8953219681986151437803149974110692608867429622675756052317
2777520353613936, which we approximate to 1.61803.

Now if we take the reciprocal of $\phi = \frac{1+\sqrt{5}}{2}$ to get $\frac{1}{\phi} = \frac{2}{1+\sqrt{5}}$ and then multiply this fraction by 1 in the form of $1 = \frac{1-\sqrt{5}}{1-\sqrt{5}}$, we get

$\frac{2}{1+\sqrt{5}} \cdot \frac{1-\sqrt{5}}{1-\sqrt{5}} = \frac{\sqrt{5}-1}{2} = \frac{1}{\phi}$, which then gives us the approximate value[3]:

$$\frac{1}{\phi} = \frac{\sqrt{5}-1}{2}$$

≈.618033988749894848204586834365638117720309179805762862135448622705260462818902449707207204189391137484754088075386891752126633862223536931793180060766726354433389086595939582905638322661319928290267880675208766892501711696207032221043216269548626296313614438149758701220340805887954454749246185695364864449241044320771344947049565846788509874339442212544877066478091588460749988712400765217057517978834166256249407589069704000281210427621771117778053153171410117046665991466979873176135600670874807101317952368942752194843530567830022878569978297783478458782289110976250030269615617002504643382437764861028383126833037242926752631165339247316711121158818638513316203840052221657912866752946549068113171599343235973494985090409476213222981017261070596116456299098162905552085247903524060201727997471753427775927786256194320827505131218156285512224809394712341451702237358057727861600868838295230459264787801788992199027077690389532196819861514378031499741106926088674296226757560523172777520353613936, which we approximate to 0.61803.

We see that the value of $\phi$ has a unique characteristic. Aside from the usual fact that the product of a number and its reciprocal is 1, which, here, gives us $\phi \cdot \frac{1}{\phi}=1$, the difference of $\phi$ and its reciprocal, $\frac{1}{\phi}$, is surprisingly also 1, that is, $\phi-\frac{1}{\phi}=1$. This is the only number for which this is true!

The not-too-well-known mathematician Michael Maestlin (1550–1631), who happened to be one of Johannes Kepler's teachers and later his friend, is credited with the first expansion of the value of $\phi$ to a five-place accuracy, as $\phi \approx$ **1.61803**40, in 1597, while at the University of Tübingen (Germany). As with most famous numbers in mathematics, there is always a desire to seek greater accuracy of a value. This means calculating the value to a larger number of decimal places. Naturally, today we can use computers to facilitate this goal; here is a short history of these milestones of the recent past.

| Year | Number of places of the value of $\phi$ | Mathematician |
|---|---|---|
| 1966 | 4,599 | M. Berg |
| 1976 | 10,000 | J. Shallit |
| 1996 | 10,000,000 | G. J. Fee and S. Plouffe |
| 2000 | 1,500,000,000 | X. Gourdon and P. Sebah |
| 2007 | 5,000,000,000 | A. Irlande |
| 2008 | 17,000,000,000 | A. Irlande |
| 2008 | 31,415,927,000 | X. Gourdon and P. Sebah |
| 2008 | 100,000,000,000 | S. Kondo and S. Pagliarulo |
| 2010 | 1,000,000,000,000 | A. Yee |

Having now established the numerical value of the golden ratio, let us inspect some of the properties of this most unusual number. We begin by considering the irrationality of $\phi$. To do this, we will embark on a nifty little excursion through some simple number theory. The realm of *real* numbers is composed of *rational* and *irrational* numbers. They can be either positive or negative. When expressed in decimal form, the rational numbers are either terminating decimals or repeating decimals, while the irrational numbers do not repeat with any repeating pattern and continue indefinitely. Another way of distinguishing these numbers is that only the rational numbers can be expressed as quotients of integers.

Here are some examples:

Rational numbers: $3$;

$$-\frac{1}{2} = -0.5000\ldots = -0.5\overline{0} = 0.5;$$

$$\frac{2}{3} = 0.666\ldots = 0.\overline{6}.$$

Irrational numbers: $\sqrt{2} = 1.414213562\ldots$;

$$\pi = 3.141592653\ldots;$$

$$e = 2.718281828\ldots.$$

We claim that the number $\phi$ is an irrational number—one that has an unending decimal value—one that has no repeating pattern. We can establish that $\sqrt{5}$ is irrational and therefore

$$\frac{\sqrt{5}+1}{2}=\phi$$

would also be irrational. To prove that $\sqrt{5}$ is irrational, we begin by supposing the contrary, namely that $\sqrt{5}$ is rational, implying that $\sqrt{5}=\frac{p}{q}$, a fraction that may be assumed to be in lowest terms. Squaring both sides and clearing denominators, we get $5q^2=p^2$. Thus the left-hand side is divisible by 5 and therefore so is the right-hand side. But 5 is a prime. Therefore, since 5 divides $p^2$, it must also divide $p$. Thus, $p=5r$, for some $r$. Then we have $5q^2=p^2=25r^2$, so that $q^2=5r^2$. Repeating the previous argument, we find that $q$ is also divisible by 5, contradicting our assumption that the fraction was in lowest terms. Therefore, $\sqrt{5}$ is not rational. Thus, $\phi$ must then be an irrational number.

As we will see in chapter 4, the irrationality of $\phi$ will be realized by the fact that the diagonal of a regular pentagon is incommensurate with a side of the pentagon, which means they have no common measure. Similarly, the irrationality of $\pi$ is seen by the fact that the diameter of a circle and its circumference have no common measure.

We continue by considering the powers of $\phi$. To do so, we first must find the value of $\phi^2$ in terms of $\phi$.
Since

$$\phi=\frac{\sqrt{5}+1}{2},$$

$$\phi^2=\left(\frac{\sqrt{5}+1}{2}\right)^2=\frac{5+2\sqrt{5}+1}{4}=$$

$$\frac{2\sqrt{5}+6}{4}=\frac{\sqrt{5}+3}{2}=\frac{\sqrt{5}+1}{2}+1=\phi+1.$$

You may find this equation $\phi^2=\phi+1$ somewhat familiar, as we have encountered it at several points already. From this equation, we can generate an interesting mathematical expression to express the value of

$\phi$ in a very unusual fashion. By taking the square root of both sides, this equation can be rewritten as $\phi=\sqrt{\phi+1}$, or $\phi=\sqrt{1+\phi}$. We will now replace $\phi$ under the radical sign with its equivalent $\phi=\sqrt{1+\phi}$ to get

$$\phi=\sqrt{1+\left(\sqrt{1+\phi}\right)}=\sqrt{1+\sqrt{1+\phi}}.$$

Then, repeating this process (i.e., replacing the last $\phi$ with $\phi=\sqrt{1+\phi}$), we get

$$\phi=\sqrt{1+\sqrt{1+\sqrt{1+\phi}}}.$$

Continuing this process gives us

$$\phi=\sqrt{1+\sqrt{1+\sqrt{1+\sqrt{1+\phi}}}},$$

and so on, until you realize that this will go on to infinity and look something like

$$\phi=\sqrt{1+\sqrt{1+\sqrt{1+\sqrt{1+\sqrt{1+\sqrt{1+\sqrt{1+\sqrt{1+\sqrt{1+\sqrt{1+\sqrt{1+\sqrt{1+\sqrt{1+\sqrt{1+\sqrt{1+\sqrt{1+\ldots}}}}}}}}}}}}}}}}$$

Suppose we now consider the following analogous nest of radicals:

$$x=\sqrt{1-\sqrt{1-\sqrt{1-\sqrt{1-\sqrt{1-\sqrt{1-\sqrt{1-\sqrt{1-\sqrt{1-\sqrt{1-\sqrt{1-\sqrt{1-\sqrt{1-\sqrt{1-\sqrt{1-\sqrt{1-\ldots}}}}}}}}}}}}}}}}$$

We can evaluate this value for $x$ by using the following technique: There is an infinite number of radicals in this nest. Without loss of accuracy, we can temporarily "ignore" the outermost radical and see that the remaining expression is actually the same as the original one:

$$x=\sqrt{1-\sqrt{1-\sqrt{1-\sqrt{1-\sqrt{1-\sqrt{1-\sqrt{1-\sqrt{1-\sqrt{1-\sqrt{1-\sqrt{1-\sqrt{1-\sqrt{1-\sqrt{1-\sqrt{1-\sqrt{1-\ldots}}}}}}}}}}}}}}}}$$

So if we substitute this value of $x$ into the original expression, we get $x=\sqrt{1-x}$. By squaring both sides, we get the following quadratic equation: $x^2=1-x$, or $x^2+x-1=0$. When we apply the quadratic formula to this equation, we get (ignoring the negative root)

$$x=\frac{\sqrt{5}-1}{2}\approx 0.61803,$$

which is $\frac{1}{\phi}$. Again, we have a most unusual relationship between $\phi$ and $\frac{1}{\phi}$:

$$\phi=\sqrt{1+\sqrt{1+\sqrt{1+\sqrt{1+\sqrt{1+\sqrt{1+\sqrt{1+\sqrt{1+\sqrt{1+\sqrt{1+\sqrt{1+\sqrt{1+\sqrt{1+\sqrt{1+\sqrt{1+\sqrt{1+\ldots}}}}}}}}}}}}}}}}$$

and

$$\frac{1}{\phi}=\sqrt{1-\sqrt{1-\sqrt{1-\sqrt{1-\sqrt{1-\sqrt{1-\sqrt{1-\sqrt{1-\sqrt{1-\sqrt{1-\sqrt{1-\sqrt{1-\sqrt{1-\sqrt{1-\sqrt{1-\sqrt{1-\ldots}}}}}}}}}}}}}}}}$$

Let us now investigate powers of $\phi$. In order to inspect the successive powers of $\phi$, we will break them down into their component parts. It may at first appear more complicated than it really is. You should try to follow each step (it's really not difficult—and yet very rewarding!) and then extend it to further powers of $\phi$.

$\phi=\phi$
$\phi^2=\phi+1$
$\phi^3=\phi\cdot\phi^2=\phi(\phi+1)=\phi^2+\phi=(\phi+1)+\phi=2\phi+1$
$\phi^4=\phi^2\cdot\phi^2=(\phi+1)(\phi+1)=\phi^2+2\phi+1=(\phi+1)+2\phi+1=3\phi+2$
$\phi^5=\phi^3\cdot\phi^2=(2\phi+1)(\phi+1)=2\phi^2+3\phi+1=2(\phi+1)+3\phi+1=5\phi+3$
$\phi^6=\phi^3\cdot\phi^3=(2\phi+1)(2\phi+1)=4\phi^2+4\phi+1=4(\phi+1)+4\phi+1=8\phi+5$
$\phi^7=\phi^4\cdot\phi^3=(3\phi+2)(2\phi+1)=6\phi^2+7\phi+2=6(\phi+1)+7\phi+2=13\phi+8$
$\phi^8=\phi^4\cdot\phi^4=(3\phi+2)(3\phi+2)=9\phi^2+12\phi+4=9(\phi+1)+12\phi+4=21\phi+13$
$\phi^9=\phi^5\cdot\phi^4=(5\phi+3)(3\phi+2)=15\phi^2+19\phi+6=15(\phi+1)+19\phi+6=34\phi+21$
$\phi^{10}=\phi^5\cdot\phi^5=(5\phi+3)(5\phi+3)=25\phi^2+30\phi+9=25(\phi+1)+30\phi+9=55\phi+34$
and so on ... .

By this point, you should be able to see a pattern emerging. As we take further powers of $\phi$, the end result of each power of $\phi$ is actually equal to a multiple of $\phi$ plus a constant. Further inspection shows that the coefficients of $\phi$ and the constants follow the pattern 1, 1, 2, 3, 5, 8, 13, 21, 34, 55, 89, 144, .... This sequence of numbers is famous and is known as the *Fibonacci sequence*.[4] Beginning with two 1s, each successive number is the sum of the two preceding numbers. The Fibonacci numbers are perhaps the most ubiquitous numbers in all of mathematics; they come up in just about every field of the subject. Yet, as we mentioned earlier, they only made their "debut" in the Western world in chapter 12 of a 1202 publication, *Liber Abaci*, by Leonardo of Pisa, most commonly known today as Fibonacci (ca. 1175–after 1240), in the solution of a simple problem about the breeding of rabbits.

We recently discovered that the Fibonacci numbers were described in early Indian mathematics writings.[5] The earliest appearance can be found under the name *mātrāmeru* (Mountain of Cadence), which appeared in *Chandahsūtras* (*Art of Prosody*) by the Sanskrit grammarian Pingala (between the fifth and second century BCE). In a more complete fashion were the writings of Virahānka (sixth century CE) and Ācārya Hemacandra (1089–1172), who cites the Fibonacci numbers. It is speculated that Fibonacci may have come to these numbers from his Arabic sources, which exposed him to these Indian writings.

Sometime before his death in 1564, the German calculation master Simon Jacob[6] made the first published connection between the golden ratio and the Fibonacci series, but it appears to have been something of a side note.[7] Jacob had published a numerical solution for the golden ratio. In the margin of the page discussing the Euclidean algorithm from the second proposition of book 7 of Euclid's *Elements*, he wrote the first twenty-eight terms of the Fibonacci sequence and stated:

> In following this sequence one comes nearer and nearer to that proportion described in the 11th proposition of the 2nd book and the 30th of the 6th book of Euclid, and though one comes nearer and nearer to this proportion it is impossible to reach or to overcome it.

We will use the symbol $F_7$ to represent the seventh Fibonacci number, and $F_n$ to represent the *n*th Fibonacci number, or as we say, the general Fibonacci number, that is, any Fibonacci number. Therefore, in general terms, we would write the rule of the Fibonacci numbers as $F_{n+2}=F_n+F_{n+1}$ with $n \geq 1$, and $F_1=F_2=1$.

Let us look at the first thirty Fibonacci numbers.

| | | |
|---|---|---|
| $F_1 = 1$ | $F_{11} = 89$ | $F_{21} = 10{,}946$ |
| $F_2 = 1$ | $F_{12} = 144$ | $F_{22} = 17{,}711$ |
| $F_3 = 2$ | $F_{13} = 233$ | $F_{23} = 28{,}657$ |
| $F_4 = 3$ | $F_{14} = 377$ | $F_{24} = 46{,}368$ |
| $F_5 = 5$ | $F_{15} = 610$ | $F_{25} = 75{,}025$ |
| $F_6 = 8$ | $F_{16} = 987$ | $F_{26} = 121{,}393$ |
| $F_7 = 13$ | $F_{17} = 1{,}597$ | $F_{27} = 196{,}418$ |
| $F_8 = 21$ | $F_{18} = 2{,}584$ | $F_{28} = 317{,}811$ |
| $F_9 = 34$ | $F_{19} = 4{,}181$ | $F_{29} = 514{,}229$ |
| $F_{10} = 55$ | $F_{20} = 6{,}765$ | $F_{30} = 832{,}040$ |

The list of powers of $\phi$ can easily be extended by using the Fibonacci numbers directly in the pattern we developed above.

$$
\begin{aligned}
\phi &= 1\phi + 0\\
\phi^2 &= 1\phi + 1\\
\phi^3 &= 2\phi + 1\\
\phi^4 &= 3\phi + 2\\
\phi^5 &= 5\phi + 3\\
\phi^6 &= 8\phi + 5\\
\phi^7 &= 13\phi + 8\\
\phi^8 &= 21\phi + 13\\
\phi^9 &= 34\phi + 21\\
\phi^{10} &= 55\phi + 34\\
\phi^{11} &= 89\phi + 55\\
\phi^{12} &= 144\phi + 89\\
\phi^{13} &= 233\phi + 144\\
\phi^{14} &= 377\phi + 233\\
&\ldots
\end{aligned}
$$

Since the Fibonacci numbers appear as the coefficients of $\phi$, as well as the constants, we can write all powers of $\phi$ in a linear form: $\phi^n = a\phi + b$, where $a$ and $b$ are consecutive Fibonacci numbers. In the general case, we can write this as: $\phi^n = F_n\phi + F_{n-1}$, with $n \geq 1$ and $F_0 = 0$. (See the appendix for a proof of this statement.)

You should also take note that each power of $\phi$ is the sum of the two immediately preceding powers of $\phi$. We can develop another amazing pattern involving the Fibonacci numbers and the golden ratio. This will involve a structure that is called a *continued fraction*. We will begin with a brief introduction to continued fractions.[8] A continued fraction is a fraction in which the denominator contains a mixed number (a whole number and a proper fraction). We can take an improper fraction such as $\frac{13}{7}$ and express it as a mixed number: $1\frac{6}{7} = 1 + \frac{6}{7}$. Without changing the value, we could then write this as

$$1 + \frac{6}{7} = 1 + \cfrac{1}{\frac{7}{6}},$$

which in turn could be written (again without any value change) as

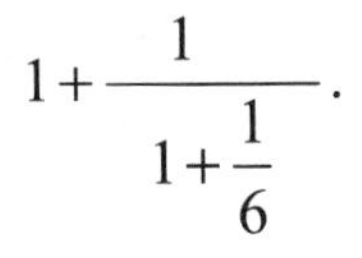

This is a continued fraction. We could have continued this process, but when we reach a unit fraction[9] (as in this case, the unit fraction is $\frac{1}{6}$), we are essentially finished.

To enable you to get a better grasp of this technique, we will create another continued fraction. We will convert $\frac{12}{7}$ to a continued fraction form. Notice that, at each stage, when a proper fraction is reached, we take the reciprocal of the reciprocal (e.g., change

$$\frac{2}{5} \text{ to } \frac{1}{\frac{5}{2}}$$

as we will do in the example that follows), which does not change its value:

$$\frac{12}{7} = 1 + \frac{5}{7} = 1 + \frac{1}{\frac{7}{5}} = 1 + \frac{1}{1 + \frac{2}{5}} = 1 + \frac{1}{1 + \frac{1}{\frac{5}{2}}} = 1 + \frac{1}{1 + \frac{1}{2 + \frac{1}{2}}}.$$

If we break up a continued fraction into its component parts (called *convergents*),[10] we get closer and closer to the actual value of the original fraction.

First convergent of $\frac{12}{7}$: $1.$

Second convergent of $\frac{12}{7}$: $1 + \frac{1}{1} = 2.$

Third convergent of $\frac{12}{7}$: $1 + \frac{1}{1 + \frac{1}{2}} = 1 + \frac{2}{3} = 1\frac{2}{3} = \frac{5}{3}.$

Fourth convergent of $\frac{12}{7}$: $1 + \frac{1}{1 + \frac{1}{2 + \frac{1}{2}}} = \frac{12}{7}.$

The above examples are all *finite* continued fractions, which are equivalent to rational numbers (those that can be expressed as simple fractions). It would then follow that an irrational number would result in an *infinite* continued fraction. And that is exactly the case. A simple example of an infinite continued fraction is that of $\sqrt{2}$.

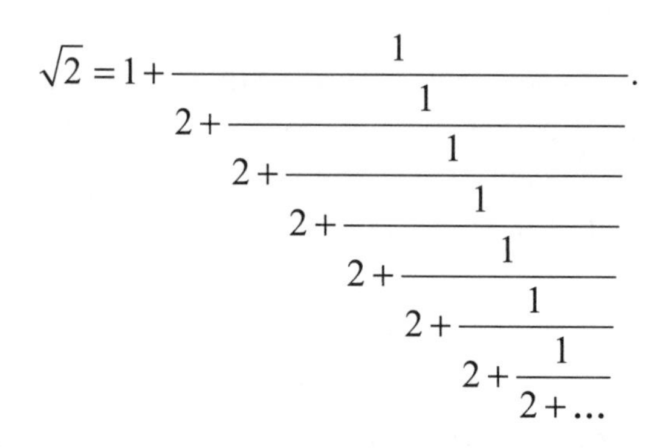

We have a short way to write a long (in this case infinitely long) continued fraction: [1; 2,2,2,2,2,2,2,...], or when there are these endless repetitions, we can write it in an even shorter form as $[1;\overline{2}]$, where the bar over the 2 indicates that the 2 repeats endlessly.

In general, we can represent a continued fraction as

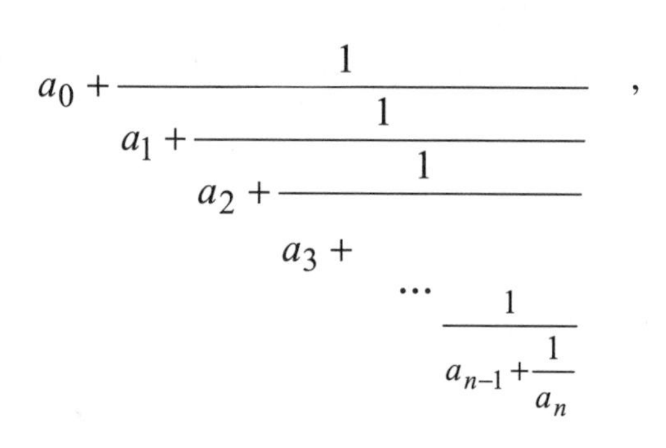

where $a_i$ are real numbers and $a_i \neq 0$ for $i > 0$. We can write this in short fashion as $[a_0; a_1, a_2, a_3, \ldots, a_{n-1}, a_n]$.

Now that the concept of a continued fraction has been described, we can apply it to the golden ratio. We begin with the equation of the golden ratio: $\phi=1+\frac{1}{\phi}$. If we substitute $1+\frac{1}{\phi}$ for the $\phi$ in the denominator of the fraction of this equation, we get

$$\phi=1+\cfrac{1}{1+\cfrac{1}{\phi}}=[1;\,1,\phi],$$

and then continue this process by substituting the value $\phi=1+\frac{1}{\phi}$ in each case for the last denominator of the previous equation, we will get the following:

$$\phi=1+\cfrac{1}{1+\cfrac{1}{\left(1+\cfrac{1}{\phi}\right)}}=[1;\,1,\,1,\,\phi].$$

Repeating this procedure, we get an *infinite* continued fraction that looks like this:

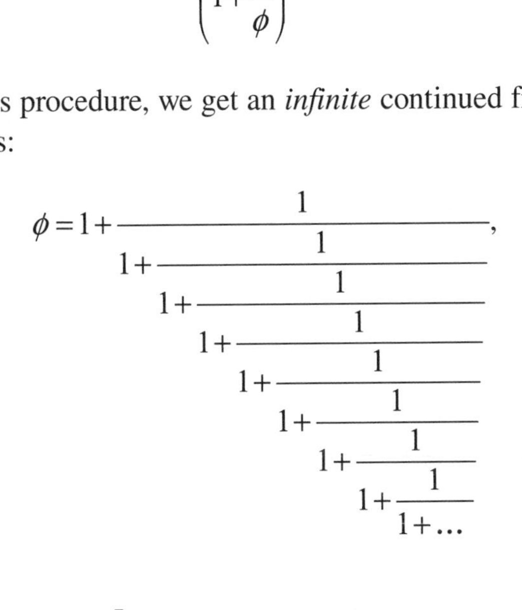

$$\phi=1+\cfrac{1}{1+\cfrac{1}{1+\cfrac{1}{1+\cfrac{1}{1+\cfrac{1}{1+\cfrac{1}{1+\cfrac{1}{1+\cfrac{1}{1+\dots}}}}}}}},$$

or $\phi=[1;\,1,\,1,\,1,\ldots]=[\bar{1}]$. (See the appendix.)

This gives our now already famous $\phi$ another unique characteristic, namely that it is equal to the most primitive infinite continued fraction—one with all 1s.

Let us take the value of this continued fraction in successive parts (which are called *convergents*), each of which will successively bring us closer to the value of the infinite continued fraction. The successive convergents are as follows:

$$1 = \frac{F_2}{F_1} = \frac{1}{1} = [1] = 1$$

$$1 + \frac{1}{1} = 2$$

$$= \frac{F_3}{F_2} = \frac{2}{1} = [1;\ 1] = 2$$

$$1 + \cfrac{1}{1 + \cfrac{1}{1}} = 1 + \frac{1}{2} = \frac{3}{2}$$

$$= \frac{F_4}{F_3} = \frac{3}{2} = [1;\ 1,1] = 1.5$$

$$1 + \cfrac{1}{1 + \cfrac{1}{1 + \cfrac{1}{1}}} = 1 + \cfrac{1}{1 + \cfrac{1}{2}} = 1 + \cfrac{1}{\cfrac{3}{2}} = \frac{5}{3}$$

$$= \frac{F_5}{F_4} = \frac{5}{3} = [1;\ 1,1,1] = 1.\overline{6}$$

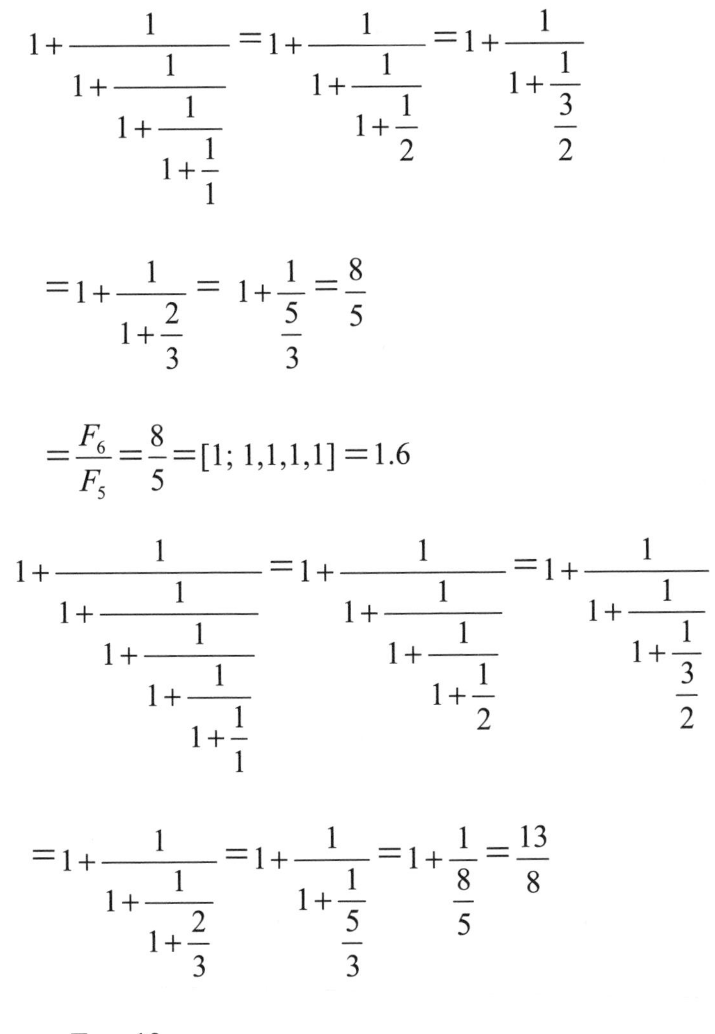

$$1+\cfrac{1}{1+\cfrac{1}{1+\cfrac{1}{1+\cfrac{1}{1}}}}=1+\cfrac{1}{1+\cfrac{1}{1+\cfrac{1}{2}}}=1+\cfrac{1}{1+\cfrac{1}{\frac{3}{2}}}$$

$$=1+\cfrac{1}{1+\cfrac{2}{3}}=1+\cfrac{1}{\frac{5}{3}}=\frac{8}{5}$$

$$=\frac{F_6}{F_5}=\frac{8}{5}=[1;\ 1,1,1,1]=1.6$$

$$1+\cfrac{1}{1+\cfrac{1}{1+\cfrac{1}{1+\cfrac{1}{1+\cfrac{1}{1}}}}}=1+\cfrac{1}{1+\cfrac{1}{1+\cfrac{1}{1+\cfrac{1}{2}}}}=1+\cfrac{1}{1+\cfrac{1}{1+\cfrac{1}{\frac{3}{2}}}}$$

$$=1+\cfrac{1}{1+\cfrac{1}{1+\cfrac{2}{3}}}=1+\cfrac{1}{1+\cfrac{1}{\frac{5}{3}}}=1+\cfrac{1}{\frac{8}{5}}=\frac{13}{8}$$

$$=\frac{F_7}{F_6}=\frac{13}{8}=[1;\ 1,1,1,1,1]=1.625$$

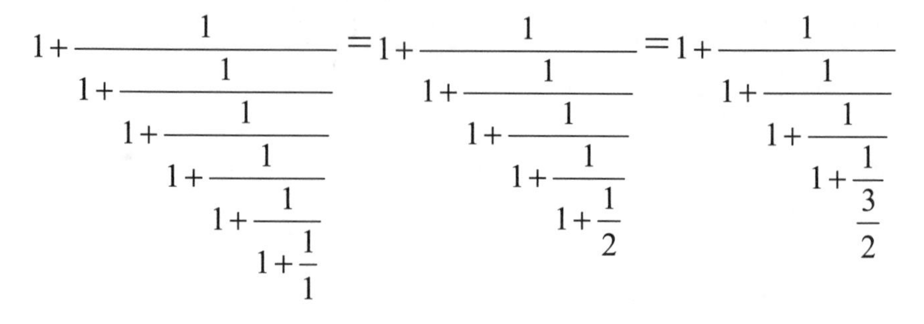

$$1+\cfrac{1}{1+\cfrac{1}{1+\cfrac{1}{1+\cfrac{1}{1+\cfrac{1}{1+\cfrac{1}{1}}}}}}=1+\cfrac{1}{1+\cfrac{1}{1+\cfrac{1}{1+\cfrac{1}{1+\cfrac{1}{2}}}}}=1+\cfrac{1}{1+\cfrac{1}{1+\cfrac{1}{1+\cfrac{1}{\cfrac{3}{2}}}}}$$

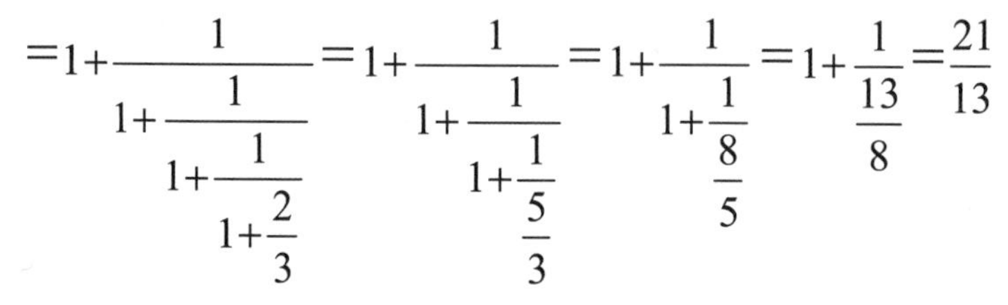

$$=1+\cfrac{1}{1+\cfrac{1}{1+\cfrac{1}{1+\cfrac{2}{3}}}}=1+\cfrac{1}{1+\cfrac{1}{1+\cfrac{1}{\cfrac{5}{3}}}}=1+\cfrac{1}{1+\cfrac{1}{\cfrac{8}{5}}}=1+\cfrac{1}{\cfrac{13}{8}}=\frac{21}{13}$$

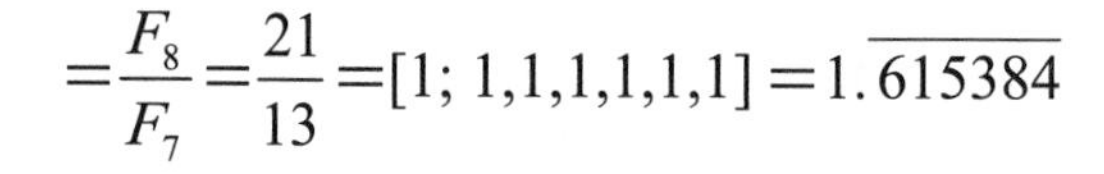

$$=\frac{F_8}{F_7}=\frac{21}{13}=[1;\ 1,1,1,1,1,1]=1.\overline{615384}$$

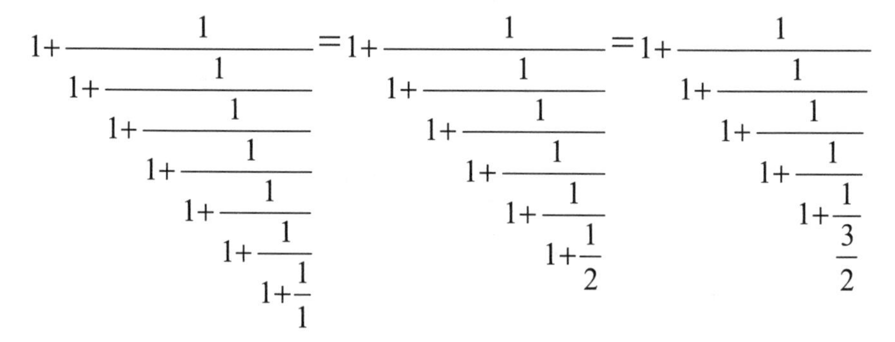

$$1+\cfrac{1}{1+\cfrac{1}{1+\cfrac{1}{1+\cfrac{1}{1+\cfrac{1}{1+\cfrac{1}{1+\cfrac{1}{1}}}}}}}=1+\cfrac{1}{1+\cfrac{1}{1+\cfrac{1}{1+\cfrac{1}{1+\cfrac{1}{1+\cfrac{1}{2}}}}}}=1+\cfrac{1}{1+\cfrac{1}{1+\cfrac{1}{1+\cfrac{1}{1+\cfrac{1}{\cfrac{3}{2}}}}}}$$

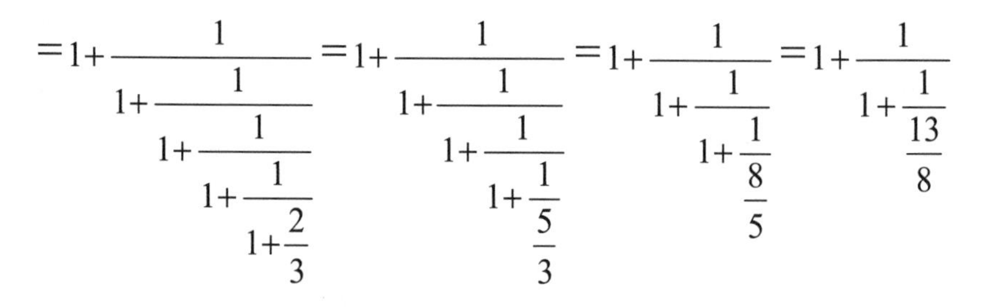

$$=1+\cfrac{1}{1+\cfrac{1}{1+\cfrac{1}{1+\cfrac{1}{1+\cfrac{2}{3}}}}}=1+\cfrac{1}{1+\cfrac{1}{1+\cfrac{1}{1+\cfrac{1}{\cfrac{5}{3}}}}}=1+\cfrac{1}{1+\cfrac{1}{1+\cfrac{1}{\cfrac{8}{5}}}}=1+\cfrac{1}{1+\cfrac{1}{\cfrac{13}{8}}}$$

$$=1+\cfrac{1}{\cfrac{21}{13}}=\frac{34}{21}=\frac{F_9}{F_8}=\frac{34}{21}=[1;\ 1,1,1,1,1,1,1]=1.\overline{619047}$$

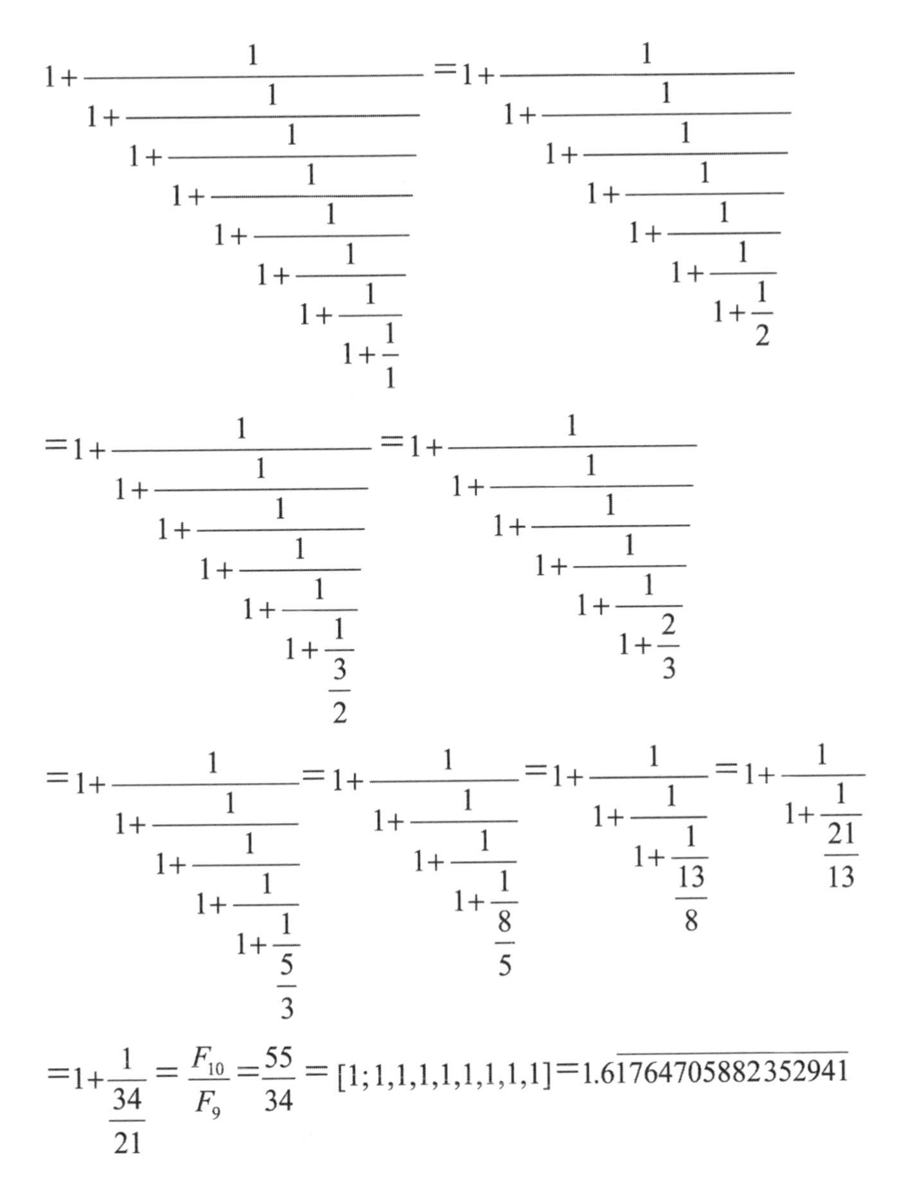

$$1+\cfrac{1}{1+\cfrac{1}{1+\cfrac{1}{1+\cfrac{1}{1+\cfrac{1}{1+\cfrac{1}{1+\cfrac{1}{1+\cfrac{1}{1}}}}}}}}=1+\cfrac{1}{1+\cfrac{1}{1+\cfrac{1}{1+\cfrac{1}{1+\cfrac{1}{1+\cfrac{1}{1+\cfrac{1}{2}}}}}}}$$

$$=1+\cfrac{1}{1+\cfrac{1}{1+\cfrac{1}{1+\cfrac{1}{1+\cfrac{1}{1+\cfrac{1}{\frac{3}{2}}}}}}}=1+\cfrac{1}{1+\cfrac{1}{1+\cfrac{1}{1+\cfrac{1}{1+\cfrac{2}{3}}}}}$$

$$=1+\cfrac{1}{1+\cfrac{1}{1+\cfrac{1}{1+\cfrac{1}{1+\cfrac{1}{\frac{5}{3}}}}}}=1+\cfrac{1}{1+\cfrac{1}{1+\cfrac{1}{1+\cfrac{1}{\frac{8}{5}}}}}=1+\cfrac{1}{1+\cfrac{1}{1+\cfrac{1}{\frac{13}{8}}}}=1+\cfrac{1}{1+\cfrac{1}{\frac{21}{13}}}$$

$$=1+\cfrac{1}{\frac{34}{21}}=\frac{F_{10}}{F_9}=\frac{55}{34}=[1;1,1,1,1,1,1,1,1]=1.6\overline{1764705882352941}$$

As they progress, you will notice how these convergents seem to "sandwich in," or converge, to the value of $\phi$, with which we are now quite familiar, approximately 1.618034. What also emerges from these continually increasing convergents is that the final simple fractional values of these convergents happen to be composed of the Fibonacci numbers.

Aside from the continued fraction

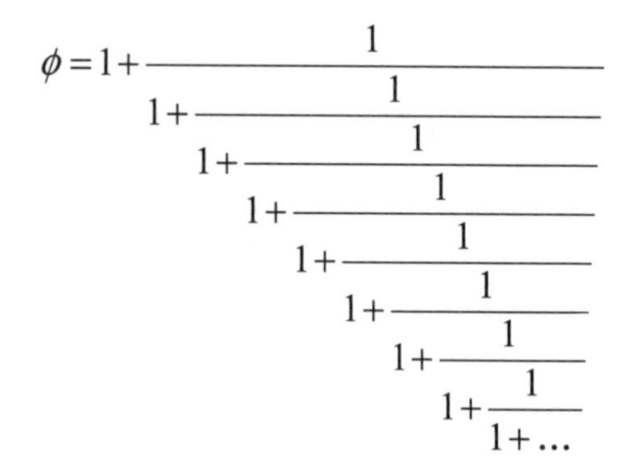

getting ever closer to the value of $\phi$ as we increase its length, we shall now see another surprising relationship of $\phi$ and the Fibonacci numbers.

In the following chart, we can see that the ratios of consecutive members of the Fibonacci sequence also approach the value of $\phi$. In mathematical terms, we say that the limit of the quotient/ratio of two consecutive Fibonacci numbers,

$$\frac{F_{n+1}}{F_n}$$

is the value of $\phi$. Mathematicians typically write this as

$$\lim_{n \to \infty} \frac{F_{n+1}}{F_n} = \phi .$$

The famous Scottish mathematician Robert Simson (1687–1768), who wrote an English-language book based on Euclid's *Elements*, which is largely responsible for the development of the foundation of the high school geometry course taught in the United States, was the first to popularize the notion that the ratio

$$\frac{F_{n+1}}{F_n}$$

of two consecutive Fibonacci numbers will approach the value $\phi$ of the golden ratio. Yet it was Johannes Kepler (1571–1630) whom we credit with discovering that the reciprocal quotient

$$\frac{F_n}{F_{n+1}}$$

of two consecutive Fibonacci numbers approaches the reciprocal of the golden ratio $\frac{1}{\phi}$.

We can see this in the left column of the following table, where $F_n$ represents the $n$th Fibonacci number and $F_{n+1}$ the next, or ($n$+1)st, Fibonacci number.

**The Ratios of Consecutive Fibonacci Numbers**[11]

| $\frac{F_{n+1}}{F_n}$ | $\frac{F_n}{F_{n+1}}$ |
|---|---|
| $\frac{1}{1}=1.000000000$ | $\frac{1}{1}=1.000000000$ |
| $\frac{2}{1}=2.000000000$ | $\frac{1}{2}=0.500000000$ |
| $\frac{3}{2}=1.500000000$ | $\frac{2}{3}\approx 0.666666667$ |
| $\frac{5}{3}\approx 1.666666667$ | $\frac{3}{5}=0.600000000$ |
| $\frac{8}{5}=1.600000000$ | $\frac{5}{8}=0.625000000$ |
| $\frac{13}{8}=1.625000000$ | $\frac{8}{13}\approx 0.615384615$ |
| $\frac{21}{13}\approx 1.615384615$ | $\frac{13}{21}\approx 0.619047619$ |
| $\frac{34}{21}\approx 1.619047619$ | $\frac{21}{34}\approx 0.617647059$ |
| $\frac{55}{34}\approx 1.617647059$ | $\frac{34}{55}\approx 0.618181818$ |
| $\frac{89}{55}\approx 1.618181818$ | $\frac{55}{89}\approx 0.617977528$ |
| $\frac{144}{89}\approx 1.617977528$ | $\frac{89}{144}\approx 0.618055556$ |
| $\frac{233}{144}\approx 1.618055556$ | $\frac{144}{233}\approx 0.618025751$ |
| $\frac{377}{233}\approx 1.618025751$ | $\frac{233}{377}\approx 0.618037135$ |
| $\frac{610}{377}\approx 1.618037135$ | $\frac{377}{610}\approx 0.618032787$ |
| $\frac{987}{610}\approx 1.618032787$ | $\frac{610}{987}\approx 0.618034448$ |

By taking the reciprocals of each of the fractions on the left side, we get the column on the right side—also, as expected, approaching the value of $\frac{1}{\phi} \approx 0.618034$.[12] Once again we notice this most unusual relationship between $\phi$ and $\frac{1}{\phi}$, namely that $\phi = \frac{1}{\phi} + 1$—this time via the Fibonacci numbers.

## THE BINET FORMULA

Until now, we accessed the Fibonacci numbers as members of their sequence. If we wish to find a specific Fibonacci number without listing all of its predecessors, we have a general formula to do just that. In other words, if you would like to find the thirtieth Fibonacci number without writing the sequence of Fibonacci numbers up to the twenty-ninth member ($F_{29}$) (which is a procedure that is somewhat cumbersome), you would use the Binet formula. In 1843 the French mathematician Jacques-Philippe Marie Binet[13] (1786–1856) developed this formula, which allows us to find any Fibonacci number without actually listing the sequence as we would otherwise have to do.

The *Binet formula*[14] is as follows:

$$F_n = \frac{1}{\sqrt{5}}\left[\phi^n - \left(-\frac{1}{\phi}\right)^n\right],$$

or without using the $\phi$, we have

$$F_n = \frac{1}{\sqrt{5}}\left[\left(\frac{1+\sqrt{5}}{2}\right)^n - \left(\frac{1-\sqrt{5}}{2}\right)^n\right],$$

which will give us the Fibonacci number ($F_n$) for any natural number $n$ (a proof of this formula can be found in the appendix).

As is often the case in mathematics when a formula is named after a mathematician, controversies arise as to who was actually the first to discover it. Even today, when a mathematician comes up with what appears to be a new idea, others are usually hesitant to attribute the work to that person. They often say something like: "It looks original, but how do we know it wasn't done by someone else earlier?" Such is the case with the Binet formula. When he publicized his work, there were no challenges to Binet, but in the course of time, some claims have surfaced that Abraham de Moivre (1667–1754) was aware of it in 1718, Nicolaus Bernoulli (1687–1759) knew it in 1728, and his cousin Daniel Bernoulli (1700–1782)[15] also seems to have known the formula before Binet. Also, the prolific mathematician Leonhard Euler (1707–1783) is said to have known it in 1765. Nevertheless, it is still known today as the *Binet formula*.

Let's stop and marvel at this wonderful formula. For any natural number $n$, the irrational numbers in the form of $\sqrt{5}$ seem to disappear in the calculation, and a Fibonacci number appears. In other words, the Binet formula delivers the possibility of obtaining any Fibonacci number, and can also be expressed in terms of the golden ratio, $\phi$.

So, now we shall use this formula. Let's try using it to find a Fibonacci number, say, the 128th Fibonacci number. We would ordinarily have a hard time getting to this Fibonacci number—that is, by writing out the Fibonacci sequence with 128 terms until we arrive at it.

Applying the Binet formula, and using a calculator of course, for $n=128$ we get:

$$F_{128}=\frac{1}{\sqrt{5}}\left(\phi^{128}-\left(-\frac{1}{\phi}\right)^{128}\right)=\frac{1}{\sqrt{5}}\left(\left(\frac{1+\sqrt{5}}{2}\right)^{128}-\left(\frac{1-\sqrt{5}}{2}\right)^{128}\right)$$

$$=251{,}728{,}825{,}683{,}549{,}488{,}150{,}424{,}261.$$

As we claimed earlier, we can also express the Fibonacci numbers (in Binet form) exclusively in terms of the golden ratio, $\phi$, as

$$F_n=\frac{\phi^n-\left(-\frac{1}{\phi}\right)^n}{\phi+\frac{1}{\phi}},\text{ where } n\geq 1.$$

Familiarity with the Fibonacci numbers reminds us of their recursive definition: $F_{n+2}-F_{n+1}-F_n=0$, which comes from the original definition of the Fibonacci numbers: $F_{n+2}=F_n+F_{n+1}$, where $n\geq 1$ and $F_1=F_2=1$. Recall the Fibonacci number sequence:

| $n$ | 1 | 2 | 3 | 4 | 5 | 6 | 7 | 8 | 9 | 10 | 11 | 12 | 13 | 14 | 15 | 16 |
|---|---|---|---|---|---|---|---|---|---|---|---|---|---|---|---|---|
| $F_n$ | 1 | 1 | 2 | 3 | 5 | 8 | 13 | 21 | 34 | 55 | 89 | 144 | 233 | 377 | 610 | 987 |

Rather than beginning with 1 and 1, suppose we were to begin with 1 and 2. Then we would still generate a similar sequence, except we would be missing the first 1. Edouard Lucas[16] (1842–1891), the French mathematician who is largely responsible for bringing the Fibonacci numbers to light in recent years, suggested an analogous sequence; however, this time beginning with 1 and 3. That is, for the (now-called) *Lucas numbers*: $L_{n+2}=L_n+L_{n+1}$, when $n\geq 1$, and $L_1=1$ and $L_2=3$. The sequence looks like this:

| $n$ | 1 | 2 | 3 | 4 | 5 | 6 | 7 | 8 | 9 | 10 | 11 | 12 | 13 | 14 | 15 | 16 |
|---|---|---|---|---|---|---|---|---|---|---|---|---|---|---|---|---|
| $L_n$ | 1 | 3 | 4 | 7 | 11 | 18 | 29 | 47 | 76 | 123 | 199 | 322 | 521 | 843 | 1,364 | 2,207 |

Once again, our golden ratio comes into play in that we can also express the Lucas numbers in terms of the golden ratio:

$$L_n=\phi^n+\left(-\frac{1}{\phi}\right)^n, \text{ where } n\geq 1.$$

Let's admire the continued fraction development of $\frac{L_{n+1}}{L_n}$, and notice how it differs from that of $\frac{F_{n+1}}{F_n}$.

Only the last denominator is different. It is a 3 instead of a 1—this is also the difference in the beginning of the Lucas sequence of numbers: The second number is a 3 instead of a 1, as with the Fibonacci numbers.

For example, consider the following two examples:

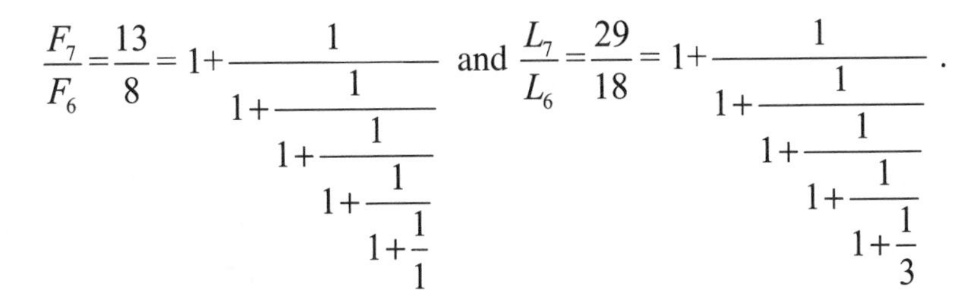

In general and in the shortened format we have the following:

$\frac{L_2}{L_1}=\frac{3}{1}=[3]=3$

$\frac{L_3}{L_2}=\frac{4}{3}=[1;\ 3]=1.\overline{3}$

$\frac{L_4}{L_3}=\frac{7}{4}=[1;\ 1,3]=1.75$

$\frac{L_5}{L_4}=\frac{11}{7}=[1;\ 1,1,3]=1.\overline{571428}$

$\frac{L_6}{L_5}=\frac{18}{11}=[1;\ 1,1,1,3]=1.\overline{63}$

$\frac{L_7}{L_6}=\frac{29}{18}=[1;\ 1,1,1,1,3]=1.6\overline{1}$

$\frac{L_8}{L_7}=\frac{47}{29}=[1;\ 1,1,1,1,1,3]=1.\overline{6206896551724137931034482758}$

$\frac{L_9}{L_8}=\frac{76}{47}=[1;\ 1,1,1,1,1,1,3]$
$=1.\overline{6170212765957446808510638297872340425531914893}$

$\frac{L_{10}}{L_9}=\frac{123}{76}=[1;\ 1,1,1,1,1,1,1,3]=1.61\overline{8421052631578947361}$, etc.

One might then ask if this can be extended to any starting pair of numbers. That is, were we to begin a Fibonacci-like sequence with other starting numbers, would we also be able to express the numbers in terms of $\phi$?

Suppose we choose the starting numbers of such a sequence to be $f_1=7$ and $f_2=13$ with the same recursive relationship as before, where $f_{n+2}=f_n+f_{n+1}$ (with $n\geq 1$). We would then have the following sequence, which does not have a particular name, as the Fibonacci or Lucas sequences do:

| $n$ | 1 | 2 | 3 | 4 | 5 | 6 | 7 | 8 | 9 | 10 | 11 | 12 | 13 | 14 | 15 |
|---|---|---|---|---|---|---|---|---|---|---|---|---|---|---|---|
| $f_n$ | 7 | 13 | 20 | 33 | 53 | 86 | 139 | 225 | 364 | 589 | 953 | 1,542 | 2,495 | 4,037 | 6,532 |

Yet the big surprise is that the ratio of consecutive members of the sequence will tend toward the golden ratio as the numbers increase—as was the case with the Fibonacci and the Lucas numbers before. In the chart below, notice how the ratio of $\frac{f_{n+1}}{f_n}$ approaches $\phi = 1.6180339887498948482\ldots$ as a limit. It is believed that the Fibonacci numbers provide the best approximation of $\phi$, though this is not easily seen from the chart.[17]

| $n$ | $\frac{F_{n+1}}{F_n}$ | $\frac{L_{n+1}}{L_n}$ | $\frac{f_{n+1}}{f_n}$ |
|---|---|---|---|
| 1 | 1 | 3 | 1.857142857 |
| 2 | 2 | 1.333333333 | 1.538461538 |
| 3 | 1.5 | 1.75 | 1.65 |
| 4 | 1.666666666 | 1.571428571 | 1.606060606 |
| 5 | 1.6 | 1.636363636 | 1.622641509 |
| 6 | 1.625 | 1.611111111 | 1.616279069 |
| 7 | 1.615384615 | 1.620689655 | 1.618705035 |
| 8 | 1.619047619 | 1.617021276 | 1.617777777 |
| 9 | 1.617647058 | 1.618421052 | 1.618131868 |
| 10 | 1.618181818 | 1.617886178 | 1.617996604 |
| 11 | 1.617977528 | 1.618090452 | 1.618048268 |
| 12 | 1.618055555 | 1.618012422 | 1.618028534 |
| 13 | 1.618025751 | 1.618042226 | 1.618036072 |
| 14 | 1.618037135 | 1.618030842 | 1.618033192 |
| 15 | 1.618032786 | 1.618035190 | 1.618034292 |
| 16 | 1.618034447 | 1.618033529 | 1.618033872 |
| 17 | 1.618033813 | 1.618034164 | 1.618034033 |
| 18 | 1.618034055 | 1.618033921 | 1.618033971 |
| 19 | 1.618033963 | 1.618034014 | 1.618033995 |
| 20 | **1.6180339**98 | **1.6180339**78 | **1.6180339**86 |
| … | | | |
| 100 | **1.618033988** | **1.618033988** | **1.618033988** |

We can see this better when we take the fifty-place approximation of the value of

$\phi = 1.61803398874989484820458683436563811772030917980 57\ldots$.

Now compare this value to the approximations below for $n = 100$:

$\frac{F_{n+1}}{F_n} = \mathbf{1.6180339887498948482045868343656381177203}127439637\ldots$

$\frac{L_{n+1}}{L_n} = \mathbf{1.6180339887498948482045868343656381177203}056156477\ldots$

$\frac{f_{n+1}}{f_n} = \mathbf{1.6180339887498948482045868343656381177203}082783971\ldots$

Curiously enough, if we take the ratio of the Lucas numbers to the Fibonacci numbers, $\frac{L_n}{F_n}$, it seems to approach $\sqrt{5} = 2.236067977\ldots$, as shown in the chart below.

| $n$ | $\frac{L_n}{F_n}$ |
|---|---|
| 1 | 1 |
| 2 | 3 |
| 3 | 2 |
| 4 | 2.333333333 |
| 5 | 2.2 |
| 6 | 2.25 |
| 7 | 2.230769230 |
| 8 | 2.238095238 |
| 9 | 2.235294117 |
| 10 | 2.236363636 |
| 11 | 2.235955056 |
| 12 | 2.236111111 |
| 13 | 2.236051502 |
| 14 | 2.236074270 |
| 15 | 2.236065573 |
| 16 | 2.236068895 |
| 17 | 2.236067626 |
| 18 | 2.236068111 |
| 19 | 2.236067926 |
| 20 | 2.236067997 |
| … | |
| 100 | **2.236067977** |

You may be impressed further by observing that if we take the ratio of alternating Fibonacci numbers, the limit as the numbers increase will approach the value $\phi+1$. Another way of saying this is that by taking increasing Fibonacci numbers for $\frac{F_{n+2}}{F_n}$, we gradually approach the value of $\phi+1$ as shown in the chart below:

| $n$ | $\frac{F_{n+2}}{F_n}$ | Approximation of $\frac{F_{n+2}}{F_n}$ | $\sqrt{\frac{F_{n+2}}{F_n}}$ | Approximation of $\sqrt{\frac{F_{n+2}}{F_n}}$ |
|---|---|---|---|---|
| 1 | $\frac{F_3}{F_1}=\frac{2}{1}$ | 2 | $\sqrt{\frac{F_3}{F_1}}=\sqrt{\frac{2}{1}}$ | 1.414213562 |
| 2 | $\frac{F_4}{F_2}=\frac{3}{1}$ | 3 | $\sqrt{\frac{F_4}{F_2}}=\sqrt{\frac{3}{1}}$ | 1.732050807 |
| 3 | $\frac{F_5}{F_3}=\frac{5}{2}$ | 2.5 | $\sqrt{\frac{F_5}{F_3}}=\sqrt{\frac{5}{2}}$ | 1.581138830 |
| 4 | $\frac{F_6}{F_4}=\frac{8}{3}$ | 2.666666666 | $\sqrt{\frac{F_6}{F_4}}=\sqrt{\frac{8}{3}}$ | 1.632993161 |
| 5 | $\frac{F_7}{F_5}=\frac{13}{5}$ | 2.6 | $\sqrt{\frac{F_7}{F_5}}=\sqrt{\frac{13}{5}}$ | 1.612451549 |
| 6 | $\frac{F_8}{F_6}=\frac{21}{8}$ | 2.625 | $\sqrt{\frac{F_8}{F_6}}=\sqrt{\frac{21}{8}}$ | 1.620185174 |
| 7 | $\frac{F_9}{F_7}=\frac{34}{13}$ | 2.615384615 | $\sqrt{\frac{F_9}{F_7}}=\sqrt{\frac{34}{13}}$ | 1.617215080 |
| 8 | $\frac{F_{10}}{F_8}=\frac{55}{21}$ | 2.619047619 | $\sqrt{\frac{F_{10}}{F_8}}=\sqrt{\frac{55}{21}}$ | 1.618347187 |
| 9 | $\frac{F_{11}}{F_9}=\frac{89}{34}$ | 2.617647058 | $\sqrt{\frac{F_{11}}{F_9}}=\sqrt{\frac{89}{34}}$ | 1.617914416 |
| 10 | $\frac{F_{12}}{F_{10}}=\frac{144}{55}$ | 2.618181818 | $\sqrt{\frac{F_{12}}{F_{10}}}=\sqrt{\frac{144}{55}}$ | 1.618079669 |
| … | | | | |
| 100 | $\frac{F_{102}}{F_{100}}$ | 2.618033988 | $\sqrt{\frac{F_{102}}{F_{100}}}$ | 1.618033988 |

Yet, if we consider the series

$$\sqrt{\frac{F_{n+2}}{F_n}},$$

it approaches the golden ratio, $\phi$, as compared to the value reached by the series $\frac{F_{n+2}}{F_n}$, which tends toward $\phi+1$. If you consider that we already established that $\phi+1=\phi^2$, the relationship above should not be completely unexpected.[18]

One more little tidbit relating the Fibonacci numbers to the golden ratio can be seen by taking the series of reciprocals of Fibonacci numbers in the position of powers of 2.

$$\frac{1}{F_1}+\frac{1}{F_2}+\frac{1}{F_4}+\frac{1}{F_8}+\frac{1}{F_{16}}+\ldots+\frac{1}{F_{2^k}}+\ldots=4-\phi\approx 2.3819660112501051517,$$

or written another way,

$$\sum_{k=0}^{\infty}\frac{1}{F_{2^k}}=4-\phi.$$

At the point at which $k=6$, we get a rather good approximation[19]:

$$\frac{1}{1}+\frac{1}{1}+\frac{1}{3}+\frac{1}{21}+\frac{1}{987}+\frac{1}{2{,}178{,}309}+\frac{1}{10{,}610{,}209{,}857{,}723}$$
$$\approx \mathbf{2.3819660112501051517954131}6166,$$

which you can appreciate when comparing it to the value $4-\phi \approx \mathbf{2.3819660112501051517954131}6563$.

There are many numerical expressions that characterize the golden ratio, but none as simply as the unique relationship between the golden ratio and its reciprocal: $\phi\cdot\frac{1}{\phi}=1$ and $\phi-\frac{1}{\phi}=1$. For no other number is this true!

If we look at $\phi$ with the above relationships in mind, there are a number of variations that can result. For example, which positive number is 1 greater than its reciprocal? Yes, by now you probably guessed it: $\phi$.

The question yields the following equation: $x = \frac{1}{x} + 1$, which can then be written as $x^2 = 1 + x$, or $x^2 - x - 1 = 0$. This has its positive root:

$$x = \frac{1+\sqrt{5}}{2} = \phi.$$

Yes, the golden ratio!

One might also say that $\phi$ is the only number that is 1 less than its square. That is, $x = x^2 - 1$, which leads us back to the previous equation, $x^2 - x - 1 = 0$, which when substituting for $x$ gives us $\phi = \phi^2 - 1$. This is a relationship that we saw earlier as $\phi^2 = \phi + 1$, when we then used it to generate the powers of $\phi$.

As we further investigate the representation of the golden ratio, we can ask: What is the solution to each of the following equations?

$$\begin{aligned} x^2 - x - 1 &= 0 \\ x^3 - x^2 - x &= 0 \\ x^4 - x^3 - x^2 &= 0 \\ x^5 - x^4 - x^3 &= 0 \\ &\ldots \\ x^{n+2} - x^{n+1} - x^n &= 0 \end{aligned}$$

If we divide each of these equations by $x^n$, where $n$ is the power of the third term of the equation, we get the following equation: $x^2 - x - 1 = 0$, a solution with which we are by now quite familiar; namely $\phi$, and $-\frac{1}{\phi}$, hence the golden ratio!

Furthermore, when we consider the equation in the form of $ax^2 + bx + c = 0$, where $a$, $b$, and $c$ take on values 1 and –1, we get the golden ratio, $\phi$, in a number of ways, as long as the roots are not complex numbers.[20]

The chart below shows the various solutions to these equations.

| *a* | *b* | *c* | The Roots | The Roots Elaborated |
|---|---|---|---|---|
| 1 | 1 | 1 | $-\frac{1}{2} \pm \frac{\sqrt{3}}{2} \cdot i$ | Complex roots |
| 1 | 1 | –1 | $-\frac{1}{2} \pm \frac{\sqrt{5}}{2}$ | $-\phi = -\frac{\sqrt{5}+1}{2}; \frac{1}{\phi} = \frac{\sqrt{5}-1}{2}$ |
| 1 | –1 | 1 | $\frac{1}{2} \pm \frac{\sqrt{3}}{2} \cdot i$ | Complex roots |
| 1 | –1 | –1 | $\frac{1}{2} \pm \frac{\sqrt{5}}{2}$ | $\phi = \frac{\sqrt{5}+1}{2}; -\frac{1}{\phi} = -\frac{\sqrt{5}-1}{2}$ |
| –1 | 1 | 1 | $\frac{1}{2} \pm \frac{\sqrt{5}}{2}$ | $\phi = \frac{\sqrt{5}+1}{2}; -\frac{1}{\phi} = -\frac{\sqrt{5}-1}{2}$ |
| –1 | 1 | –1 | $\frac{1}{2} \pm \frac{\sqrt{3}}{2} \cdot i$ | Complex roots |
| –1 | –1 | 1 | $-\frac{1}{2} \pm \frac{\sqrt{5}}{2}$ | $-\phi = -\frac{\sqrt{5}+1}{2}; \frac{1}{\phi} = \frac{\sqrt{5}-1}{2}$ |
| –1 | –1 | –1 | $-\frac{1}{2} \pm \frac{\sqrt{3}}{2} \cdot i$ | Complex roots |

As we search for ways to express the value of $\phi$, we cannot neglect the value of $\pi$. We can take an approximation to make this comparison. We can show that the circumference of a circle with radius of length $\sqrt{\phi}$ is approximately equal to the perimeter of a square with side length 2. That is, the circumference of this circle is $2\pi\sqrt{\phi} \approx 7.992 \approx 8$, and the perimeter of the square is $4 \cdot 2 = 8$.

This can give us an approximate value for $\pi$ in terms of $\phi$; since we have $\pi\sqrt{\phi} \approx 4$, this then can be written as $\sqrt{\phi} \approx \frac{4}{\pi}$, or $\phi \approx \frac{16}{\pi^2} \approx 1.621$, a very close approximation of $\phi \approx 1.618$.

Irrational numbers can only be *approximated* by fractions in the decimal system. For example, Archimedes (287–212 BCE) found an approximation for the irrational number $\pi = 3.141592653\ldots$, namely $\frac{223}{71} < \pi < \frac{22}{7}$.

Inspecting these two limiting fractions, we find that $\frac{223}{71}$ has a period of length 35 (i.e., after which the decimal begins to repeat itself) as shown here: 3.**14084507042253521126760563380281690**1408450… . And the fraction $\frac{22}{7}$ has a period of length 6, as: 3.**142857**142857 142… . Yet we notice that both fractions establish the value of $\pi$ to two-decimal-place accuracy. On the other hand, the fraction $\frac{355}{113}$ approximates the value of $\pi$ to an accuracy of six decimal places as: $\frac{355}{113} =$ **3.141592**920….

For the golden ratio, $\phi = 1.618033988\ldots$, we have, for example, the following approximation (using Fibonacci numbers) correct to five decimal places: $\frac{987}{610} =$ 1.**61803**2786… .

That should not come as a surprise since we already saw that the quotient of consecutive Fibonacci numbers yields ever closer approximations of the golden ratio, $\phi$.

The best approximations of $\phi$, when both numerator and denominator have the same number of digits, is achieved with the Fibonacci numbers, as seen, for example, with single-digit fractions as $\frac{8}{5} =$ 1.**6** and with double-digit fractions as $\frac{89}{55} =$ 1.**618**181818… .

From our study of continued fractions, we saw that $\phi$ and $\frac{1}{\phi}$ can be represented by the simplest of all continued fractions, since they consisted of all 1s. The sad news here is that despite their "simplicity," they require one of the largest numbers of fractions to reach their convergence at infinity. We might then say that the golden ratio and its reciprocal are the most irrational numbers, because they require the most fractions to reach their best approximation. We will see later, however, that despite this rather sad assessment, the golden ratio will show its beauty in art and nature well above all other numbers.

There are many other numerical representations of $\phi$. Some of these involve trigonometric functions. We'll show a few of these here. You may wish to verify their (correct) values.

$$\phi = 2\sin\frac{3\pi}{10} = 2\sin 54^\circ, \qquad \phi = 2\cos\frac{\pi}{5} = 2\cos 36^\circ,$$

$$\phi = \frac{3-\tan^2\frac{\pi}{5}}{1+\tan^2\frac{\pi}{5}} = \frac{3-\tan^2 36^\circ}{1+\tan^2 36^\circ}, \qquad \frac{1}{\phi} = 2\cos\frac{2\pi}{5} = 2\cos 72^\circ,$$

$$\frac{1}{\phi} = 2\sin\frac{\pi}{10} = 2\sin 18^\circ, \qquad \phi = \frac{1}{2\cos\frac{2\pi}{5}} = \frac{1}{2\cos 72^\circ},$$

and $\phi = \dfrac{1}{2\sin\frac{\pi}{10}} = \dfrac{1}{2\sin 18^\circ}$.

See the appendix for more trigonometric relationships that result in the golden ratio.

While on the topic of the trigonometric functions, it might be interesting and noteworthy to see that through trigonometry we can connect the value of $\pi$ to the golden ratio as in the following, where we can express the value of $\pi$ in terms of $\phi$.

$$\pi = 2(\arctan\frac{1}{\phi^5} + \arctan\phi^5), \text{ or}$$

$$\pi = 6\arctan\frac{1}{\phi} - 2\arctan\frac{1}{\phi^5}.$$

The justification for these representations can be found in the appendix.

As you can see, the golden ratio, $\phi$, can be represented in a number of ways. In chapter 5, we will provide some surprising appearances of this apparently ubiquitous number.

# Chapter 4
# Golden Geometric Figures

The term *golden*, as noted earlier, has come, in large measure, from the belief that the rectangle with dimensions in the golden ratio is the most pleasing of all rectangles to the eye, and therefore is called the *golden rectangle*.[1] It is then only fitting that we now embark on a careful inspection of the golden rectangle and other geometric figures that possess the golden ratio among their dimensions.

## THE GOLDEN RECTANGLE

We recall that the golden rectangle has dimensions $L$ (length) and $W$ (width) such that the golden ratio appears as $\frac{L}{W}=\frac{W+L}{L}$, which we conveniently call $\phi$. We can consider the golden rectangle $ABCD$ (fig. 4-1), where the length $L=a+b$ and the width $W=a$ and the golden ratio $\frac{L}{W}=\frac{a+b}{a}=\phi$. It then follows that $\frac{a}{b}=\phi$. (As a refresher, see pp. 17–18.)

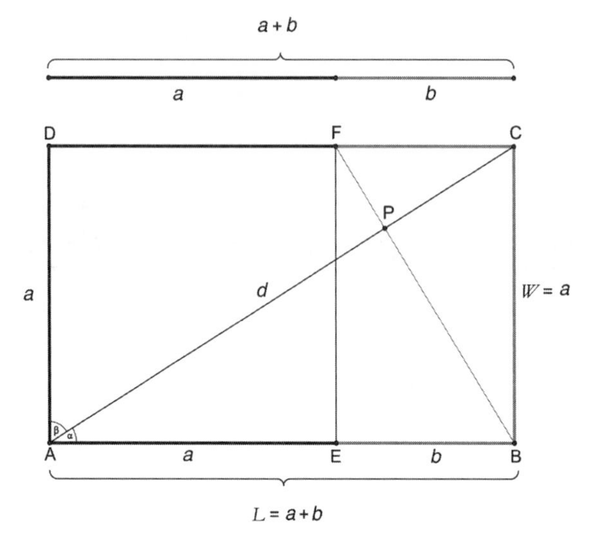


Figure 4-1

In figure 4-1, if we cut off a square (*AEFD*) from the golden rectangle *ABCD*, we remain with rectangle *EBCF*, which is itself a golden rectangle, since its dimensions are in the ratio $\frac{L}{W}=\frac{a}{b}=\phi$.

The point *E* on a segment *AB* is sometimes called the *goldpoint*, or *golden point*, of *AB*, if $\frac{AE}{EB}$ is equal to the golden ratio, or $\phi$.

Further describing the golden rectangle, we can use simple trigonometry to determine the measure of the angles in which the diagonals partition the rectangle's right angles: In figure 4-1, the angle marked $\alpha=\angle BAC$ can be found from $\tan\alpha=\frac{BC}{AB}=\frac{a}{a+b}$, which then gives us the measure of $\alpha\approx 31.71747441°$. Rounding off to $\alpha\approx 31.72°$, we can get its complementary angle, $\beta=\angle CAD\approx 58.28°$.

Now to find the length of the diagonal *AC* of the golden rectangle, the Pythagorean theorem applied to triangle *ACD* gives us

$$d^2=(a+b)^2+a^2=2a^2+2ab+b^2$$

$$\frac{d^2}{a^2}=2+2\cdot\frac{b}{a}+\frac{b^2}{a^2}=2+2\cdot\frac{1}{\phi}+\frac{1}{\phi^2}=2+2(\phi-1)+(\phi-1)^2=\phi^2+1$$

$$\frac{d^2}{(a+b)^2}=\frac{(a+b)^2+a^2}{(a+b)^2}=1+\frac{a^2}{a^2+b^2}=1+\frac{1}{\phi^2}$$

To get a "feel" for the relationship between the sides and the diagonal of a golden rectangle, we find that $\frac{d}{a}=\sqrt{\frac{5+\sqrt{5}}{2}}=\sqrt{\phi^2+1}$, and $\frac{d}{a+b}=\frac{d}{a\phi}=\frac{d}{a}\cdot\frac{1}{\phi}=\sqrt{\phi^2+1}\cdot\frac{1}{\phi}=\frac{\sqrt{\phi^2+1}}{\phi}=\sqrt{\frac{5-\sqrt{5}}{2}}$.

Combining these, we obtain the ratio of the width, length, and diagonal of a golden rectangle as $d : (a+b) : a=\sqrt{\phi^2+1} : \phi : 1$. Notice how the golden ratio appears to be omnipresent throughout the golden rectangle!

Our fascination with this ratio continues; for when we inspect the area of the golden rectangle *ABCD*, we find surprisingly that once again it reveals the golden ratio. This can be seen by considering its relationship to the area of square *AEFD* (with side length *a*): that is,

$$\frac{A_{ABCD}}{A_{AEFD}}=\frac{a(a+b)}{a^2}=\frac{a+b}{a}=\phi.$$

Interestingly, the ratio of the area of square *AEFD* (with side length *a*) to the area of the golden rectangle *EBCF* (with side lengths *a* and *b*) also exposes the golden ratio:

$$\frac{A_{AEFD}}{A_{EBCF}}=\frac{a^2}{ab}=\frac{a}{b}=\phi.$$

Now comparing the areas of the two golden rectangles, *ABCD* and *EBCF* (with sides of length *a* and *b*), we get

$$\frac{A_{ABCD}}{A_{EBCF}}=\frac{a(a+b)}{ab}=\frac{a+b}{b}=\frac{a+b}{\frac{a}{\phi}}=\frac{a+b}{a}\cdot\phi=\phi^2.$$

Again the golden ratio appears—albeit this time squared.

Since all golden rectangles (fig. 4-1) have the same shape, golden rectangle *ABCD* is similar to golden rectangle *EBCF*. This implies that $\Delta BCF \sim \Delta ABC$. Therefore, $\angle CFB$ is congruent to $\angle BCA$ ($\beta$). And $\angle BCA$ is complementary to $\angle FCA$ ($\alpha$). Therefore, $\angle CFB$ is complementary to $\angle FCA$. Thus, $\angle FPC$ must be a right angle, or $AC \perp BF$. We will be using this relationship of perpendicularity between the diagonals of consecutive golden rectangles. In general, if the width of one rectangle is the length of the other and the rectangles are similar, then the rectangles are said to be *reciprocal rectangles*. In this case, the ratio between the corresponding sides of the similar rectangles—called the ratio of similitude—is $\phi$. We also can show that reciprocal rectangles have perpendicular corresponding diagonals, as we have shown for these special reciprocal rectangles—namely the golden rectangles.

In figure 4-2, we see that rectangle *ABCD* and rectangle *EBCF* are reciprocal rectangles. Furthermore, we have just established that reciprocal rectangles have corresponding diagonals that are perpendicular.

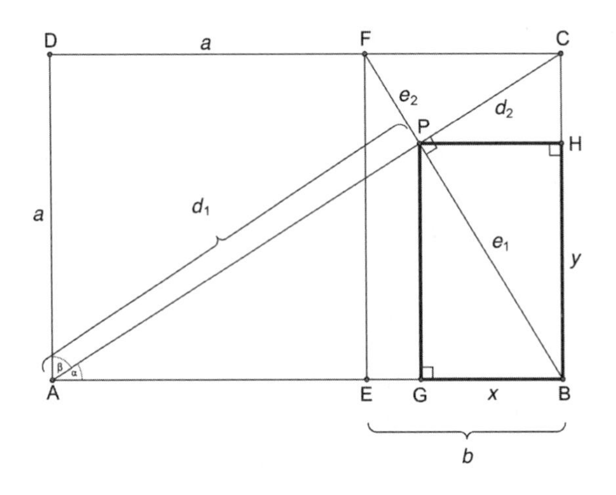


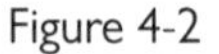
Figure 4-2

As you will see, there are practically endless manifestations of the golden ratio to be found on the golden rectangle. Take, for instance, the rectangle formed by dropping perpendiculars from the intersection *P* of the two diagonals *AC* and *BF* to the two sides *AB* and *BC* as shown in figure 4-2. By using the lengths shown in figure 4-2, we get the following:

$$\frac{a+b}{a}=\frac{y}{x}=\frac{a-y}{b-x}=\phi=\frac{\sqrt{5}+1}{2}.$$

We can take this a step further with another comparison to get

$$\frac{d_1}{d_2} = \frac{e_1}{e_2} = \phi^2 = \phi + 1 = \frac{\sqrt{5}+3}{2}.$$

This shows us that the diagonals intersect each other in a ratio involving the golden ratio. (The proof of this can be found in the appendix.)

Let us once again consider a square with side of length $a$, with a rectangle whose length is also $a$ and whose width is $b$ and which is placed as in figure 4-3 with the diagonals of the two rectangles intersecting as shown. We will now determine that the ratio of the two diagonals is—as you can guess by now—once again the golden ratio.

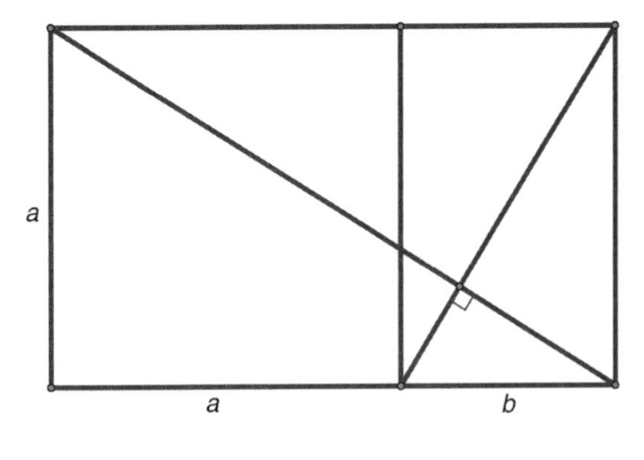


Figure 4-3

We shall now consider the two diagonals (fig. 4-4) along with another line segment, $HKJ \perp FJE$. Recall also that the diagonals $DE$ and $BF$ are perpendicular at point $G$.

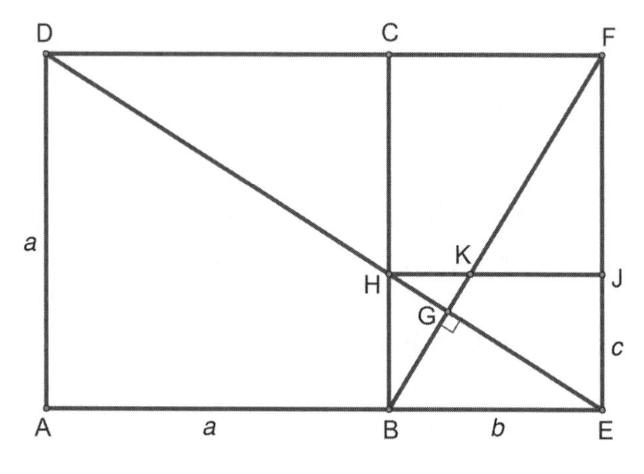


Figure 4-4

We now get lots of similar right triangles: $\Delta ADE \sim \Delta BCF \sim \Delta BEF \sim \Delta BEG \sim \Delta BEH \sim \Delta BGH \sim \Delta CDH \sim \Delta DEF \sim \Delta DFG \sim \Delta EFG \sim \Delta EHJ \sim \Delta FJK \sim \Delta GHK$.

The first two ($\Delta ADE \sim \Delta BCF$) give us the following: $\frac{AD}{AE} = \frac{CF}{BC}$, which can also be expressed as $\frac{a}{a+b} = \frac{b}{a}$, which then yields:

$$a^2 = b(a+b) = ab + b^2.$$

If we divide each term of the previous equation by $b^2$, we get $\frac{a^2}{b^2} = \frac{ab}{b^2} + \frac{b^2}{b^2}$, and when we let $x = \frac{a}{b}$, we get the equation of the golden section with which we are already familiar, $x^2 - x - 1 = 0$.

Because $x$ is positive, we conclude that

$$x = \frac{a}{b} = \frac{\sqrt{5}+1}{2} = \phi.$$

This also can be written as $b = \frac{a}{\phi}$. Now to consider our diagonals, $DE$ and $BF$. From $\Delta ADE \sim \Delta BEF$, it follows that $\frac{DE}{BF} = \frac{AE}{EF}$, whereupon $\frac{DE}{BF} = \frac{a+b}{a} = \frac{a}{b} = \phi$. The diagonals also form the golden ratio.

There are still lots of manifestations of the golden ratio. Using figure 4-4 you might want to justify the following:

$$\frac{AE}{AB}=\frac{a+b}{a}=\phi,\ \frac{AB}{BE}=\frac{a}{b}=\phi;$$

$$\frac{CH}{BH}=\frac{a-c}{c}=\phi,\text{ then } c=BH=\frac{a}{\phi+1}=\frac{a}{\phi^2}=\frac{b}{\phi};$$

$$\frac{BF}{EH}=\frac{BC}{BE}=\frac{a}{b}=\phi;$$

$$\frac{EH}{BK}=\frac{HJ}{BH}=\frac{b}{c}=\frac{\frac{a}{\phi}}{\frac{a}{\phi^2}}=\phi.$$

Segments along the diagonals also fit the golden ratio:

$$\frac{DH}{FK}=\frac{CD}{FJ}=\frac{a}{a-c}=\frac{a}{a-\frac{a}{\phi^2}}=\frac{1}{1-\frac{1}{\phi^2}}=\frac{\phi^2}{\phi^2-1}=\frac{\phi^2}{\phi+1-1}=\phi;$$

$$\frac{DH}{EH}=\frac{CD}{BE}=\frac{a}{b}=\phi;$$

$$\frac{EG}{BG}=\frac{BE}{BH}=\frac{b}{c}=\frac{\frac{a}{\phi}}{\frac{a}{\phi^2}}=\phi;$$

$$\frac{GH}{GK}=\frac{BH}{HK}=\frac{AE}{AD}=\frac{a+b}{a}=\frac{a}{b}=\phi.$$

This should convince you that the golden ratio can be found throughout the golden rectangle and its related parts.

Let's now take another interesting situation for the golden rectangle. Consider one that is inscribed in a square so that the golden rectangle's vertices on the sides of the square divide the sides into the golden ratio, as shown in figure 4-5. That is, the points $P$, $Q$, $R$, and $S$ divide the sides $AB$, $BC$, $CD$, and $AD$ into the golden ratio, respectively.

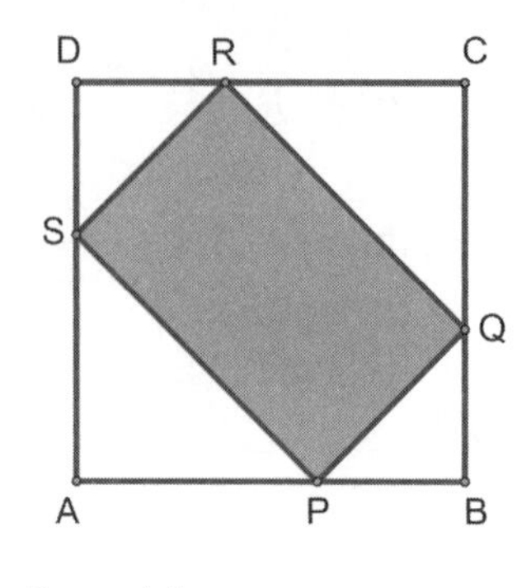


Figure 4-5

In figure 4-6, we can establish that the line segments *PS*, *BD*, and *QR* are parallel, since from the golden ratio we have the following proportions: $\frac{AP}{BP}=\frac{AS}{DS}(=\phi)$ and $\frac{CQ}{BQ}=\frac{CR}{DR}(=\phi)$. It then, similarly, follows that since $\frac{AP}{BP}=\frac{CQ}{BQ}(=\phi)$ and $\frac{AS}{DS}=\frac{CR}{DR}(=\phi)$, $PQ \parallel AC \parallel RS$.

This allows us to establish that the quadrilateral *PQRS* is a parallelogram. Furthermore, this parallelogram is a rectangle, since the sides are respectively parallel to the perpendicular diagonals of square *ABCD*. Now it remains for us to show that this rectangle (*PQRS*) is a golden rectangle.

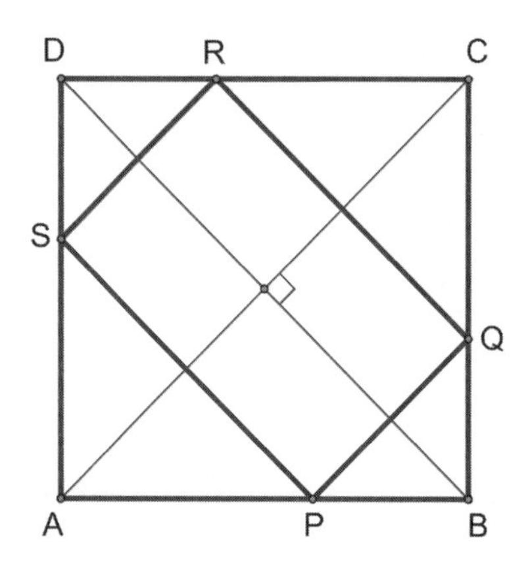


Figure 4-6

From the golden ratios $\frac{AP}{BP}=\frac{AS}{DS}=\frac{CQ}{BQ}=\frac{CR}{DR}=\phi$, we can conclude that the triangles *APS*, *BPQ*, *CQR*, and *DRS* are similar to each other. From

$\Delta APS \sim \Delta BPQ$ and $\frac{AP}{BP} = \frac{AS}{BQ} = \phi$ and also that $\frac{PS}{PQ} = \phi$, we thus establish that rectangle *PQRS* is a golden rectangle.

We shall now consider a special type of right triangle—one where the product of the length of a leg and the hypotenuse equals the square of the other leg. In figure 4-7, we have a right triangle with sides *a*, *b*, and *c*, where $a \leq b \leq c$ and $ac = b^2$. Which relationship, might you guess, exists among the sides and the areas of the quadrilaterals shown in figure 4-7? Read on, and you will see that your assumption is probably correct.

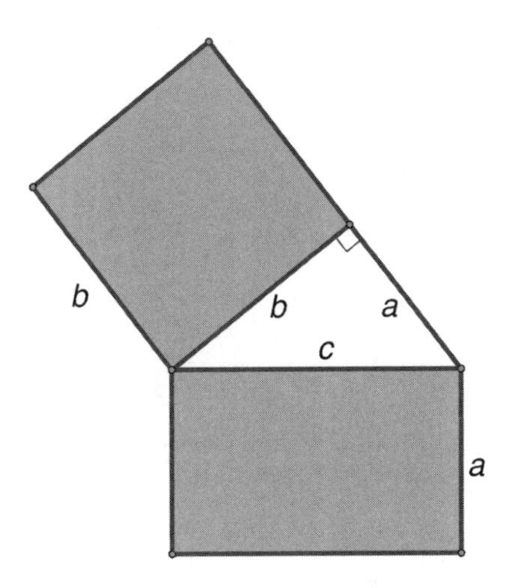


Figure 4-7

Applying the Pythagorean theorem to the right triangle, we get $a^2 + b^2 = c^2$, and then using the given relationship, $ac = b^2$, we have the following result: $a^2 + ac = c^2$, or put another way, $a^2 + ac - c^2 = 0$. Dividing both sides of the equation by $c^2$ and substituting $x = \frac{a}{c}$, we get our now-familiar equation exhibiting the golden ratio: $x^2 + x - 1 = 0$, whose solution is $x = \frac{1}{\phi} = \frac{a}{c}$, or $c = a \cdot \phi$ (the negative root $-\phi$ is disregarded). In other words, this tells us that the ratio of the hypotenuse to the shorter leg is the golden ratio! (Incidentally, a right triangle whose hypotenuse and shorter leg are in the golden ratio is called a *Kepler triangle*, after the famous mathematician Johannes Kepler [1571–1630].)

We can also compare the hypotenuse to the longer leg. From

$ac = \frac{c}{\phi} \cdot c = b^2$, we have $c^2 = \phi \cdot b^2$, which then gives us $c = \sqrt{\phi} \cdot b$. This tells us that the ratio of the hypotenuse, $c$, to the longer leg, $b$, is $\sqrt{\phi} : 1$. This now allows us to establish the ratio of the legs, in this case

$$\frac{b}{a} = \frac{\frac{c}{\sqrt{\phi}}}{\frac{c}{\phi}} = \frac{\phi}{\sqrt{\phi}} = \sqrt{\phi}.$$

This allows us to easily compare the areas of the squares on the legs of this right triangle, since the ratio of the areas of two similar figures (here they are both squares) is the square of the ratio of corresponding sides. Thus,

$$c = \phi \cdot a \quad \text{gives us} \quad \frac{c^2}{a^2} = \phi^2,$$

$$c = \sqrt{\phi} \cdot b \quad \text{gives us} \quad \frac{c^2}{b^2} = \phi,$$

$$\frac{b}{a} = \sqrt{\phi} \quad \text{gives us} \quad \frac{b^2}{a^2} = \phi.$$

Furthermore, in figure 4-8 we have a golden rectangle (with side lengths $a$ and $c$) along the hypotenuse of a right triangle. We can summarize the various manifestations of the golden ratio in this configuration in the following way: $c : b : a = \phi : \sqrt{\phi} : 1$, or represented another way: $c^2 : b^2 : a^2 = \phi^2 : \phi : 1$.

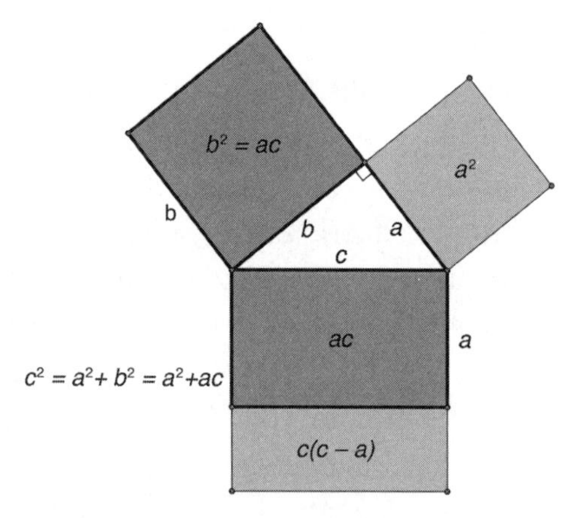


Figure 4-8

Let us now see if we can set up a method for constructing line segments of lengths $\phi, \phi^2, \phi^3, \phi^4, \ldots, \phi^n$, given a segment of length 1 unit. We begin by setting up an iterative process—using straightedges and compasses. Applying what we have previously established, that the golden ratio, $\phi$, satisfies $\phi^2 = \phi + 1$, we can begin the process.[2]

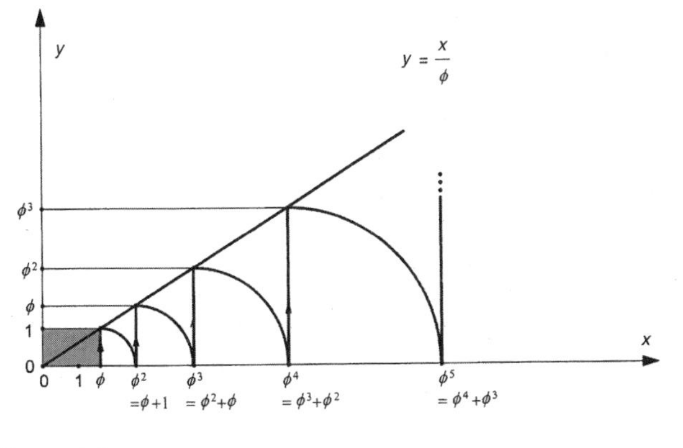


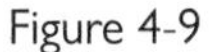
Figure 4-9

We begin with a golden rectangle whose vertices have coordinates (0, 0), ($\phi$, 0), ($\phi$, 1), and (0, 1). The rectangle's diagonal determines the line, whose equation is $y = f(x) = \frac{x}{\phi}$, and which we then draw through the point ($\phi$, 1).

The circle with center at point ($\phi$, 0) and with radius $y = 1$ cuts the $x$-axis at the point ($\phi + 1$, 0), since $\phi^2 = \phi + 1$, this point is ($\phi^2$, 0).

We continue this process. The circle with center at ($\phi^2$, 0) and with radius $f(\phi^2) = \phi$ cuts the $x$-axis at point ($\phi^2 + \phi$, 0). However, since $\phi^2 + \phi = \phi(\phi + 1) = \phi\phi^2 = \phi^3$, we arrive at the point ($\phi^3$, 0). In general, the circles with center ($\phi^n$, 0) and radius $f(\phi^n) = \phi^{n-1}$ will produce the point ($\phi^{n+1}$, 0) on the $x$-axis.

Recall that in chapter 3 (p. 58) we developed the following sequence of powers of $\phi$:

$$\begin{aligned}
\phi &= \phi\phi^0 = 1\phi + 0 & \phi^9 &= \phi\phi^8 = 34\phi + 21\\
\phi^2 &= \phi\phi^1 = 1\phi + 1 & \phi^{10} &= \phi\phi^9 = 55\phi + 34\\
\phi^3 &= \phi\phi^2 = 2\phi + 1 & \phi^{11} &= \phi\phi^{10} = 89\phi + 55\\
\phi^4 &= \phi\phi^3 = 3\phi + 2 & \phi^{12} &= \phi\phi^{11} = 144\phi + 89\\
\phi^5 &= \phi\phi^4 = 5\phi + 3 & \phi^{13} &= \phi\phi^{12} = 233\phi + 144\\
\phi^6 &= \phi\phi^5 = 8\phi + 5 & \phi^{14} &= \phi\phi^{13} = 377\phi + 233\\
\phi^7 &= \phi\phi^6 = 13\phi + 8 & \phi^{15} &= \phi\phi^{14} = 610\phi + 377\\
\phi^8 &= \phi\phi^7 = 21\phi + 13 & & \ldots.
\end{aligned}$$

This sequence of numbers, $1, \phi, \phi^2, \phi^3, \phi^4, \phi^5, \phi^6, \ldots$, for which we now have a method of getting the geometric representation, is sometimes referred to as the *golden sequence*.

While we are discussing the diagonals of the golden rectangle, we can expect the golden ratio to appear along the diagonals as well. Of course, we know that the sides of a golden rectangle are in the golden ratio. This will allow us to find the point along the diagonal that cuts it into a ratio related to the golden ratio (see point $H$ in figure 4-4). It is only because of the unique properties of this special rectangle that we can do this so easily.

Consider the golden rectangle $ABCD$, whose sides $AB=a$ and $BC=b$, so that $\frac{a}{b}=\phi$. As shown in figure 4-10, two semicircles are drawn on the sides $AB$ and $BC$ to intersect at $S$. If we now draw segments $SA$, $SB$, and $SC$, we find that $\angle ASB$ and $\angle BSC$ are right angles (since they are each inscribed in a semicircle). Therefore, $AC$ is a straight line, namely the diagonal of the rectangle. We can now show, rather elegantly, that point $S$ divides the diagonal in the golden section.

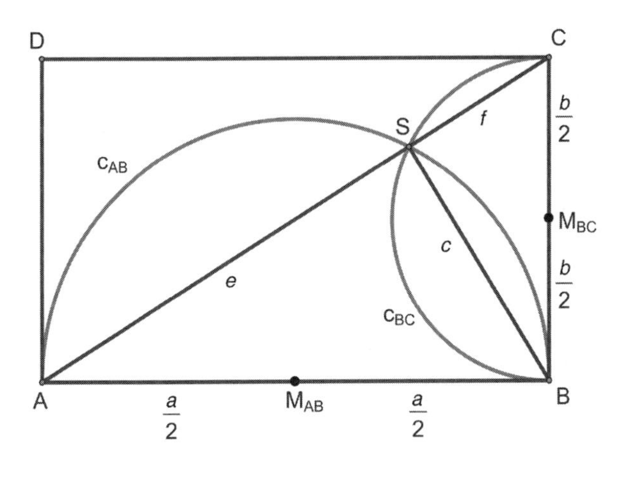


Figure 4-10

From the mean proportional ratios (obtained from similar triangles *ABC*, *ASB*, and *BSC*) we get the following:

We use the segment length markings in figure 4-10; for each of the right triangles the mean proportional relationship gives us

$$\Delta ASB: a^2 = e(e+f)$$
$$\Delta BSC: b^2 = f(e+f).$$

Therefore, $\frac{a^2}{b^2} = \frac{e}{f}$.

But, since $\frac{a}{b} = \phi$, then $\frac{e}{f} = \frac{a^2}{b^2} = \phi^2 = \phi + 1 = \frac{\sqrt{5}+3}{2}$.

Thus the point *S* divides the diagonal of the golden rectangle in a ratio involving the golden section in the following way:

$$e : f = \phi^2 : 1,$$
$$\text{or } e : f = (\phi+1) : 1.$$

Let's exploit the diagonal of a golden rectangle just a bit more. Suppose we draw perpendiculars to a diagonal of a golden rectangle as shown in figure 4-11.

(Remember that when $AD=1$, $AB=\phi=\frac{\sqrt{5}+1}{2}$.)

Surprisingly, we can show that $AE=EF=FC$, or $x=y=z$.

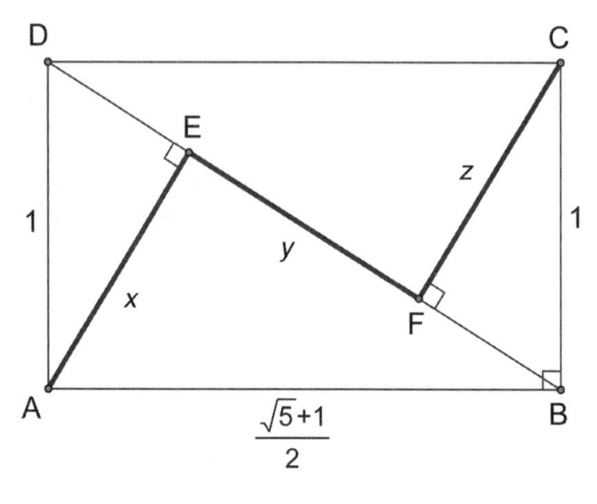


Figure 4-11

Let's consider an auxiliary line segment, the other diagonal $AC$, to help us establish this interesting equality. (See fig. 4-12.)

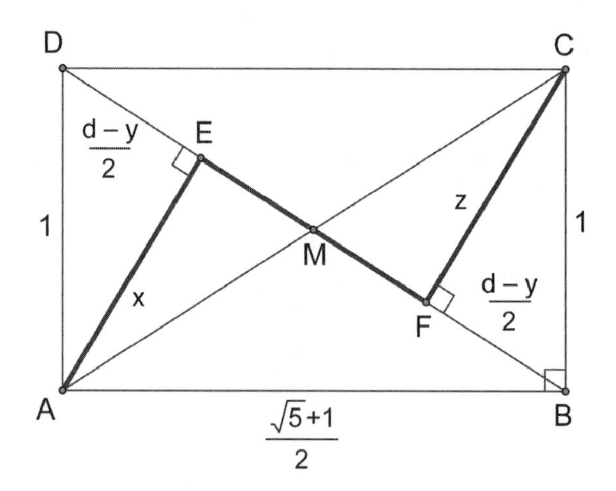


Figure 4-12

For expediency we will accept some relationships on the basis of symmetry: $AE = CF$, or $x = z$. Applying the Pythagorean theorem to triangle $ABC$, we get

$$AB^2 + BC^2 = AC^2, \text{ or } AC^2 = \left(\frac{\sqrt{5}+1}{2}\right)^2 + 1^2 = \frac{5+\sqrt{5}}{2}.$$

Therefore, $d = AC = BD = \sqrt{\phi^2 + 1} = \sqrt{\frac{\sqrt{5}+5}{2}}$.

We can also apply the Pythagorean theorem to triangles $AED$ and $AEM$, which gives us

$$AE^2 + DE^2 = AD^2, \text{ and } AE^2 + EM^2 = AM^2, \text{ or}$$

$$x^2 + \left(\frac{d-y}{2}\right)^2 = 1, \text{ and } x^2 + \left(\frac{y}{2}\right)^2 = \left(\frac{d}{2}\right)^2.$$

By subtracting the two equations, we get $\left(\frac{d-y}{2}\right)^2 - \left(\frac{y}{2}\right)^2 = 1 - \left(\frac{d}{2}\right)^2$, which, when we solve for $y$, gives us

$$y = \sqrt{\frac{\sqrt{5}+5}{10}} \approx 0.8506508083.$$

By substituting the values for $d$ and $y$ in the equation that we got above, $x^2 + \left(\frac{y}{2}\right)^2 = \left(\frac{d}{2}\right)^2$, we find

$$x = y = \sqrt{\frac{\sqrt{5}+5}{10}} \approx 0.8506508083.$$

Therefore, $x = y = z = \sqrt{\frac{\sqrt{5}+5}{10}} \approx 0.8506508083$,

which is what we originally wanted to show.

Our next manifestation of the golden rectangle comes from a fascinating property that is embedded in a somewhat unusual pattern. We begin with two congruent rectangles perpendicular to each other and inscribed in a circle, as shown in figure 4-13. The shaded region formed by these two rectangles is of interest to us here. Curiously, it turns out that this shaded region is greatest when the two rectangles are golden rectangles, that is, when $\frac{a}{b} = \phi$. To show that this is actually true requires nothing more than a bit of high school mathematics. For interested readers, we provide this proof in the appendix.

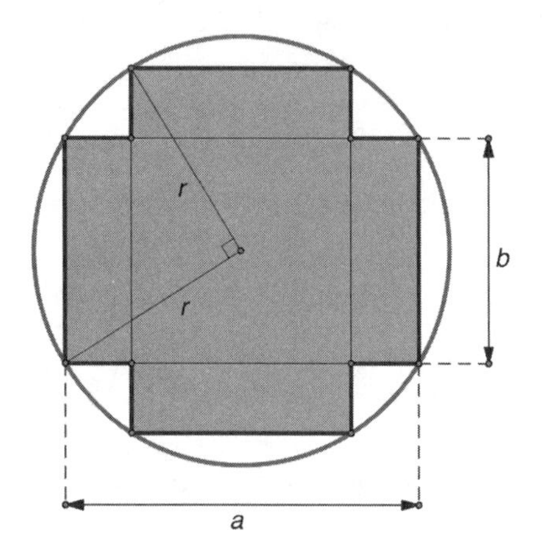


Figure 4-13

Another interesting appearance—when one might least expect it—can be found in three dimensions. If we have a sphere (with radius $R$) and a right circular cylinder with maximum surface area inscribed in the sphere, then the ratio of the cylinder's diameter $d$ and its height $h$ are in the golden ratio.

In figure 4-14, $\frac{AB}{BC} = \frac{d}{h} = \frac{2r}{h} = \phi$. In other words, the cross section ($ABCD$) through the cylinder's axis is a golden rectangle.

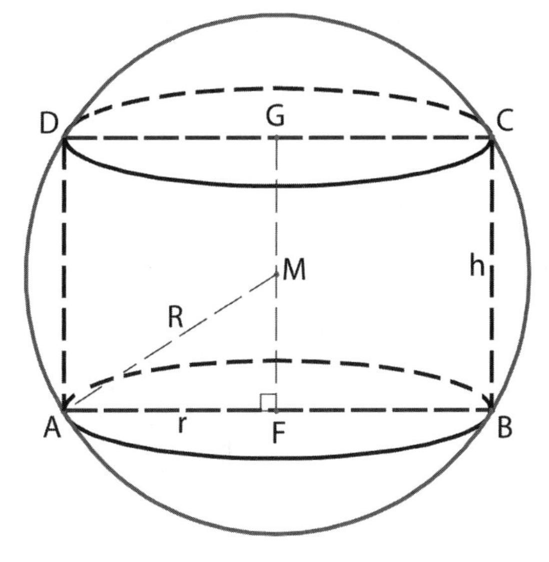


Figure 4-14

We shall now consider a rectangle with length 1 and width $\frac{1}{\phi}$, which is, in fact, a golden rectangle because $1 : \frac{1}{\phi} = \phi$. Suppose we cut off a square with side length $\frac{1}{\phi}$, as shown in figure 4-15. The remaining rectangle is also a golden rectangle with side lengths $\frac{1}{\phi}$ and $1 - \frac{1}{\phi} = \frac{1}{\phi^2}$.

The area of the square in figure 4-15 is

$$\frac{1}{\phi} \cdot \frac{1}{\phi} = \frac{1}{\phi^2} = \frac{3-\sqrt{5}}{2},$$

while the area of the smaller rectangle is

$$\left(1 - \frac{1}{\phi}\right) \cdot \frac{1}{\phi} = \frac{1}{\phi^2} \cdot \frac{1}{\phi} = \frac{1}{\phi^3} = \sqrt{5} - 2.$$

The area of the entire (larger) rectangle is

$$1 \cdot \frac{1}{\phi} = \frac{1}{\phi} = \frac{\sqrt{5}-1}{2} = \frac{1}{\phi}.$$

That is, we have now shown *geometrically* the following relationship:

$$\frac{1}{\phi}=\frac{1}{\phi^2}+\frac{1}{\phi^3}.$$

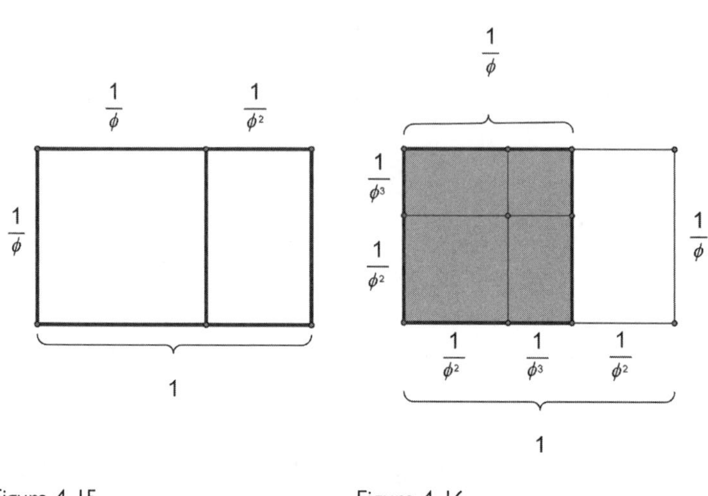


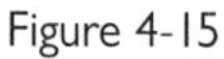
Figure 4-15

Figure 4-16

The square with side length $\frac{1}{\phi}$ will now be further partitioned into the golden ratio (fig. 4-16) so that we get two line segments of lengths $\frac{1}{\phi^2}$ and $\frac{1}{\phi^3}$.

Another square is then constructed with the longer of these segments as a side. As shown in figure 4-16, we now have three squares of different sizes in the shaded region. We shall now calculate the areas of each of these squares.

The area of the largest square is

$$\frac{1}{\phi}\cdot\frac{1}{\phi}=\frac{1}{\phi^2}=\frac{3-\sqrt{5}}{2}.$$

The next smaller square has an area

$$\frac{1}{\phi^2}\cdot\frac{1}{\phi^2}=\frac{7-3\sqrt{5}}{2}=\frac{1}{\phi^4}.$$

The smallest square's area is

$$\frac{1}{\phi^3} \cdot \frac{1}{\phi^3} = 9 - 4\sqrt{5} = \frac{1}{\phi^6}.$$

Each of the rectangles (which are not squares) has an area

$$\frac{1}{\phi^2} \cdot \frac{1}{\phi^3} = \frac{5\sqrt{5} - 11}{2} = \frac{1}{\phi^5}.$$

The area of the large square of the shaded region is then equal to the areas of the two smaller squares and twice the area of the smaller rectangle inside the square, which gives us

$$\frac{1}{\phi^2} = \frac{1}{\phi^4} + \frac{1}{\phi^6} + \frac{1}{\phi^5} + \frac{1}{\phi^5} = \frac{1}{\phi^4} + \frac{1}{\phi^6} + \frac{2}{\phi^5}.$$

We could continue this process by subdividing the segments into the golden ratio to obtain further representations of $\frac{1}{\phi}$. Look at the square with side length $\phi$ shown in figure 4-17. If we partition the square as we did above and continue the process successively so that the segments partition as

$$\phi \rightarrow (1, \frac{1}{\phi}),\ 1 \rightarrow (\frac{1}{\phi}, \frac{1}{\phi^2}),\ \frac{1}{\phi} \rightarrow (\frac{1}{\phi^2}, \frac{1}{\phi^3}),\ \ldots, \text{ and so on,}$$

we get the following sequence: $1, \frac{1}{\phi}, \frac{1}{\phi^2}, \frac{1}{\phi^3}, \frac{1}{\phi^4}, \ldots$ .

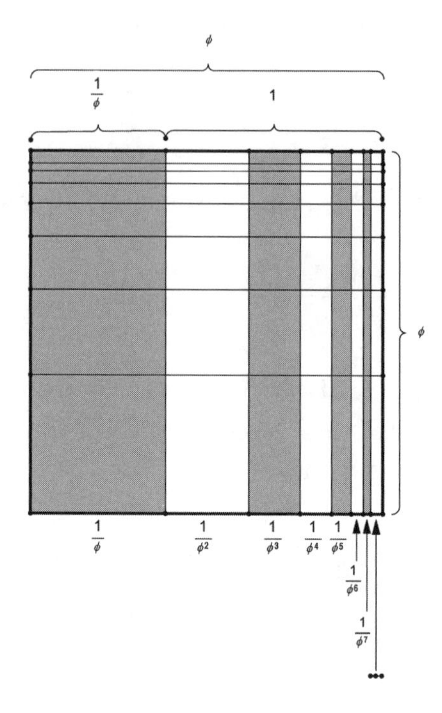


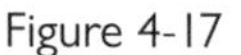

Figure 4-17

The area of the large square is $A=\phi^2$. The areas of the rectangles with sides of length $\phi$ and $\frac{1}{\phi^i}$ (where $i=1, 2, 3, \ldots$) will completely fill the large square. So if we continue this process indefinitely, we will get

$$\phi^2=\phi\cdot\frac{1}{\phi}+\phi\cdot\frac{1}{\phi^2}+\phi\cdot\frac{1}{\phi^3}+\phi\cdot\frac{1}{\phi^4}+\phi\cdot\frac{1}{\phi^5}+\phi\cdot\frac{1}{\phi^6}+\phi\cdot\frac{1}{\phi^7}+\ldots$$
$$=1+\frac{1}{\phi}+\frac{1}{\phi^2}+\frac{1}{\phi^3}+\frac{1}{\phi^4}+\frac{1}{\phi^5}+\frac{1}{\phi^6}+\ldots.$$

Because, as we know, $\phi^2=\phi+1$, we get the following:

$$\phi=\frac{1}{\phi}+\frac{1}{\phi^2}+\frac{1}{\phi^3}+\frac{1}{\phi^4}+\frac{1}{\phi^5}+\frac{1}{\phi^6}+\ldots.$$

Suppose we now inspect this same square with side length $\phi$. But rather than partition it into the rectangles with dimensions $(\phi, \frac{1}{\phi^i})$ as we did above, we will consider its area as the sum of the tiling shown in figure 4-18.

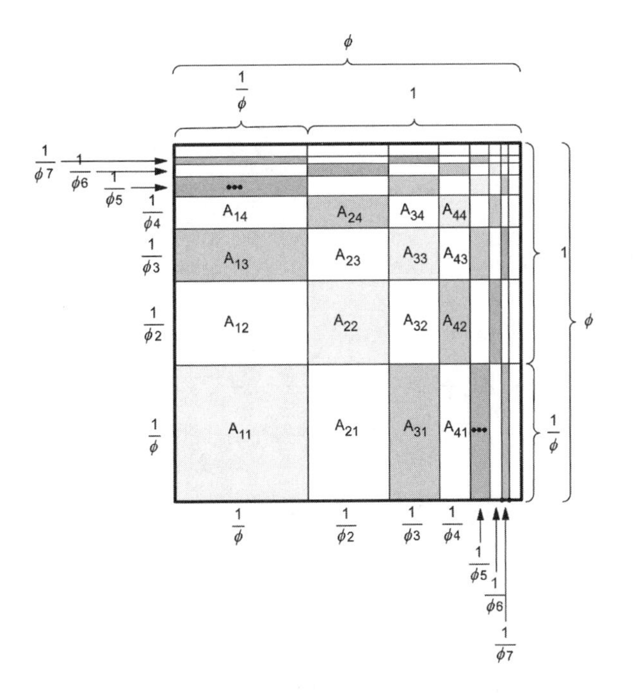

Figure 4-18

The index $i$ provides the exponents of the powers of $\frac{1}{\phi^i}=\phi^{-i}$. That is, $A_{ij}$ refers to the area of the rectangle with sides of length $\frac{1}{\phi^i}$ and $\frac{1}{\phi^j}$, also

$A_{ij}=\frac{1}{\phi^i}\cdot\frac{1}{\phi^j}=\frac{1}{\phi^{i+j}}=\phi^{-(i+j)}$, as shown in figure 4-18.

We then have:

$$A=(A_{-1-1}=)\,\phi^2$$

$$A_{11}=\frac{1}{\phi}\cdot\frac{1}{\phi}=\frac{1}{\phi^2}$$

$$A_{21}=\frac{1}{\phi^2}\cdot\frac{1}{\phi}=\frac{1}{\phi^3}=A_{12}$$

$$A_{22}=\frac{1}{\phi^2}\cdot\frac{1}{\phi^2}=\frac{1}{\phi^4}$$

$$A_{31}=\frac{1}{\phi^3}\cdot\frac{1}{\phi}=\frac{1}{\phi^4}=A_{13}$$

$$A_{32}=\frac{1}{\phi^3}\cdot\frac{1}{\phi^2}=\frac{1}{\phi^5}=A_{23}$$

$$A_{33}=\frac{1}{\phi^3}\cdot\frac{1}{\phi^3}=\frac{1}{\phi^6}$$

$$A_{41}=\frac{1}{\phi^4}\cdot\frac{1}{\phi}=\frac{1}{\phi^5}=A_{14},\ldots,$$

therefore the sum is

$$\begin{aligned}A&=A_{11}+A_{21}+A_{22}+A_{12}+A_{31}+A_{32}+A_{33}+A_{23}+A_{13}+A_{41}+\ldots\\&=A_{11}+2A_{21}+A_{22}+2A_{31}+2A_{32}+A_{33}+2A_{41}+\ldots\\&=\frac{1}{\phi^2}+2\frac{1}{\phi^3}+\frac{1}{\phi^4}+2\frac{1}{\phi^4}+2\frac{1}{\phi^5}+\frac{1}{\phi^6}+2\frac{1}{\phi^5}+\ldots\\&=1\frac{1}{\phi^2}+2\frac{1}{\phi^3}+3\frac{1}{\phi^4}+4\frac{1}{\phi^5}+5\frac{1}{\phi^6}+6\frac{1}{\phi^7}+\ldots,\end{aligned}$$

which is[3] $\frac{1}{\phi^2}+\frac{2}{\phi^3}+\frac{3}{\phi^4}+\frac{4}{\phi^5}+\frac{5}{\phi^6}+\frac{6}{\phi^7}+\ldots+\frac{n}{\phi^{n+1}}+\ldots=\phi^2$.[3]

If we consider the squares or partial rectangles, then a slew of other relationships with powers of $\frac{1}{\phi}$ emerge:[4]

$$\frac{1}{\phi}+\frac{1}{\phi^3}+\frac{1}{\phi^5}+\frac{1}{\phi^7}+\ldots+\frac{1}{\phi^{2n-1}}+\ldots=1,$$

$$\frac{1}{\phi^2}+\frac{1}{\phi^4}+\frac{1}{\phi^6}+\frac{1}{\phi^8}+\ldots+\frac{1}{\phi^{2n}}+\ldots=\frac{1}{\phi},$$

$$\frac{1}{\phi^2}+\frac{2}{\phi^3}+\frac{3}{\phi^4}+\frac{4}{\phi^5}+\ldots+\frac{n}{\phi^{n+1}}+\ldots=\phi^2 \quad \text{(see above)},$$

$$\frac{1}{\phi^4}+\frac{2}{\phi^6}+\frac{3}{\phi^8}+\frac{4}{\phi^{10}}+\ldots+\frac{n}{\phi^{2n+2}}+\ldots=\frac{1}{\phi^2},$$

$$\frac{1}{\phi^3}+\frac{2}{\phi^5}+\frac{3}{\phi^7}+\frac{4}{\phi^9}+\ldots+\frac{n}{\phi^{2n+1}}+\ldots=\frac{1}{\phi},$$

$$\frac{1}{\phi}+\frac{1}{\phi^2}+\frac{1}{\phi^3}+\frac{1}{\phi^4}+\ldots+\frac{1}{\phi^n}+\ldots=\phi.$$

Let us continue our discussion with golden rectangle *ABCD*, but now in a very curious way. We will admire some of the beautiful geometric manifestations of this golden section. We shall begin with the golden rectangle *ABCD* with side lengths $a+b$ and $a$ (fig. 4-19).

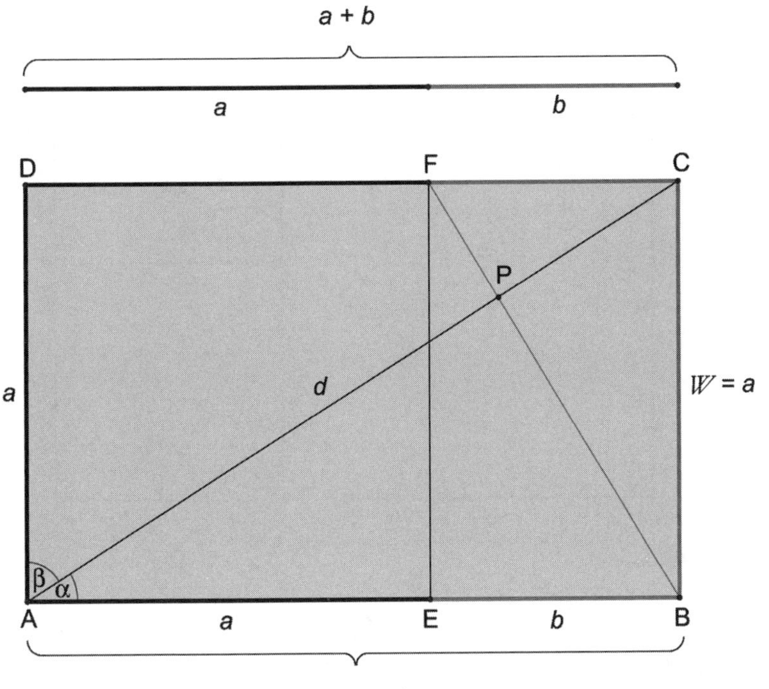


Figure 4-19

By cutting off square *AEFD*, we get the golden rectangle *EBCF*, with side lengths $a$ and $b$. As we have shown earlier, we know that $\frac{a+b}{a} = \phi$, as well as $\frac{a}{b} = \phi$. If we continue this process of cutting off a square (side length $b$) from the smaller golden rectangle (*EBCF*), we once again end up with a golden rectangle. We can continue this process indefinitely. In figure 4-20, you can see how this will look.

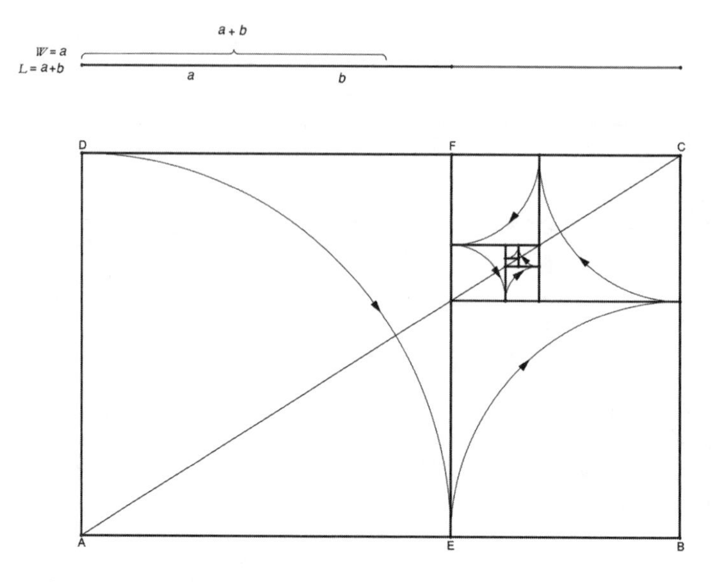

Figure 4-20

If we draw quarter-circle arcs in each of the successive squares, we get a spiral that approximates the golden spiral, as you can see in figure 4-21.

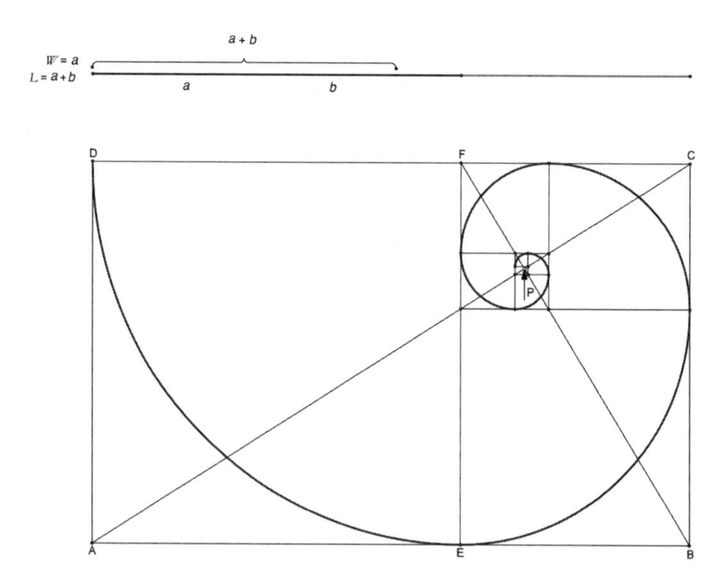

Figure 4-21

We established earlier that the two rectangles *AEFD* and *EBCF* are reciprocal rectangles (fig. 4-21) and that their diagonals are perpendicular. This point of intersection is the limiting point for the spiral.

If we consider a rectangle (fig. 4-22) with sides of length 55 and 34 (two Fibonacci numbers), we have a rectangle that is very close to being a golden rectangle, since the ratio of the sides $\frac{55}{34} = 1.6\overline{1764705882352941}$, which is very close to the golden ratio, $\phi \approx 1.618033988$. By cutting off successive squares that are Fibonacci numbers, we also get a spiral, often referred to as the *Fibonacci spiral*—one very similar to the *golden spiral*.

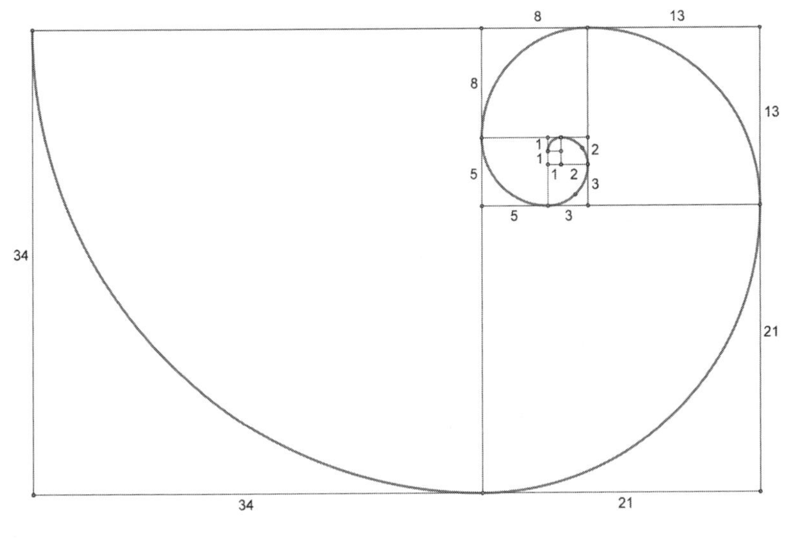


Figure 4-22

The actual golden spiral does not evolve from these quarter circles. This demonstration merely gives us a good and easily understood approximation. Referring to figure 4-23, the table in figure 4-24 provides a progressive calculation of the length of the spiral.

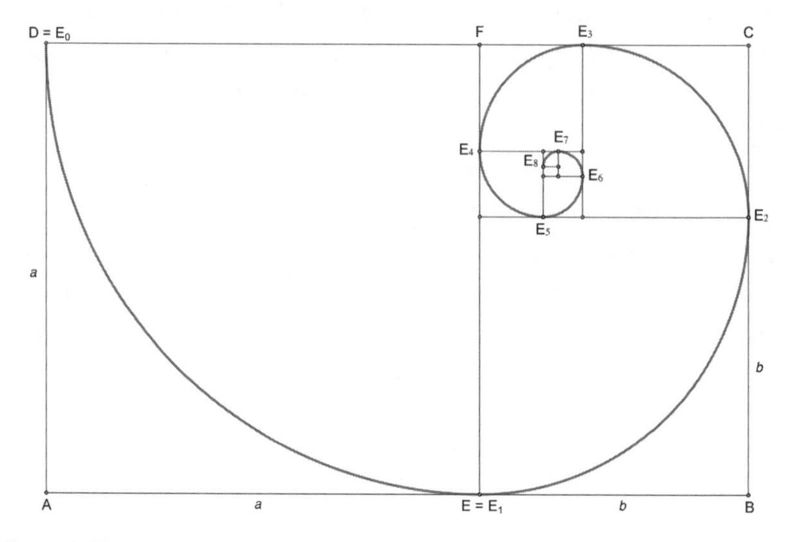


Figure 4-23

The side of a square is also the length of the quarter-circle radius $r$. The length of a quarter-circle arc is $\frac{2\pi \cdot r}{4} = \frac{\pi \cdot r}{2}$.

| $n$ | Square with diagonal $E_nE_{n+1}$ | Side of the square | Length of the quarter-circle arc | | |
|---|---|---|---|---|---|
| 0 | $E_0E_1 = DE_1$ | $\frac{a}{\phi^0} = a$ | $\frac{\pi \cdot a}{2} \cdot \frac{1}{\phi^0}$ | $= \frac{\pi \cdot a}{2}$ | $= -\frac{0\sqrt{5}-2}{2} \cdot \frac{\pi \cdot a}{2}$ |
| 1 | $E_1E_2$ | $\frac{a}{\phi^1} = b$ | $\frac{\pi \cdot a}{2} \cdot \frac{1}{\phi^1}$ | $= \frac{\pi \cdot a}{2} \cdot \frac{\sqrt{5}-1}{2}$ | $= \frac{1\sqrt{5}-1}{2} \cdot \frac{\pi \cdot a}{2}$ |
| 2 | $E_2E_3$ | $\frac{a}{\phi^2}$ | $\frac{\pi \cdot a}{2} \cdot \frac{1}{\phi^2}$ | $= \frac{\pi \cdot a}{2} \cdot \frac{3-\sqrt{5}}{2}$ | $= -\frac{1\sqrt{5}-3}{2} \cdot \frac{\pi \cdot a}{2}$ |
| 3 | $E_3E_4$ | $\frac{a}{\phi^3}$ | $\frac{\pi \cdot a}{2} \cdot \frac{1}{\phi^3}$ | $= \frac{\pi \cdot a}{2} \cdot \frac{2\sqrt{5}-4}{2}$ | $= \frac{2\sqrt{5}-4}{2} \cdot \frac{\pi \cdot a}{2}$ |
| 4 | $E_4E_5$ | $\frac{a}{\phi^4}$ | $\frac{\pi \cdot a}{2} \cdot \frac{1}{\phi^4}$ | $= \frac{\pi \cdot a}{2} \cdot \frac{7-3\sqrt{5}}{2}$ | $= -\frac{3\sqrt{5}-7}{2} \cdot \frac{\pi \cdot a}{2}$ |
| 5 | $E_5E_6$ | $\frac{a}{\phi^5}$ | $\frac{\pi \cdot a}{2} \cdot \frac{1}{\phi^5}$ | $= \frac{\pi \cdot a}{2} \cdot \frac{5\sqrt{5}-11}{2}$ | $= \frac{5\sqrt{5}-11}{2} \cdot \frac{\pi \cdot a}{2}$ |
| 6 | $E_6E_7$ | $\frac{a}{\phi^6}$ | $\frac{\pi \cdot a}{2} \cdot \frac{1}{\phi^6}$ | $= \frac{\pi \cdot a}{2} \cdot \frac{18-8\sqrt{5}}{2}$ | $= -\frac{8\sqrt{5}-18}{2} \cdot \frac{\pi \cdot a}{2}$ |
| 7 | $E_7E_8$ | $\frac{a}{\phi^7}$ | $\frac{\pi \cdot a}{2} \cdot \frac{1}{\phi^7}$ | $= \frac{\pi \cdot a}{2} \cdot \frac{13\sqrt{5}-29}{2}$ | $= \frac{13\sqrt{5}-29}{2} \cdot \frac{\pi \cdot a}{2}$ |
| ... | | | | | |

Figure 4-24

Amazingly, as you inspect the results, you will see progressively the Fibonacci numbers, $F_n$ (0, 1, 1, 2, 3, 5, 8, 13, ...), as the product with $\sqrt{5}$, followed by the constants that represent the Lucas numbers, $L_n$ (2, 1, 3, 4, 7, 11, 18, 29, ...). This could justify, perhaps, calling this spiral the *Fibonacci-Lucas spiral.*

The length of this spiral in figure 4-23 from point $E_0$ to point $E_8$ is

$$\text{Length} \left( \overset{\frown}{E_0E_1} + \overset{\frown}{E_1E_2} + \overset{\frown}{E_2E_3} + \overset{\frown}{E_3E_4} + \overset{\frown}{E_4E_5} + \overset{\frown}{E_5E_6} + \overset{\frown}{E_6E_7} + \overset{\frown}{E_7E_8} \right)$$

$$= \frac{\pi \cdot a}{2} \cdot \left( \frac{1}{\phi^0} + \frac{1}{\phi^1} + \frac{1}{\phi^2} + \frac{1}{\phi^3} + \frac{1}{\phi^4} + \frac{1}{\phi^5} + \frac{1}{\phi^6} + \frac{1}{\phi^7} \right)$$

$$= \frac{\pi \cdot a}{2} \cdot \frac{9\sqrt{5} - 15}{2} \approx 4.024860693 \cdot a.$$

As a matter of fact, the spiral has a definite length ($L$), which can be obtained by continuing this process indefinitely.

$$L = \frac{\pi \cdot a}{2} \cdot \left( \frac{1}{\phi^0} + \frac{1}{\phi^1} + \frac{1}{\phi^2} + \ldots + \frac{1}{\phi^n} + \ldots \right)$$

$$= \frac{\pi \cdot a}{2} \cdot \phi^2 = \frac{\pi \cdot a}{2} \cdot \frac{\sqrt{5} + 3}{2} \approx 4.112398172 \cdot a < 5a.$$

We established earlier (p. 100) $\frac{1}{\phi^1} + \frac{1}{\phi^2} + \ldots + \frac{1}{\phi^n} + \ldots = \phi$
Therefore,

$$\frac{1}{\phi^0} + \frac{1}{\phi^1} + \frac{1}{\phi^2} + \ldots + \frac{1}{\phi^n} + \ldots = \frac{1}{\phi^0} + \left( \frac{1}{\phi^1} + \frac{1}{\phi^2} + \ldots + \frac{1}{\phi^n} + \ldots \right) = \frac{1}{\phi^0} + \phi = 1 + \phi = \phi^2.$$

The symmetric parts of this complex-looking figure are the squares, with *ABCD* as the golden rectangle. Suppose we locate the center of each of these squares. We can draw arcs through each of these points and then see that the centers of these squares lie in another approximation of a logarithmic spiral (fig. 4-25).

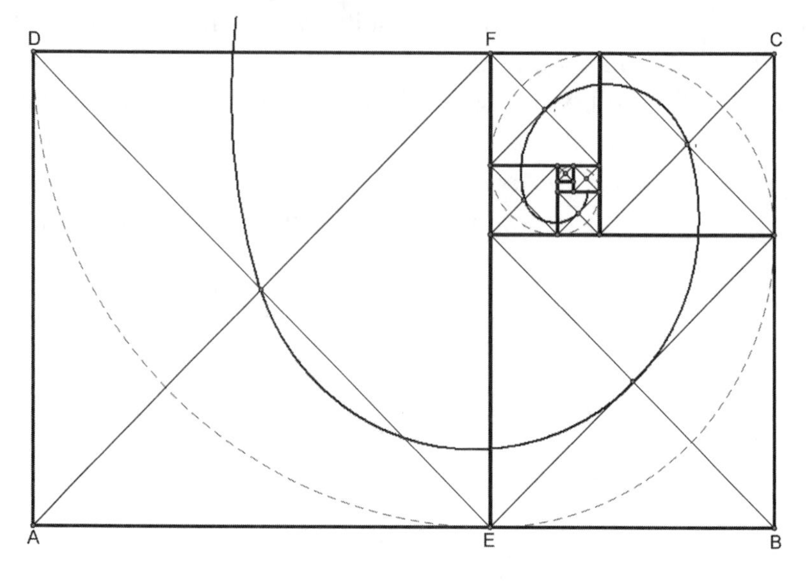


Figure 4-25

There are several similar spirals that approximate, but do not exactly equal, a *golden spiral.*

## THE GOLDEN SPIRAL

The *golden spiral* is essentially a *logarithmic spiral*, which is one that has a tangent to it at any point forming equal angles with the radius from the origin. In figure 4-26, we marked two such equal angles that two randomly selected tangent lines make with the radius of the spiral. You should also note that the spiral looks like it eventually reaches point *P*, but, in fact, it does not!

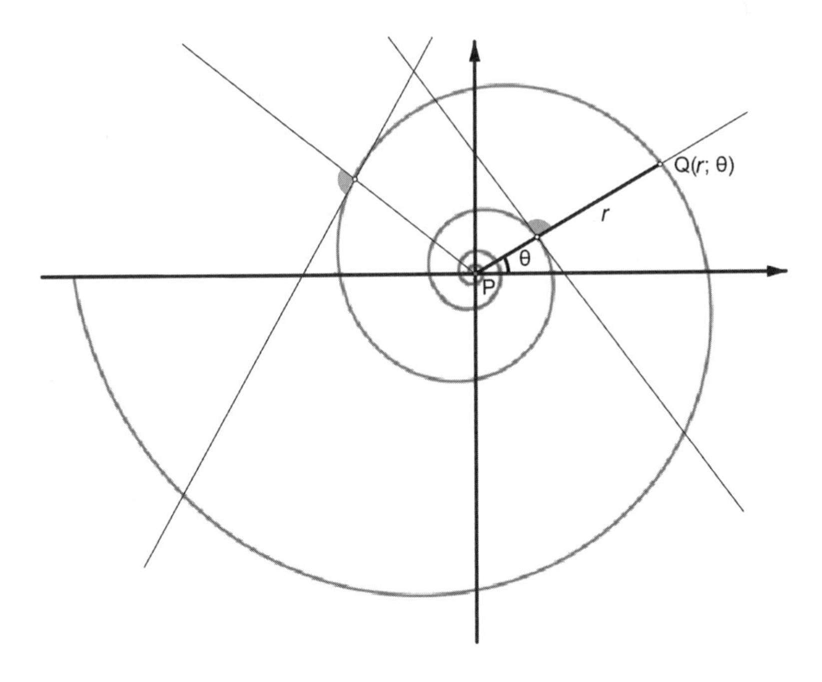


Figure 4-26

The length of the radius, $r$, grows exponentially with the polar angle, $\theta$. With the polar coordinates $(r, \theta)$, we get the equation $r(\theta) = ae^{k\theta}$, where (in fig. 4-26) $\theta$ is the measure of the angle that the radius makes with the $x$-axis and $a$ and $k$ are positive real numbers.[5]

Since every line from the origin, $P$, to the spiral creates an equal angle with the tangent to the curve at that point, the famous French mathematician René Descartes (1596–1650)[6] called this an *equiangular spiral.*

Incidentally, the name *logarithmic spiral* is credited to the Swiss mathematician Jakob Bernoulli (1654–1705), who also called it the *spira mirabilis* (Latin for "miraculous spiral"). He was so enchanted with this spiral and its properties that he requested it as his epitaph with the words *Eadem mutata resurgo* ("although changed, I shall arise the same"). However, as fate would have it, the sculptor who then created the epitaph chiselled not a logarithmic spiral but rather an Archimedean one!

In figure 4-27, we try to show how close the golden spiral[7] is to the spiral formed by successive quarter circles in squares cut off from successive golden rectangles—yes, they are almost indistinguishable! The discrepancy between the spirals is actually only quite noticeable in the larger squares, whereas in the smaller squares it is not easily noticed. The origin of the spiral is at the intersection of the two diagonals shown in figure 4-27.

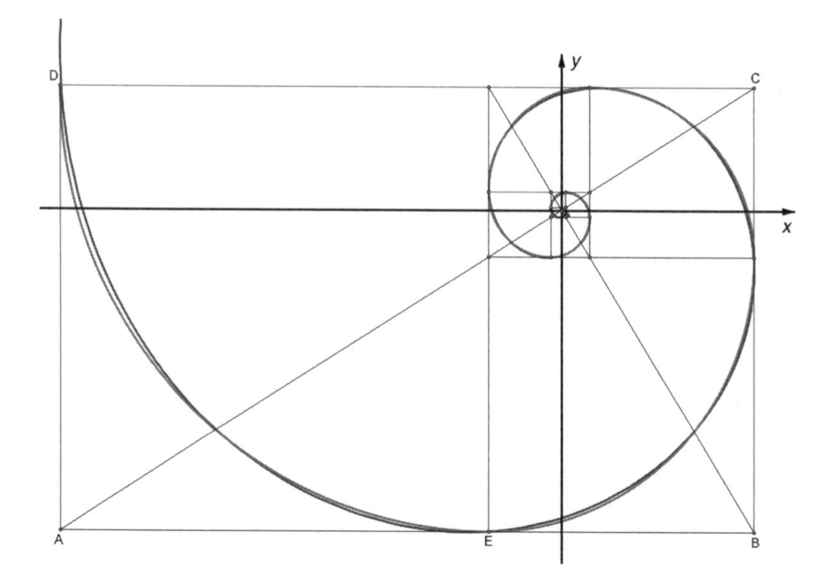


Figure 4-27

As we try to show in figure 4-28, the golden spiral is not tangent to the sides of the golden rectangles, yet cuts the sides at very small angles, whereas the approximation spiral (which consists of quarter circles) is tangent to the sides. So the sides of the golden rectangles are not tangents to this golden spiral (as in the case of the approximation). They are each cut twice, one of these points of intersection being: $E_0$, $E_1$, $E_2$, $E_3$, . . . . (See p. 106 figure 4-23, where $D = E_0$.)

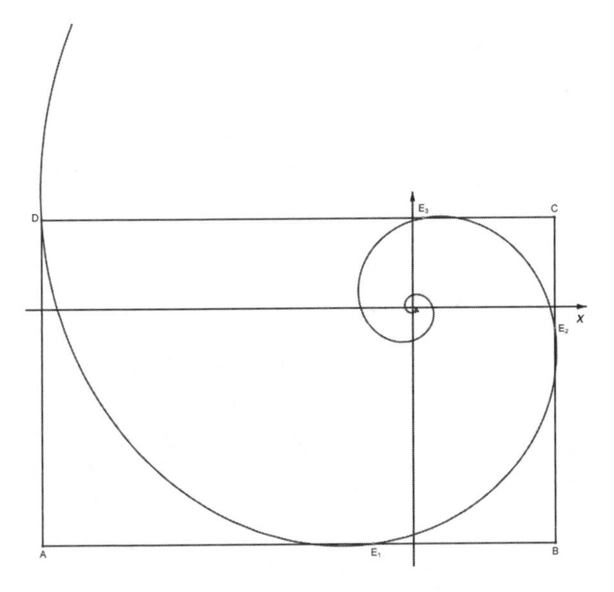


Figure 4-28

In nature, the nautilus shell (fig. 4-29) exhibits such a golden spiral, as does the snail in figure 4-30.

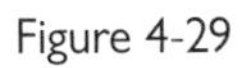
Figure 4-29

This is not surprising, because the curve tends to a logarithmic spiral as it expands.

Figure 4-30

The interested reader may wish to find other manifestations of the golden spiral in nature.

## THE GOLDEN RHOMBUS

You will recall from high school mathematics that a parallelogram is a quadrilateral with opposite sides parallel. Opposite sides of a parallelogram are also equal in length. When the adjacent sides of a parallelogram are also equal, then the parallelogram is called a *rhombus*.

We show a rhombus in figure 4-31, where $AB \parallel CD$, $BC \parallel AD$, and $AB = BC = CD = AD$, or $a = b = c = d$. Furthermore, the diagonals of a rhombus are perpendicular. In figure 4-31, $AC \perp BD$, or $e \perp f$.

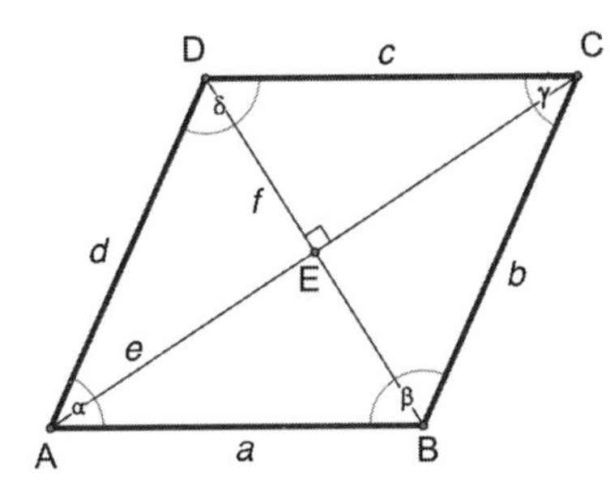


Figure 4-31

When the lengths of the diagonals are in the ratio of $\phi$ : 1, we call the rhombus a *golden rhombus* (fig. 4-32).

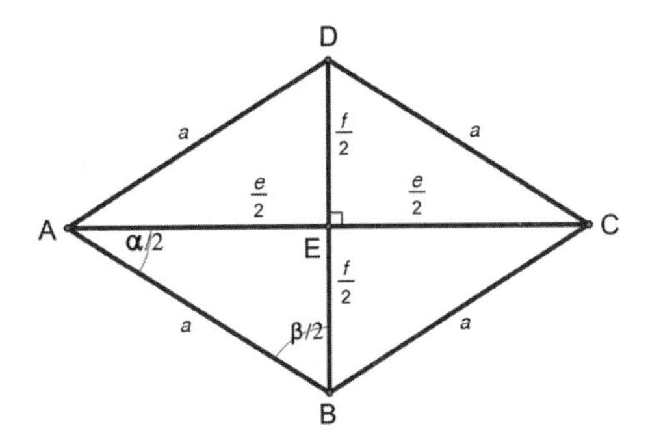


Figure 4-32

Since we just defined the golden rhombus as having its diagonals in the ratio of $\phi$ : 1, we can restate this as $\frac{e}{f}=\frac{\phi}{1}$, or put another way, $e=\phi f$.

If we apply the Pythagorean theorem to triangle $ABE$, we get

$$a = AB = \sqrt{AE^2 + BE^2} = \sqrt{\frac{e^2}{4}+\frac{f^2}{4}} = \frac{f}{2}\sqrt{\phi^2+1} = \frac{1}{2}\sqrt{\frac{5+\sqrt{5}}{2}}\cdot f.$$

This then allows us to get

$$f = BD = \frac{2}{\sqrt{\phi^2+1}}\cdot a = \frac{\sqrt{10}}{5}\sqrt{5-\sqrt{5}}\cdot a.$$

Then

$$e = \phi \cdot f = \frac{2\phi}{\sqrt{\phi^2 + 1}} \cdot a,$$

which gives us

$$a = \frac{\sqrt{\phi^2 + 1}}{2\phi} \cdot e.$$

Since we know that the triangles *ABE*, *BCE*, *CDE*, and *ADE* are congruent, we can express the area of the golden rhombus in a number of different ways, each in terms of the lengths of some of its parts.

In terms of the shorter diagonal $f$:

$$\text{Area}_{\text{Rhombus}} = 4 \cdot \frac{1}{2} \cdot \frac{e}{2} \cdot \frac{f}{2} = \frac{e \cdot f}{2} = \frac{\phi}{2} \cdot f^2 = \frac{\sqrt{5}+1}{4} \cdot f^2.$$

In terms of the longer diagonal $e$:

$$\text{Area}_{\text{Rhombus}} = \frac{e \cdot f}{2} = \frac{e \cdot e}{2\phi} = \frac{1}{2\phi} \cdot e^2 = \frac{\sqrt{5}-1}{4} \cdot e^2.$$

In terms of the side length $a$:

$$\text{Area}_{\text{Rhombus}} = \frac{e \cdot f}{2} = \frac{\phi}{2} \cdot f^2 = \frac{2\phi}{\phi^2 + 1} \cdot a^2 = \frac{2\sqrt{5}}{5} \cdot a^2.$$

Every rhombus contains an inscribed circle. In figure 4-33, we will call the inscribed circle's radius $r_i$. We can insert the radius $r_i$ to the point of tangency with side *AD*, which then gives us some similar triangles: $\Delta ADE$, $\Delta AEF$, and $\Delta DEF$.

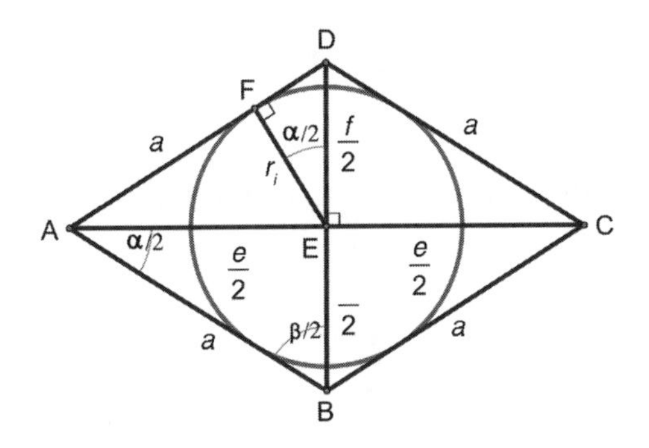


Figure 4-33

Using the fact that $\cos\angle DEF = \cos\frac{\alpha}{2} = \frac{r_i}{\frac{f}{2}}$,

we can once again express the radius of the inscribed circle in terms of each of the diagonals and the side length.

To get the radius in terms of diagonal $f$, we have

$$r_i = \frac{f}{2}\cos\frac{\alpha}{2} = \frac{\phi}{2\sqrt{\phi^2+1}} \cdot f = \frac{1}{2}\sqrt{\frac{5+\sqrt{5}}{10}} \cdot f.$$

To get the radius in terms of a side length $a$, we find that

$$r_i = \frac{f}{2}\cos\frac{\alpha}{2} = \frac{f}{2} \cdot \frac{\phi}{\sqrt{\phi^2+1}} = \frac{\phi}{\phi^2+1} \cdot a = \frac{\sqrt{5}}{5} \cdot a.$$

Finally, we can get the radius in terms of the other diagonal $e$ as

$$r_i = \frac{f}{2} \cdot \frac{\phi}{\sqrt{\phi^2+1}} = \frac{1}{2\sqrt{\phi^2+1}} \cdot e = \frac{1}{2}\sqrt{\frac{5-\sqrt{5}}{10}} \cdot e.$$

We can obtain the area of the inscribed circle in terms of the side length $a$ as

$$\text{Area}_{\text{Inscribed circle}} = \pi \cdot r_i{}^2 = \pi \cdot \frac{5}{25} \cdot a^2 = \frac{\pi}{5} \cdot a^2.$$

This now enables us to find the ratio of the area of the golden rhombus to its inscribed circle as

$$\frac{\text{Area}_{\text{Rhombus}}}{\text{Area}_{\text{Inscribed circle}}} = \frac{\frac{2\sqrt{5}}{5} \cdot a^2}{\frac{\pi}{5} \cdot a^2} = \frac{2\sqrt{5}}{\pi} \approx 1.423525086.$$

Now that we have bestowed the title *golden rhombus* on a rhombus that exhibits the golden ratio with the lengths of its diagonals, we can show how we can create such a golden rhombus by beginning with a golden rectangle *ABCD* (as in fig. 4-34), with side lengths $a$ and $b$, where $\frac{a}{b} = \phi$. By constructing lines through the four vertices of the rectangle and parallel to the diagonals, we have a rhombus, *EFGH*, since the diagonals of a rectangle are equal.

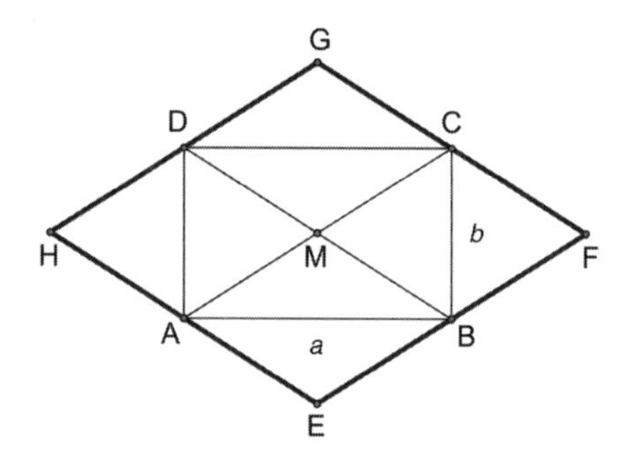


Figure 4-34

To show that this is, in fact, a golden rhombus, we would need to demonstrate that the diagonals (*HF* and *GE*) are in the golden ratio. In figure 4-35, we note that *AMGD* and *AEMD* are rhombuses.
Thus, $GE = 2 \cdot AD$. Similarly, $HF = 2 \cdot DC$. Therefore, since $\frac{DC}{AD} = \frac{\phi}{1}$, so too $\frac{HF}{GE} = \frac{2 \cdot DC}{2 \cdot AD} = \frac{\phi}{1} = \phi$, and so we have established the rhombus *EFGH* to be golden.

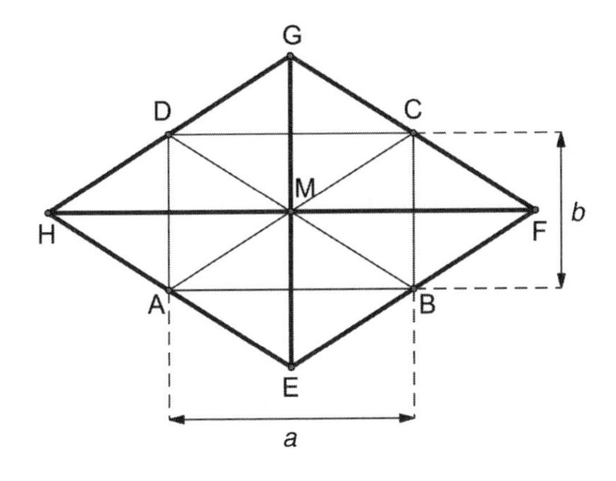


Figure 4-35

The golden rhombus also appears in the well-known *Penrose tessellations*,[8] which stem from the work of the English mathematician Roger Penrose (1931–) and the amateur mathematician Robert Ammann (1946–1994), who in 1974 developed a tessellation of the plane without a periodic repetition. The Penrose tessellations involve a two-dimensional Fibonacci lattice. There are individual small groups of symmetric regular decagons, yet the Penrose tessellations in total are not symmetric or periodic. There are various streams of Penrose tiles. In figure 4-36, we see streams of rhombuses with sides of equal length but with different-size angles.

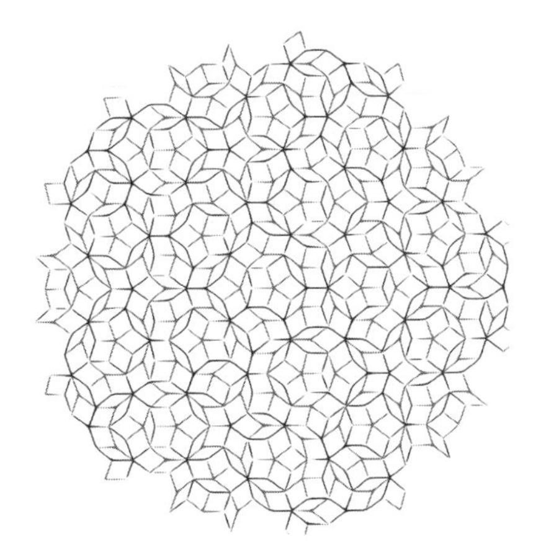

Figure 4-36

The narrow rhombus (rhombus 1, fig. 4-37) has side length 1 and angles of measures $\alpha = 36°$ and $\beta = 144°$, while the wider rhombus (rhombus 2, fig. 4-38) has side length 1 and angles of measures $\alpha = 72°$ and $\beta = 108°$. We notice that the angles are all multiples of 36°. By drawing the other diagonal in each of the rhombuses and then considering the right triangles formed, we can use trigonometry to determine the lengths of a diagonal.

For rhombus 1,

$$\sin 18° = \frac{\frac{BD}{2}}{AB}$$

and then $BD = \frac{1}{\phi}$.

For rhombus 2,

$$\cos 36° = \frac{\frac{AC}{2}}{AB},$$

which gives us $AC = \phi$. We find, once again, the emergence of the golden ratio where it was not necessarily expected. One diagonal of the narrow rhombus is $\frac{1}{\phi}$, while one diagonal of the wider rhombus is of length $\phi$. We get their respective areas as follows:

$$Area_{\text{Rhombus 1}} = 2 \cdot Area_{\Delta ABD} = 2 \cdot \frac{1}{2} \cdot AB \cdot AD \cdot \sin 36°$$

$$= 1 \cdot 1 \cdot 1 \cdot \frac{1}{2}\sqrt{\frac{5-\sqrt{5}}{2}} = \frac{\sqrt{\phi^2+1}}{2\phi}.$$

$$Area_{\text{Rhombus 2}} = 2 \cdot Area_{\Delta ABD} = 2 \cdot \frac{1}{2} \cdot AB \cdot AD \cdot \sin 72°$$

$$= 1 \cdot 1 \cdot 1 \cdot \frac{1}{2}\sqrt{\frac{5+\sqrt{5}}{2}} = \frac{\sqrt{\phi^2+1}}{2}.$$

The ratio of their areas is—as by now you might expect:

$$\frac{Area_{\text{Rhombus2}}}{Area_{\text{Rhombus1}}}=\frac{\dfrac{\sqrt{\phi^2+1}}{2}}{\dfrac{\sqrt{\phi^2+1}}{2\phi}}=\phi,$$

and once again the golden ratio emerges! It is claimed that the ratio of the number of rhombuses of each kind used for the tessellation is the golden ratio—in that $\frac{number_{wide}}{number_{narrow}}=\phi$.

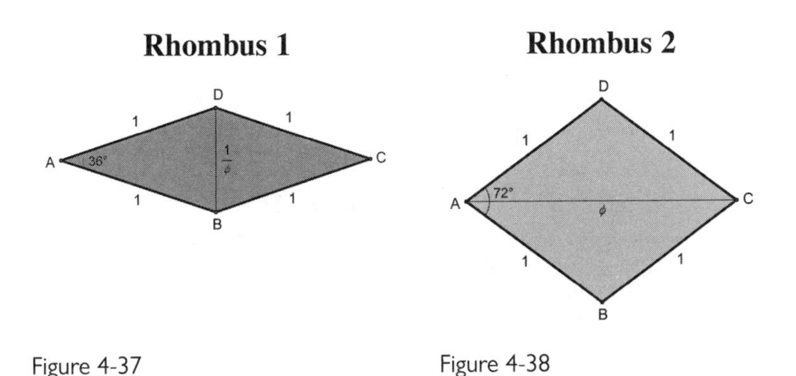


Figure 4-37

Figure 4-38

Besides these rhombus-shaped tiles, there are also other shaped tiles that produce a nonperiodic tessellation of the plane. These were named by the English mathematician John H. Conway (1937–) as *kite* and *dart*. We can see this kite and dart shape in the Penrose tessellation, when we partition the diagonal into the golden section (figs. 4-39 and 4-40). If we take the longer diagonal, this partitioning is seen as $AE : CE=\phi : 1$. Furthermore, the ratio of the longer side to the shorter side of both the kite and the dart is the golden ratio, $\phi : 1$. If that is not enough, we also find that the ratio of the longer diagonal of the rhombus to the longer side is the golden ratio, since $\frac{AC}{AB}=\frac{\phi+1}{\phi}=1+\frac{1}{\phi}=\phi$.

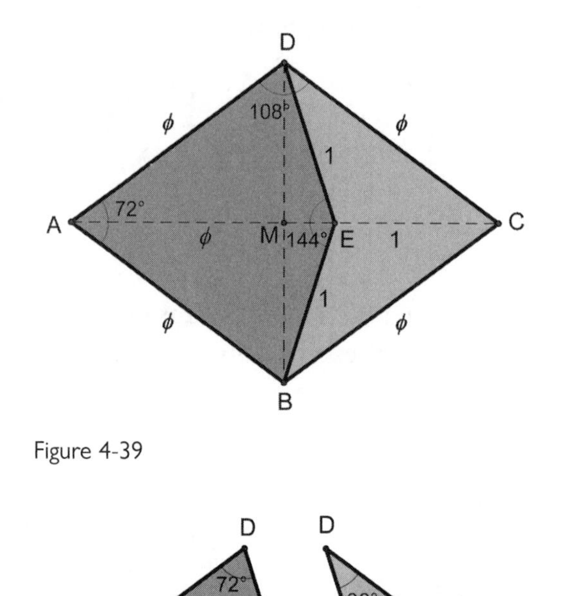


Figure 4-39

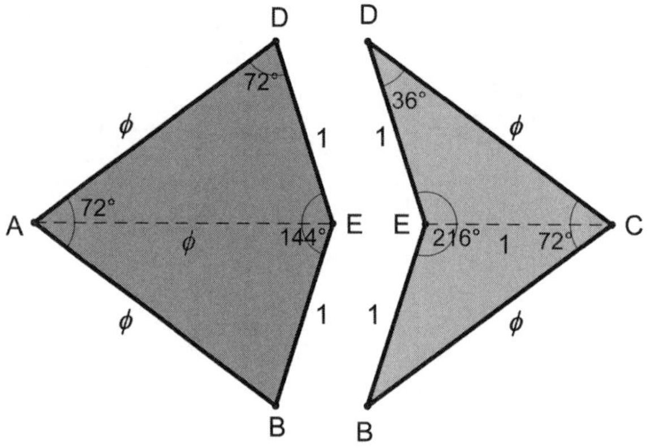

Figure 4-40

When placed together, the two shapes, the kite and the dart, produce a rhombus with side length $\phi$ and diagonal length $e=\phi+1$ and $f=\sqrt{\phi^2+1}$.

Since $\angle BDE=\angle BDC-\angle CDE=\frac{1}{2}\angle ADC-\angle CDE=54°-36°=18°$, we have for the right triangle $EDM$ the following:

$$\cos\angle BDE=\cos 18°=\frac{DM}{DE}=\frac{BD}{2DE}.$$

Therefore, $BD = 2\,DE\cos 18^\circ = 2 \cdot 1 \cdot \frac{1}{2}\sqrt{\frac{5+\sqrt{5}}{2}} = \sqrt{\phi^2+1}.$

We then get the ratio $\frac{AC}{BD} = \frac{\phi+1}{\sqrt{\phi^2+1}} = \sqrt{\frac{5+2\sqrt{5}}{5}}.$

We thus can conclude that the quadrilateral *ABCD* (fig. 4-39) is not a golden rhombus, yet the many manifestations of the golden ratio merit inclusion in our exploration of the golden section.

## THE GOLDEN TRIANGLE

The logical next step after investigating the golden rectangle is to consider the golden ratio as it pertains to a triangle—the *golden triangle*. As you would expect, much as the golden ratio is embedded in the golden rectangle, so, too, the golden triangle also exhibits the golden ratio throughout. To investigate the golden triangle, we will look at its linear relationships, its area relationships, as well as how it relates to a regular pentagon.

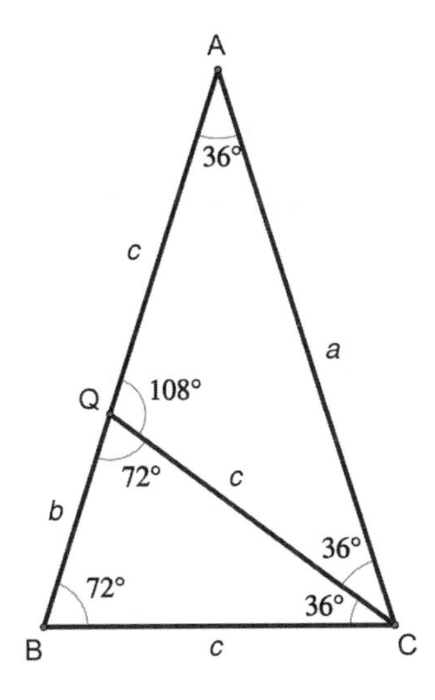


Figure 4-41

Let's begin our investigation of this special triangle, which is an isosceles $\Delta ABC$, figure 4-41, with a vertex angle of 36°, leaving 72° for each base angle. Triangle $ABC$ is called a *golden triangle.* When we draw the bisector of $\angle ACB$, we have created two similar triangles: $\Delta ABC \sim \Delta CQB$, with the angles as shown in figure 4-41. This enables us to set up the following proportion:

$$\frac{AB}{BC} = \frac{BC}{BQ}, \text{ or } \frac{b+c}{c} = \frac{c}{b}.$$

This you will, of course, recognize as the golden ratio on the side $AB$. As you can see here:

$$\frac{b+c}{c} = \frac{c}{b} = \frac{\sqrt{5}+1}{2} = \phi, \text{ or put another way, } \frac{a}{c} = \frac{c}{b} = \frac{\sqrt{5}+1}{2} = \phi.$$

## The Golden Section of the Sides of the Golden Triangle

With some algebraic manipulations, we get the following relationships; first we express $c$ and $a$ in terms of $b$:

$$c = \phi \cdot b = \frac{\sqrt{5}+1}{2}, \text{ and } a = \phi \cdot c = \phi^2 \cdot b = \left(\frac{\sqrt{5}+1}{2}\right)^2 b,$$

then $a = \dfrac{\sqrt{5}+3}{2} b$, which is the same as $\phi^2 \cdot b = (\phi + 1)b$.

Furthermore, in terms of $c$, we get

$$a = \phi \cdot c = \frac{\sqrt{5}+1}{2} \cdot c, \text{ and } b = \phi^{-1} \cdot c = \frac{1}{\phi} \cdot c = \frac{\sqrt{5}-1}{2} \cdot c;$$

then in terms of $a$, we get

$$c = \phi^{-1} \cdot a = \frac{1}{\phi} \cdot a = \frac{\sqrt{5}-1}{2} \cdot a, \text{ and } b = \phi^{-1} \cdot c = \phi^{-2} \cdot a = \frac{1}{\phi^2} \cdot a = \frac{3-\sqrt{5}}{2} \cdot a.$$

In figure 4-41, we have three golden triangles: two with an acute vertex angle, $\Delta ABC$ and $\Delta CQB$, where the ratio of the lengths of the side to the base is $\phi : 1$, or $\frac{a}{c}=\frac{\phi}{1}$; and one with an obtuse vertex angle, $\Delta ACQ$, where the ratio of the lengths of the side to the base is $1 : \phi$, or $\frac{c}{a}=\frac{1}{\phi}$.

## The Area of a Golden Triangle

To calculate the area of a golden triangle, we would use the well-known formula for the area of any triangle, namely the area is one-half the product of two sides times the sine of the included angle, or Area $= \frac{1}{2}\, ab \sin C$, where $a$ and $b$ are the lengths of the sides of a triangle and $C$ is the measure of the angle between these two sides.

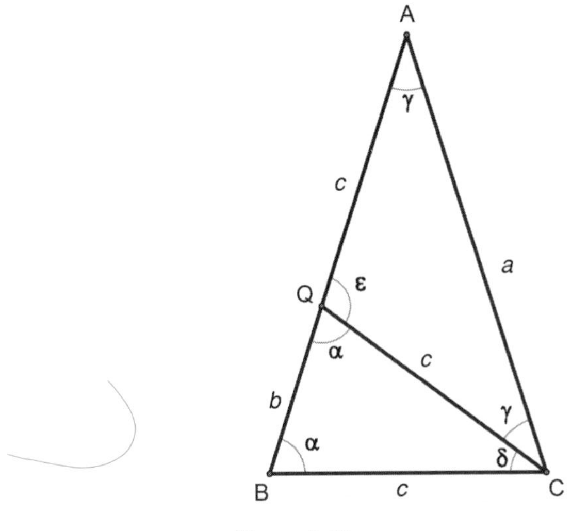


Figure 4-42

For figure 4-42, we have the following areas of the various golden triangles shown:

$$\text{Area}_{\Delta ABC}=\tfrac{1}{2}\cdot AB\cdot AC\cdot \sin\angle BAC=\tfrac{1}{2}a^2\sin\gamma=\tfrac{1}{2}a^2\sin 36^\circ$$

$$=\frac{a^2}{2}\cdot\frac{1}{2}\sqrt{\frac{5-\sqrt{5}}{2}}=\frac{a^2}{4}\cdot\frac{\sqrt{\phi^2+1}}{\phi},$$

$$\text{Area}_{\Delta ACQ} = \tfrac{1}{2} \cdot AQ \cdot CQ \cdot \sin\angle AQC = \tfrac{1}{2}c^2 \sin\varepsilon = \tfrac{1}{2}c^2 \sin 108^\circ$$

$$= \frac{c^2}{2} \cdot \frac{1}{2}\sqrt{\frac{5+\sqrt{5}}{2}} = \frac{c^2}{4} \cdot \sqrt{\phi^2+1},$$

$$\text{Area}_{\Delta BCQ} = \tfrac{1}{2} \cdot BC \cdot CQ \cdot \sin\angle BCQ = \tfrac{1}{2}c^2 \sin\delta = \tfrac{1}{2}c^2 \sin 36^\circ$$

$$= \frac{c^2}{2} \cdot \frac{1}{2}\sqrt{\frac{5-\sqrt{5}}{2}} = \frac{c^2}{4} \cdot \frac{\sqrt{\phi^2+1}}{\phi}.$$

Since $a = \phi c$ and $c = \phi b$, we find that $a = \phi^2 b$.

It follows that

$$\text{Area}_{\Delta ABC} : \text{Area}_{\Delta ACQ} = \frac{\frac{a^2}{4} \cdot \sqrt{\frac{5-\sqrt{5}}{2}}}{\frac{c^2}{4} \cdot \sqrt{\frac{5+\sqrt{5}}{2}}} = \frac{a^2}{c^2} \cdot \frac{1}{\phi} = \frac{\phi^2 \cdot c^2}{c^2} \cdot \frac{1}{\phi} = \phi,$$

$$\text{Area}_{\Delta ACQ} : \text{Area}_{\Delta BCQ} = \frac{\frac{c^2}{4} \cdot \sqrt{\frac{5+\sqrt{5}}{2}}}{\frac{c^2}{4} \cdot \sqrt{\frac{5-\sqrt{5}}{2}}} = \frac{c^2}{c^2} \cdot \phi = \phi, \text{ and}$$

$$\text{Area}_{\Delta ABC} : \text{Area}_{\Delta BCQ} = \frac{\frac{a^2}{4} \cdot \sqrt{\frac{5-\sqrt{5}}{2}}}{\frac{c^2}{4} \cdot \sqrt{\frac{5-\sqrt{5}}{2}}} = \frac{a^2}{c^2} \cdot 1 = \frac{\phi^2 \cdot c^2}{c^2} = \phi^2.$$

The areas of the three triangles *ABC*, *ACQ*, and *BCQ* are in the ratio of: $\text{Area}_{\Delta ABC} : \text{Area}_{\Delta ACQ} : \text{Area}_{\Delta BCQ} = \phi : 1 : \phi^{-1}$.

Looking back to figure 4-41, if we draw the altitudes from the vertex angle of each of the three golden triangles, we will have formed three different right triangles. Applying the Pythagorean theorem along with the trigonometric ratios, we get some interesting results—once again having the golden ratio emerge somewhat unexpectedly.

$$\sin 18° = \cos 72° = \frac{1}{2\phi}$$

$$\sin 36° = \cos 54° = \frac{1}{2} \cdot \frac{\sqrt{\phi^2 + 1}}{\phi}$$

$$\sin 54° = \cos 36° = \frac{\phi}{2}$$

$$\sin 72° = \cos 18° = \frac{\sqrt{\phi^2 + 1}}{2}$$

$$\tan 18° = \cot 72° = \frac{\sqrt{\phi^2 + 1}}{3\phi + 1}$$

$$\tan 36° = \cot 54° = \frac{\sqrt{\phi^2 + 1}}{\phi^2}$$

$$\tan 54° = \cot 36° = \frac{\phi^2}{\sqrt{\phi^2 + 1}}$$

$$\tan 72° = \cot 18° = \phi\sqrt{\phi^2 + 1}$$

This can be continued with the now-to-be-expected golden ratio present each time.

## A Golden Sequence

Figure 4-43 shows a golden triangle, $\Delta ABC$, with $BD_1$ bisecting $\angle ABC$, and then $CD_2$ bisecting $\angle BCD_1$. This continues in this figure with successive angle bisectors ($D_1D_3$, $D_2D_4$, $D_3D_5$, $D_4D_6$, …, $D_iD_{i+2}$) creating new golden triangles.

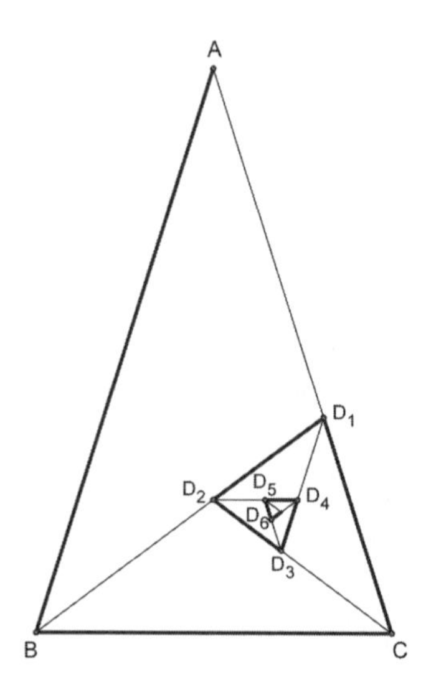


Figure 4-43

These newly formed 36°, 72°, 72° triangles, are, in fact, a series of golden triangles (see fig. 4-43): $\Delta ABC$, $\Delta BCD_1$, $\Delta CD_1D_2$, $\Delta D_1D_2D_3$, $\Delta D_2D_3D_4$, $\Delta D_3D_4D_5$, $\Delta D_4D_5D_6, \ldots, \Delta D_iD_{i+1}D_{i+2}$.

Obviously, had space permitted, we could have continued to draw angle bisectors and thereby generate more such golden triangles. Our inspection of the golden triangle will parallel that of the golden rectangle. We can thus generate a golden sequence.

Let us begin by letting $D_5D_6=1$ (fig. 4-43). Since the ratio of $\frac{\text{side}}{\text{base}}$ of a golden triangle is $\phi$, we find that for golden $\Delta D_4D_5D_6$: $\frac{D_4D_5}{D_5D_6}=\frac{\phi}{1}$, and $D_4D_5=\phi$.

Similarly, for golden $\Delta D_3D_4D_5$: $\frac{D_3D_4}{D_4D_5}=\phi$, and $D_3D_4=\phi\cdot D_4D_5=\phi\cdot\phi=\phi^2$.

In golden $\Delta D_2D_3D_4$: $\frac{D_2D_3}{D_3D_4}=\phi$, and $D_2D_3=\phi\cdot D_3D_4=\phi\cdot\phi^2=\phi^3$.

Again, for golden $\Delta D_1D_2D_3$: $\frac{D_1D_2}{D_2D_3}=\phi$, and $D_1D_2=\phi\cdot D_2D_3=\phi\cdot\phi^3=\phi^4$.

In golden $\Delta CD_1D_2$: $\frac{CD_1}{D_1D_2}=\phi$, and $CD_1=\phi\cdot D_1D_2=\phi\cdot\phi^4=\phi^5$.

Again, in golden $\Delta BCD_1$: $\frac{BC}{CD_1}=\phi$, and $BC=\phi\cdot CD_1=\phi\cdot\phi^5=\phi^6$.

Finally, in golden $\Delta ABC$: $\frac{AB}{BC}=\phi$, and $AB=\phi\cdot BC=\phi\cdot\phi^6=\phi^7$.

This can be summarized by using our knowledge of the powers of $\phi$ (developed earlier, in chap. 3) as follows, where $F_n$ are Fibonacci numbers:[9]

$$
\begin{aligned}
D_5D_6 &= \phi^0 = 0\phi+1 = F_0\phi+F_{-1}\\
D_4D_5 &= \phi^1 = 1\phi+0 = F_1\phi+F_0\\
D_3D_4 &= \phi^2 = 1\phi+1 = F_2\phi+F_1\\
D_2D_3 &= \phi^3 = 2\phi+1 = F_3\phi+F_2\\
D_1D_2 &= \phi^4 = 3\phi+2 = F_4\phi+F_3\\
CD_1 &= \phi^5 = 5\phi+3 = F_5\phi+F_4\\
BC &= \phi^6 = 8\phi+5 = F_6\phi+F_5\\
AB &= \phi^7 = 13\phi+8 = F_7\phi+F_6
\end{aligned}
$$

For the general case, we get $\phi^n = F_n\phi + F_{n-1}$ (where $n \geq 0$).

Using figure 4-44, let's now consider $\Delta ABC$ $(=\Delta D_0D_1D_2)$, which is a golden triangle with $AB = AC = a$ and $BC = c$, so that we get

$$\frac{a}{c} = \frac{\sqrt{5}+1}{2} = \phi.$$

We recall that point $D_3$ partitions the side $AC$ into the golden ratio. We also know that $AD_3 = BD_3 = BC$, or $D_0D_3 = D_1D_3 = D_1D_2$, and $\angle D_0D_3D_1 = 108°$.

As we did earlier with the golden rectangle, we can generate an approximation of a logarithmic spiral by drawing circular arcs to join the vertex angle vertices of consecutive golden triangles as shown in figure 4-44.

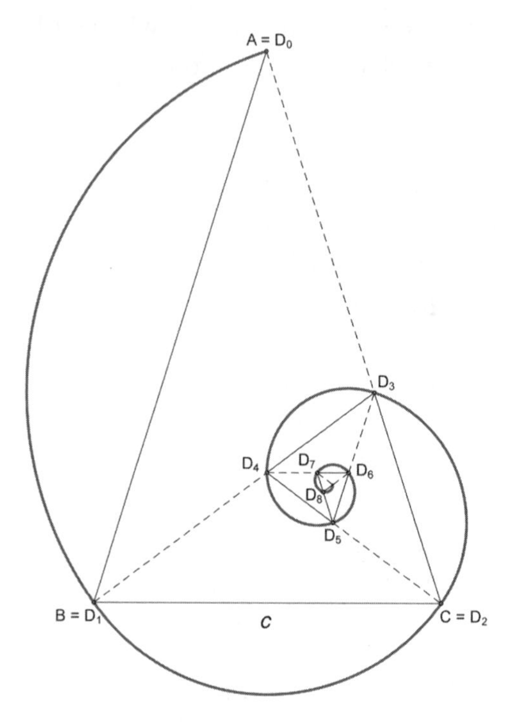


Figure 4-44

That is, we draw circular arcs as follows:

$\widehat{AB}$ with the circle center at $D_3$,

$\widehat{BC}$ with the circle center at $D_4$,

$\widehat{CD_3}$ with the circle center at $D_5$,

$\widehat{D_3D_4}$ with the circle center at $D_6$,

$\widehat{D_4D_5}$ with the circle center at $D_7$,

$\widehat{D_5D_6}$ with the circle center at $D_8$,

$\widehat{D_6D_7}$ with the circle center at $D_9$, and so on.

The base $c = BC = D_1D_2$ of the isosceles $\Delta D_0D_1D_2$ is a side ($r = D_0D_3 = D_1D_3$) in the isosceles $\Delta D_0D_1D_3$ and therefore the radius $r$

of the constructed $\frac{3}{10}$ circle arc $\widehat{AB}$ ($\angle D_0D_3D_1 = 108° = \frac{3}{10} \cdot 360°$). The length of a $\frac{3}{10}$ circle arc is

$$2\pi r \cdot \frac{3}{10} = \frac{3\pi}{5} \cdot r,$$

$$\text{therefore, } \widehat{AB} = \frac{3\pi}{5} \cdot D_1D_2 = \frac{3\pi}{5} \cdot c.$$

| $n$ | Isosceles $\Delta D_nD_{n+1}D_{n+2}$ | Base of the isosceles triangle | = Radius of the arc | Length of the $\frac{3}{10}$ circle arc |
|---|---|---|---|---|
| 0 | $\Delta D_0D_1D_2$ | $BC = D_1D_2$ | $= \frac{c}{\phi^0} = c$ | $\frac{3\pi}{5} \cdot c \cdot \frac{1}{\phi^0} = -\frac{0\sqrt{5}-2}{2} \cdot \frac{3\pi}{5} \cdot c$ |
| 1 | $\Delta D_1D_2D_3$ | $D_2D_3$ | $= \frac{c}{\phi^1}$ | $\frac{3\pi}{5} \cdot c \cdot \frac{1}{\phi^1} = \frac{1\sqrt{5}-1}{2} \cdot \frac{3\pi}{5} \cdot c$ |
| 2 | $\Delta D_2D_3D_4$ | $D_3D_4$ | $= \frac{c}{\phi^2}$ | $\frac{3\pi}{5} \cdot c \cdot \frac{1}{\phi^2} = -\frac{1\sqrt{5}-3}{2} \cdot \frac{3\pi}{5} \cdot c$ |
| 3 | $\Delta D_3D_4D_5$ | $D_4D_5$ | $= \frac{c}{\phi^3}$ | $\frac{3\pi}{5} \cdot c \cdot \frac{1}{\phi^3} = \frac{2\sqrt{5}-4}{2} \cdot \frac{3\pi}{5} \cdot c$ |
| 4 | $\Delta D_4D_5D_6$ | $D_5D_6$ | $= \frac{c}{\phi^4}$ | $\frac{3\pi}{5} \cdot c \cdot \frac{1}{\phi^4} = -\frac{3\sqrt{5}-7}{2} \cdot \frac{3\pi}{5} \cdot c$ |
| 5 | $\Delta D_5D_6D_7$ | $D_6D_7$ | $= \frac{c}{\phi^5}$ | $\frac{3\pi}{5} \cdot c \cdot \frac{1}{\phi^5} = \frac{5\sqrt{5}-11}{2} \cdot \frac{3\pi}{5} \cdot c$ |
| 6 | $\Delta D_6D_7D_8$ | $D_7D_8$ | $= \frac{c}{\phi^6}$ | $\frac{3\pi}{5} \cdot c \cdot \frac{1}{\phi^6} = -\frac{8\sqrt{5}-18}{2} \cdot \frac{3\pi}{5} \cdot c$ |
| 7 | $\Delta D_7D_8D_9$ | $D_8D_9$ | $= \frac{c}{\phi^7}$ | $\frac{3\pi}{5} \cdot c \cdot \frac{1}{\phi^7} = \frac{13\sqrt{5}-29}{2} \cdot \frac{3\pi}{5} \cdot c$ |
| ... | | | | |

Figure 4-45

Amazingly, as you inspect the results (fig. 4-45), you will again see the Fibonacci numbers, $F_n$ (0, 1, 1, 2, 3, 5, 8, 13, ...), as the product with $\sqrt{5}$, and the constants highlight the Lucas numbers, $L_n$ (2, 1, 3, 4, 7, 11, 18, 29, ...).[10]

The length of this *Fibonacci-Lucas spiral* in figure 4-44 from point $D_0$ to point $D_8$ is the sum of the following lengths:

$$\widehat{D_0D_1} + \widehat{D_1D_2} + \widehat{D_2D_3} + \widehat{D_3D_4} + \widehat{D_4D_5} + \widehat{D_5D_6} + \widehat{D_6D_7} + \widehat{D_7D_8}$$

$$=\frac{3\pi}{5}\cdot c\cdot\left(\frac{1}{\phi^0}+\frac{1}{\phi^1}+\frac{1}{\phi^2}+\frac{1}{\phi^3}+\frac{1}{\phi^4}+\frac{1}{\phi^5}+\frac{1}{\phi^6}+\frac{1}{\phi^7}\right)$$

$$=\frac{3\pi}{5}\cdot c\cdot\frac{9\sqrt{5}-15}{2}\approx 4.829832832\cdot c.$$

Again, to our amazement we find that the spiral has a definite length even though the spiral continues endlessly. Before we embark on this quest, we must recall that in chapter 3 we saw that $\frac{1}{\phi^1}+\frac{1}{\phi^2}+\ldots+\frac{1}{\phi^n}+\ldots=\phi$.

Therefore,

$$\frac{1}{\phi^0}+\frac{1}{\phi^1}+\frac{1}{\phi^2}+\ldots+\frac{1}{\phi^n}+\ldots=\frac{1}{\phi^0}+\left(\frac{1}{\phi^1}+\frac{1}{\phi^2}+\ldots+\frac{1}{\phi^n}+\ldots\right)=\frac{1}{\phi^0}+\phi=1+\phi=\phi^2.$$

We are now ready to find the spiral length as follows:

Spiral length

$$=\frac{3\pi}{5}\cdot c\cdot\left(\frac{1}{\phi^0}+\frac{1}{\phi^1}+\frac{1}{\phi^2}+\ldots+\frac{1}{\phi^n}+\ldots\right)$$

$$=\frac{3\pi}{5}\cdot c\cdot\phi^2=\frac{3\pi}{5}\cdot c\cdot\frac{\sqrt{5}+3}{2}\approx 4.934877807\cdot c<5c.$$

## The Golden Radii

You might have expected that there must also be golden radii in these configurations—these would be radii of a circle related to a triangle that will yield the golden ratio. We will use a right triangle where one leg is twice the length of the other, and where the hypotenuse is greater than the longer leg by one unit. (See fig. 4-46.)

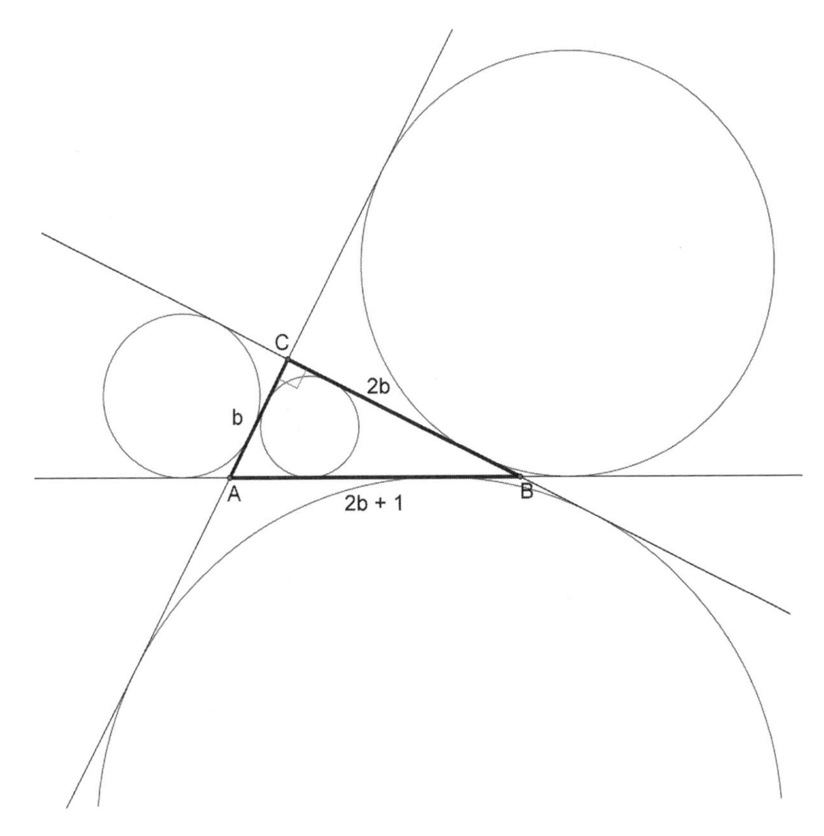


Figure 4-46

Every triangle has an *inscribed circle* (a circle inside the triangle) and three *escribed circles* (circles outside the triangle). Each of these circles is tangent to the three sides of the triangle. These are also shown in figure 4-47 for any triangle and in figure 4-46 for a right triangle. These four circles are sometimes called *equicircles*.

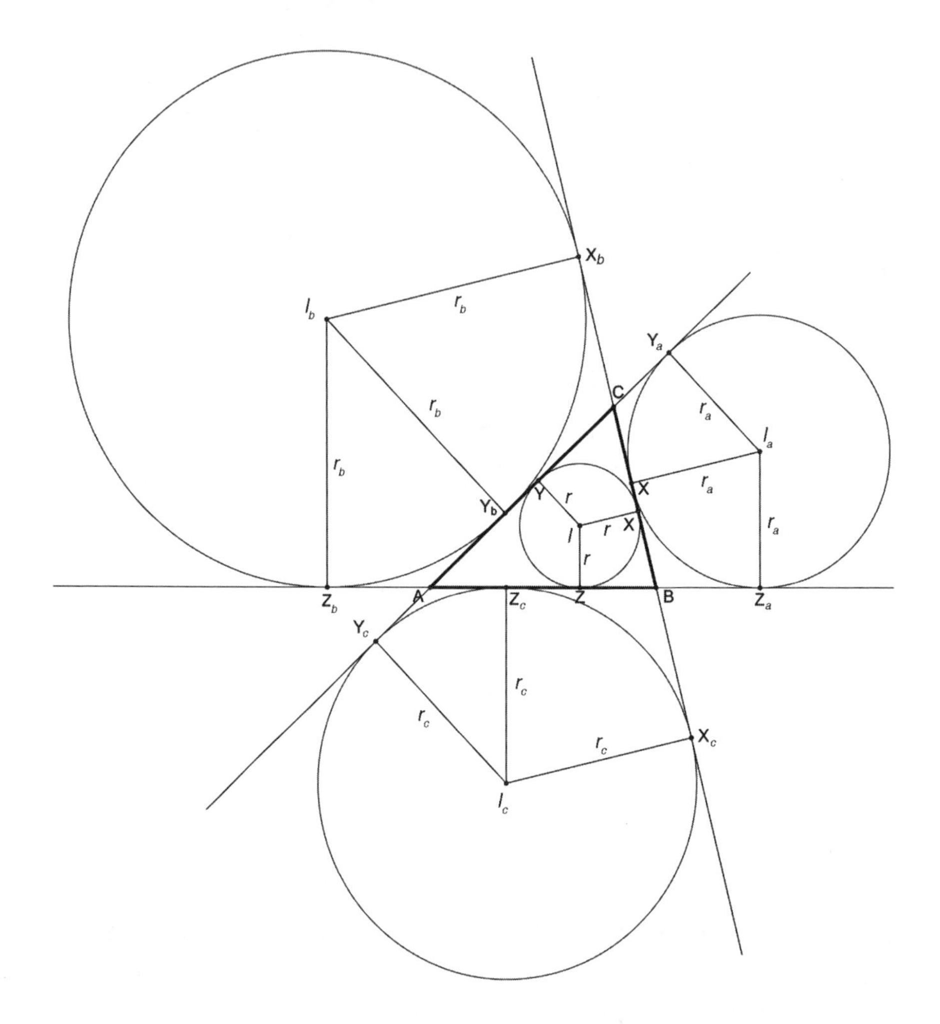


Figure 4-47

Our attention now turns to the radii of the equicircles of a triangle. We refer to their radii as *equiradii*. Let's consider first the radius of the inscribed circle. We call this the *inradius* of the triangle. We will first show that the inradius, $r$, of a triangle is equal to the ratio of the area of the triangle to its *semiperimeter*, $s$ (i.e., half of its perimeter).

In figure 4-48, we get the following equation by simply using the common formula for the area of a triangle (i.e., half the base times the height):

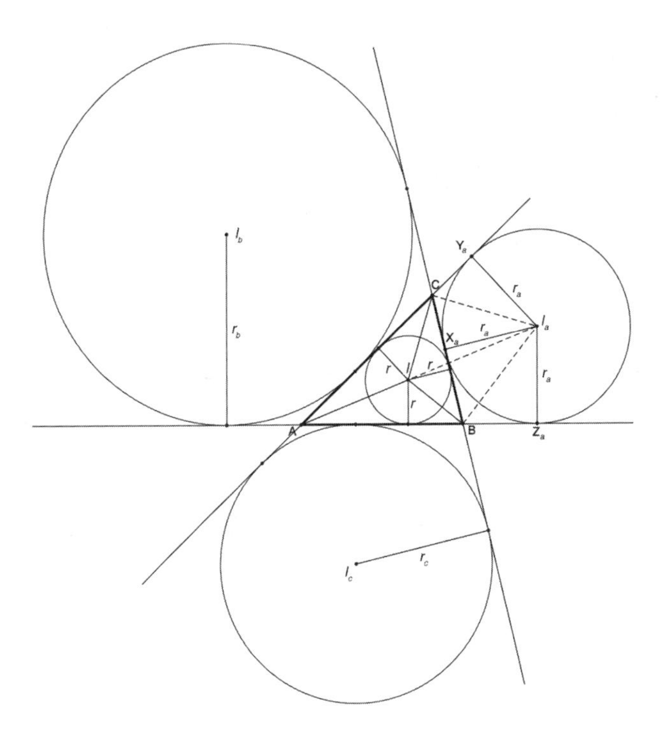


Figure 4-48

$$\text{Area}_{\Delta ABC} = \text{Area}_{\Delta BCI} + \text{Area}_{\Delta ACI} + \text{Area}_{\Delta ABI}.$$

$$\text{Area}_{\Delta ABC} = \tfrac{1}{2} IX \cdot BC + \tfrac{1}{2} IY \cdot AC + \tfrac{1}{2} IZ \cdot AB.$$

$$\text{Area}_{\Delta ABC} = \tfrac{1}{2} ra + \tfrac{1}{2} rb + \tfrac{1}{2} rc = \tfrac{1}{2} r(a+b+c) = sr, \text{ where } s = \tfrac{1}{2}(a+b+c).$$

Therefore, $r = \frac{\text{Area}_{\Delta ABC}}{s}$, which relates the area of a triangle to the radius of the inscribed circle.

We shall now examine the *exradii* (radii of the escribed circles) of a triangle in an analogous fashion.

That is, we will show that an exradius of a triangle equals the ratio of the area of the triangle to the difference between the semiperimeter and the side to which the excircle considered is internally tangent (to be internally tangent, we mean where it is tangent to the side of the triangle, rather than its extension—beyond the actual triangle).

In figure 4-48, we have the following:

$$\text{Area}_{\Delta ABC} = \text{Area}_{\Delta ABI_a} + \text{Area}_{\Delta ACI_a} - \text{Area}_{\Delta BCI_a},$$

$$\text{Area}_{\Delta ABC} = \tfrac{1}{2} I_a Z_a \cdot AB + \tfrac{1}{2} I_a Y_a \cdot AC - \tfrac{1}{2} I_a X_a \cdot BC, \text{ and}$$

$$\text{Area}_{\Delta ABC} = \tfrac{1}{2} r_a c + \tfrac{1}{2} r_a b - \tfrac{1}{2} r_a a = \tfrac{1}{2} r_a (c+b-a) = r_a (s-a);$$

therefore, $r_a = \dfrac{\text{Area}_{\Delta ABC}}{s-a}$.

In a similar manner, we can establish that $r_b = \frac{\text{Area}_{\Delta ABC}}{s-b}$ and $r_c = \frac{\text{Area}_{\Delta ABC}}{s-c}$. If we multiply these results, we get some nice outcomes (just for the reader's enjoyment):

$$r \cdot r_a \cdot r_b \cdot r_c = \frac{\text{Area}_{\Delta ABC}}{s} \cdot \frac{\text{Area}_{\Delta ABC}}{s-a} \cdot \frac{\text{Area}_{\Delta ABC}}{s-b} \cdot \frac{\text{Area}_{\Delta ABC}}{s-c}$$

$$= \frac{(\text{Area}_{\Delta ABC})^4}{s(s-a)(s-b)(s-c)}.$$

This denominator reminds us of Heron's formula for finding the area of a triangle:

$$\text{Area}_{\Delta ABC} = \sqrt{s(s-a)(s-b)(s-c)}.$$

Thus $(\text{Area}_{\Delta ABC})^2 = s(s-a)(s-b)(s-c)$.

By substitution: $r \cdot r_a \cdot r_b \cdot r_c = (\text{Area}_{\Delta ABC})^2$.

We now return to the special right triangle that we featured in figure 4-46. We have right $\Delta ABC$ ($\angle ACB = 90°$) with side lengths $b$, $2b$, and $2b+1$. Applying the Pythagorean theorem to this triangle, we get $(2b+1)^2 = 4b^2 + b^2$, which gives us $4b^2 + 4b + 1 = 4b^2 + b^2$, and further simplifies to $b^2 - 4b - 1 = 0$.[11] This quadratic equation has as its positive root $b = 2 + \sqrt{5}$ (we disregard the negative root, since for our geometric application we cannot use the negative root: $2 - \sqrt{5}$). Putting this in terms of our now-familiar $\phi$, we get $b = 2 + \sqrt{5} = \phi^3 = 2\phi + 1$.

Now that we have the lengths of the sides of this triangle (fig. 4-46), we can find its perimeter and area. But notice that we are finding them in terms of the golden ratio!

The sides of this particular right triangle $ABC$ are:

$$b = 2 + \sqrt{5} = \phi^3 = 2\phi + 1,$$

$$a = 2b = 2 \cdot (2 + \sqrt{5}) = 2\phi^3 = 4\phi + 2, \text{ and}$$

$$c = a + 1 = 5 + 2\sqrt{5} = 4\phi + 3.$$

We can get the semiperimeter as follows:

$$s = \frac{a+b+c}{2} = \frac{4\phi + 2 + 2\phi + 1 + 4\phi + 3}{2} = 5\phi + 3.$$

The area of the triangle can then be easily found:

$$\text{Area}_{\Delta ABC} = \frac{BC \cdot AC}{2} = \frac{a \cdot b}{2} = \frac{(4\phi + 2) \cdot (2\phi + 1)}{2} = \frac{8\phi^2 + 4\phi + 4\phi + 2}{2}$$

$$= 4\phi^2 + 4\phi + 1 = 8\phi + 5 = 9 + 4\sqrt{5}.$$

We are now ready to find the various equiradii of $\Delta ABC$:

$$r = \frac{A_{\Delta ABC}}{s} = \frac{8\phi + 5}{5\phi + 3} = \frac{\sqrt{5} + 1}{2} = \phi,$$

$$r_a = \frac{A_{\Delta ABC}}{s - a} = \frac{8\phi + 5}{\phi + 1} = \frac{3\sqrt{5} + 7}{2} = 3\phi + 2 = \phi^4,$$

$$r_b = \frac{A_{\Delta ABC}}{s - b} = \frac{8\phi + 5}{3\phi + 2} = \frac{\sqrt{5} + 3}{2} = \phi + 1 = \phi^2, \text{ and}$$

$$r_c = \frac{A_{\Delta ABC}}{s - c} = \frac{8\phi + 5}{\phi} = \frac{5\sqrt{5} + 11}{2} = 5\phi + 3 = \phi^5.$$

Once again, because we were able to represent these radii in terms of the golden ratio, we seem to legitimatize our referring to these radii as *golden radii*.

## THE GOLDEN ANGLE

It is now only fitting that we also identify a golden angle. As you by now suspect, we will show an angle that can be expressed in terms of the golden ratio. Let us consider an angle that partitions a circle into two arcs whose measure relate as the golden ratio. In figure 4-49, the angles $\beta$ and $\gamma$ can be shown to be in the golden ratio:

$$\frac{\text{circumference}}{\text{arc of length } a} = \frac{a+b}{a} = \frac{\text{arc of length } a}{\text{arc of length } b} = \phi.$$

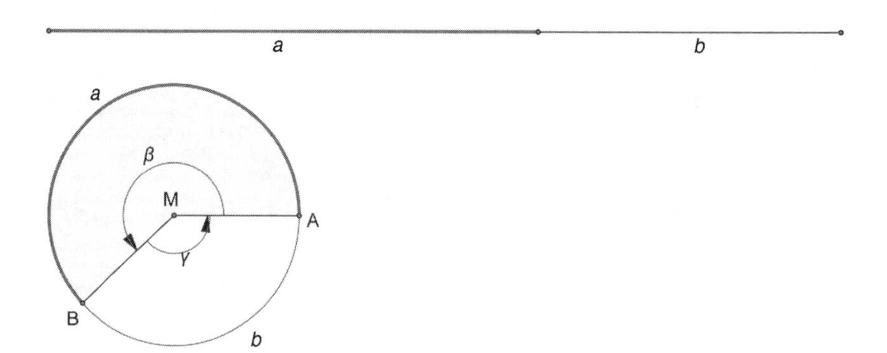


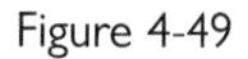
Figure 4-49

Another way of expressing this is $\frac{360°}{\phi} = \frac{\varphi}{\psi} = \phi$. Therefore, you can calculate the following:

$$\varphi = \frac{360°}{\phi} = 222.4922359\ldots° \approx 222.5° \text{ and}$$

$$\psi = 360° - \frac{360°}{\phi} = 137.5077640\ldots° \approx 137.5°.$$

This angle, $\psi$, which also manifests itself in nature, is sometimes called the *golden angle.*

## THE GOLDEN PENTAGON AND THE GOLDEN PENTAGRAM

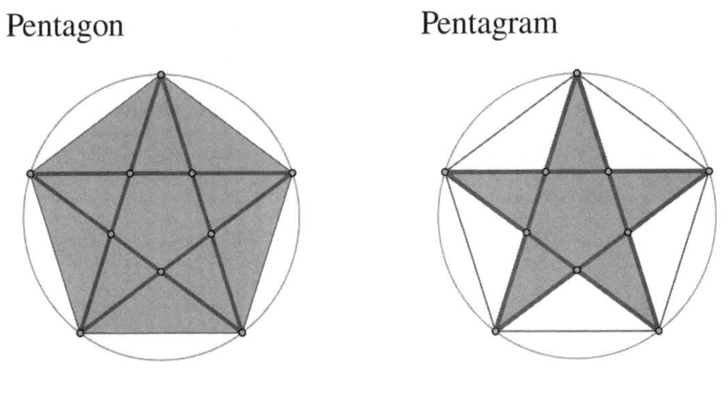


Figure 4-50

Figure 4-51

In figure 4-50, we show a regular pentagon, that is, one that has all sides and angles equal. (When we speak of a pentagon from here on, we will be referring to the *regular* pentagon—one with all sides, diagonals, and angles equal. Similarly, we will only be referring to a *regular* pentagram.) When the diagonals are drawn in the regular pentagon, a five-pointed star appears. This is called a pentagram (fig. 4-51). As you might suspect by now, this figure contains the golden ratio throughout. Yet before we explore the characteristics of the pentagon and the pentagram, we should note that the pentagram was the secret identification symbol of the Pythagoreans, a member of which was Hippasus of Metapontum (ca. 450 BCE). He presumably discovered that the ratio of the diagonal to the side of a regular pentagon could not be expressed as a fraction consisting of natural numbers. This introduced the notion of irrational numbers, which was very upsetting to the Pythagoreans, who wanted everything expressed in numbers. Their irrational number later turned out to have been based on the value of $\phi$ (see chap. 3).

Hippasus's discovery that there are lengths that do not share a common measure, or are said to be incommensurable, led to a schism within the Pythagorean community. There were the Akousmatikoi (listeners) who unquestionably accepted the word of the master, and the mathematikoi (learners), who accepted the new findings of Hippasus. According to some lore, Hippasus's drowning in a shipwreck was punishment by the gods for exposing the secret of incommensurability. There are, of course, other versions of his demise, most of which tie him to this finding of incommensurability. The pentagram has also emerged in various cultural settings. It has been a symbol that was to ward off witches and evil spirits, as it appears in Goethe's *Faust*, when it is drawn on the floor with chalk to prevent Mephisto from approaching the room.

Figure 4-52

The pentagram is omnipresent in today's society as well. It can be found on over sixty national flags—a case in point is our American flag! The European Union flag also contains the pentagram. We show a few of these in figure 4-53.

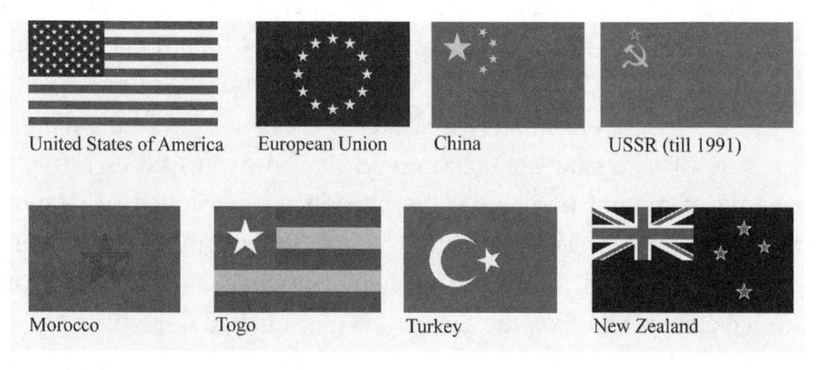


Figure 4-53

Furthermore, in addition to exhibiting the pentagram, the Togo flag is claimed to be the only flag in the world whose dimensions are also in the golden ratio.[12] Some countries whose flag dimensions approach the golden ratio—yet involve the Fibonacci numbers—in order of closeness to the golden ratio are the following:

| | |
|---|---|
| Switzerland | 1 : 1 |
| New Zealand | 1 : 2 |
| European Union | 2 : 3 |
| Turkey | 2 : 3 |
| Germany | 3 : 5 |
| Poland | 5 : 8 |
| and finally the unique flag of Togo | $1 : \phi$ |

We begin our discussion of the pentagon and the pentagram by noticing that there are five overlapping golden triangles: $\Delta ADC$, $\Delta BED$, $\Delta CAE$, $\Delta DAB$, and $\Delta EBC$. This is shown in figure 4-54 along with the inscribed and circumscribed circles. Recall that the golden triangle has a vertex angle of 36° and base angles of 72° each. The angles of the pentagon each consist of 108°. It is easy to see that the diagonals are parallel to the opposite side of the pentagon and the diagonals trisect each of the angles of the pentagon.

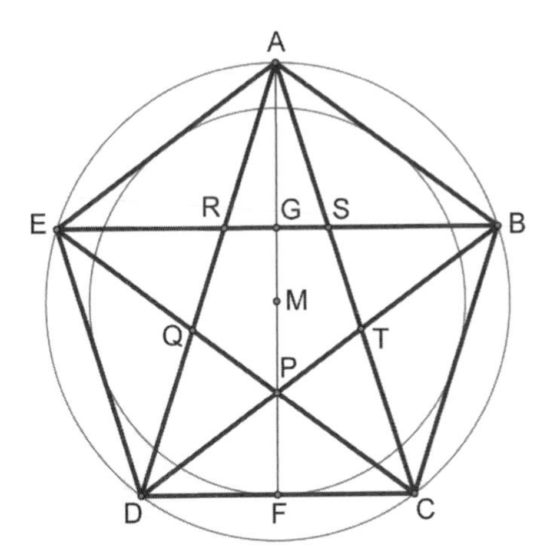


Figure 4-54

The regular pentagon contains thirty-five triangles—with six variations as depicted in the chart in figure 4-55. Furthermore, we have types IV and V, which are congruent, types II and IV, which are similar to each other, and types I, III, and V, which are also similar to each other. Surely you will have recognized the golden triangle!

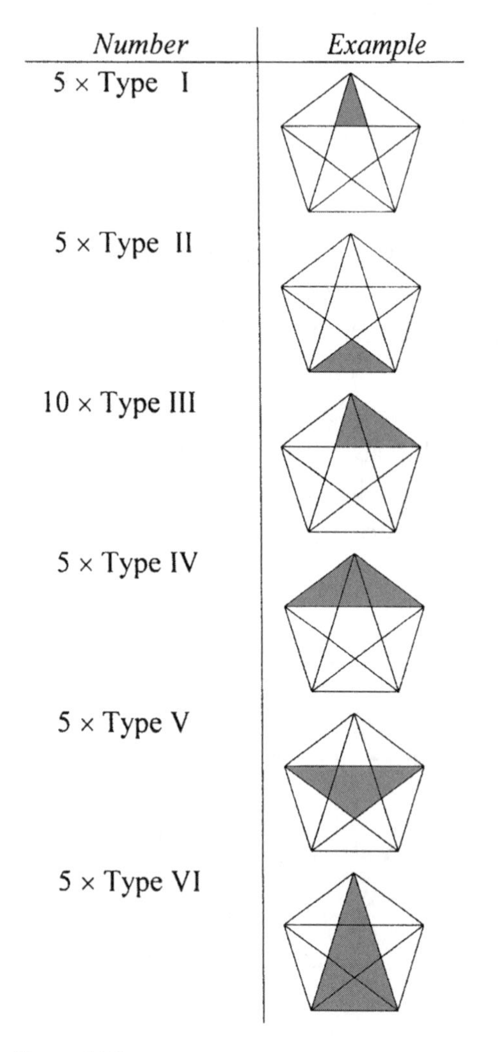

Number
Example
5 × Type I
5 × Type II
10 × Type III
5 × Type IV
5 × Type V
5 × Type VI


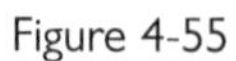
Figure 4-55

The regular pentagon also contains twenty quadrilaterals in four types and one pentagon, as shown in the chart in figure 4-56.

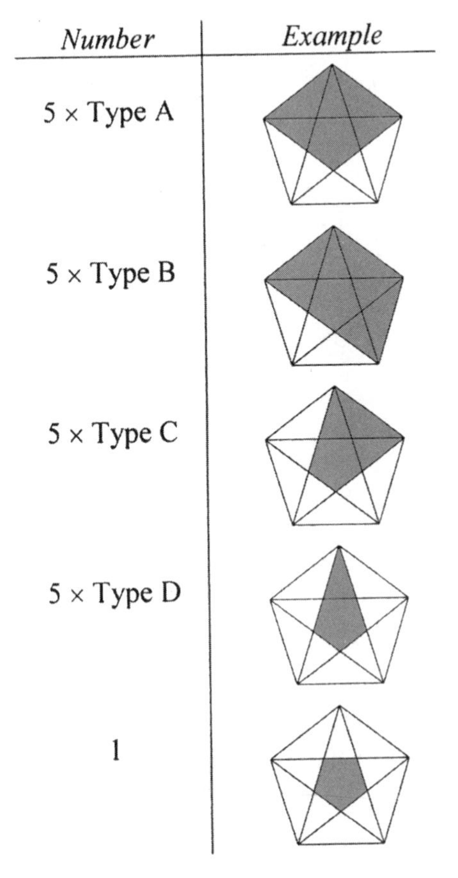

Number
Example
5 × Type A
5 × Type B
5 × Type C
5 × Type D
1

Figure 4-56

By now you should notice that the diagonals intersect each other in the golden section. Yet we shall now show the various other golden sections present in this very rich example of the golden ratio. The following are just some of the appearances of the golden ratio:

- Pentagon side **:** radius of the inscribed circle
- Pentagon side **:** radius of the circumscribed circle
- Radius of the inscribed circle **:** radius of the circumscribed circle

- Areas of various figures in the pentagon
- Area of pentagon : area of pentagram
- Areas of pentagons

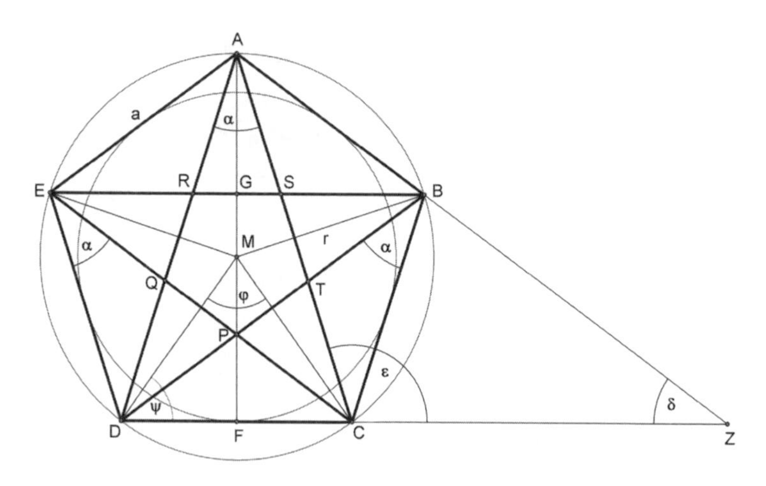


Figure 4-57

For your convenience we will identify the parts of the pentagon in figure 4-57:

*Side lengths:*

$AE = a$ (pentagon side)
$AD = d$ (diagonal of the pentagon, or side of the pentagram)
$AM = r$ (radius of the circumscribed circle)
$FM = \rho$ (radius of the inscribed circle)
$AF = b$ (pentagon height; also bisects the side and bisects the angle)
$AG = c$ (height, perpendicular bisector, and angle bisector for $\Delta ABE$ and $\Delta ARS$)
$AR = e$ (exterior portion of pentagram side)
$RS = f$ (side of smaller pentagon)

*Angle measures*:

$\angle CMD = \varphi = \frac{360°}{5} = 72°$
$\angle CDM$[13] $= \psi = 54°$
$\angle DCE = \angle ACE = \alpha = 36°$

$$\angle ACD = \angle ACE + \angle DCE = 2\alpha = 72^\circ$$

$$\angle ACZ = \varepsilon = 180^\circ - \angle ACD = 180^\circ - 72^\circ = 108^\circ$$

$$\angle AZC = \angle AZD = \delta = 36^\circ$$

$$\angle CAZ = \angle CAB = \alpha = 36^\circ$$

$$\angle AZC = 180^\circ - \angle CAB - \angle ACZ = 180^\circ - \alpha - \varepsilon = 180^\circ - 36^\circ - 108^\circ = 36^\circ$$

$$\angle CAD = \angle DBE = \angle ACE = \angle BDA = \angle BEC = 36^\circ$$

With the angle measures given above, the remaining angles in the figure can be determined, largely due to the symmetry of the pentagon or pentagram. We can see that all the triangles shown in figure 4-55 are golden triangles. Included are both acute triangles and obtuse triangles that we identified earlier as having the golden ratio along their side lengths.

The quadrilateral of type A (fig. 4-56) is a rhombus with sides of length $a$ and angles of 72° and 108°. (This is shown in fig. 4-58.)

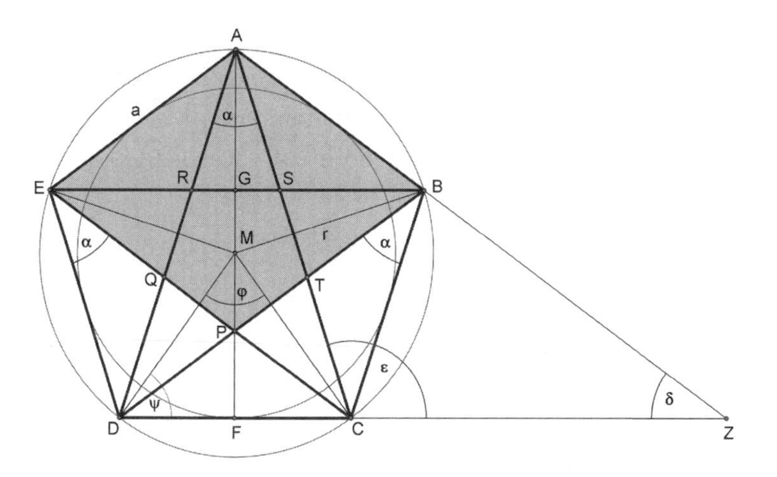


Figure 4-58

In figure 4-58, we see that the ratio of the rhombus's longer diagonal ($BE$) is in the golden ratio to the side of the pentagon. That is, when we have $BE = d$ and $AB = a$, we get $\frac{d}{a} = \phi$, and then

$$d = a\phi = a\frac{\sqrt{5}+1}{2}.$$

We can also show that the shorter diagonal of the rhombus is also related to the side of the pentagon in the following manner:

$$AP = AG + GP = c + c = 2c.$$

In $\Delta AEG$, $\sin\angle AEG = \sin\alpha = \frac{AG}{AE} = \frac{c}{a}$.

Therefore, $c = a \cdot \sin\alpha = \frac{\sqrt{5-\sqrt{5}}}{8} \cdot a = \frac{\sqrt{\phi^2+1}}{2\phi} \cdot a$.

We can write this as

$$\frac{c}{a} = \frac{\sqrt{\phi^2+1}}{2\phi}, \text{ or } \frac{2c}{a} = \frac{\sqrt{\phi^2+1}}{\phi} = \frac{AP}{AE}.$$

Again as we compare a diagonal of the rhombus to a side of the pentagon, the golden ratio appears.

Let us now focus on the two isosceles trapezoids shown as types B and C in figure 4-56, and shown in the pentagon in figures 4-59 and 4-60. Here we have the two base lengths of each of the two isosceles trapezoids in the golden ratio.

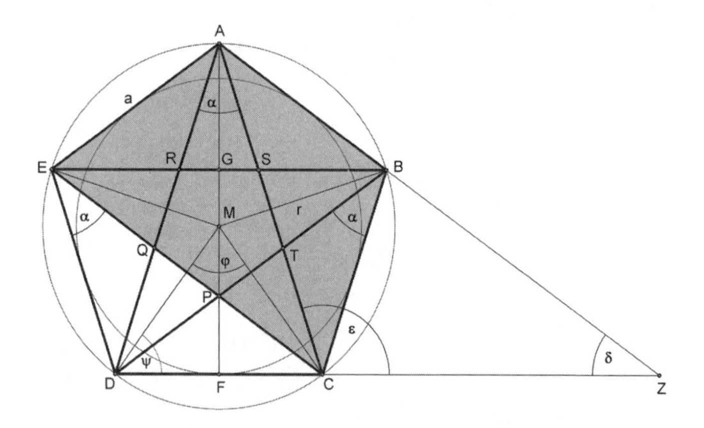


Figure 4-59

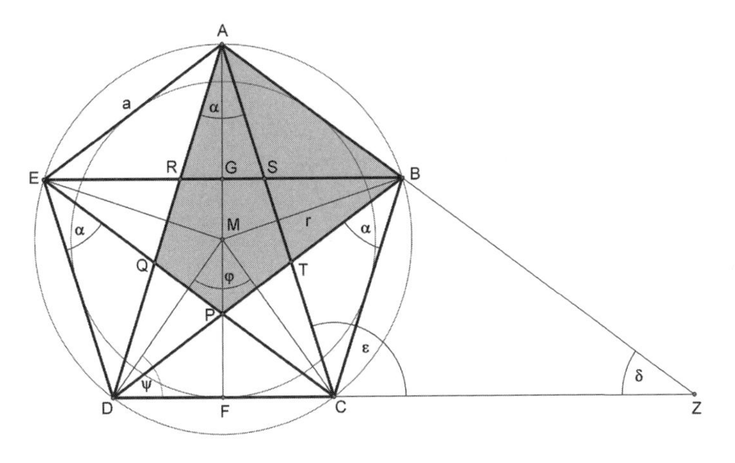


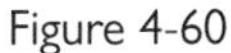
Figure 4-60

In each case you will notice that the bases are comprised of a side of the pentagon and a portion or a complete length of one of its diagonals. The shaded trapezoid in figure 4-59 has as its bases a side of the golden triangle ($EC$, a leg of $\Delta BEC$) and its base ($AB = BC$, the base of $\Delta BEC$). Therefore, given our previously determined golden triangle ratios, the ratio of the bases of the trapezoid $ABCE$ (fig. 4-59) produces the golden ratio as $\frac{EC}{AB} = \phi$.

In a similar fashion, we can show that for the trapezoid $ABPQ$ (fig. 4-60) the ratio of its two base lengths also involves the golden ratio as $\frac{AB}{QP} = \phi^2$. (A justification for this is provided in the appendix.)

The golden ratio is also present in the kite quadrilateral seen in type D of figure 4-56, and shown again in figure 4-61. Quadrilateral $QTRE$ is a parallelogram since its opposite sides are parallel. Therefore, $QT = RE = AR = e$. We know that $\frac{a}{e} = \phi$, so for one possible ratio on this kite quadrilateral, we have the ratio of a side to a diagonal, $\frac{QA}{TQ} = \phi$. (More on this can be found in the appendix.)

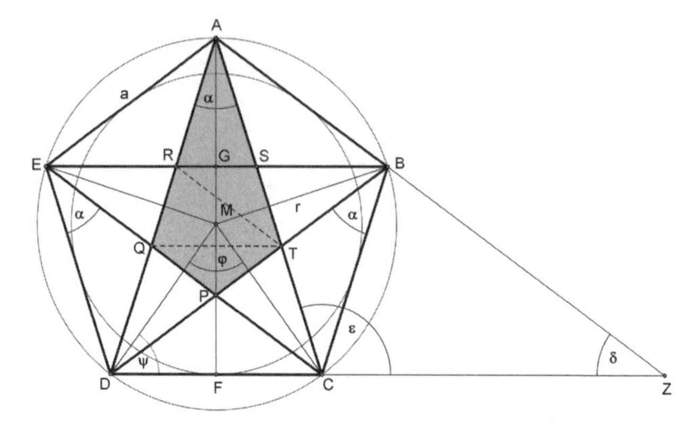


Figure 4-61

This leaves us to inspect the smaller pentagon (fig. 4-62), which we will relate to the original one to see if the golden ratio appears in such a comparison. The sides of these two pentagons are, as we might expect, also in the golden ratio, as we earlier established (in another context) as 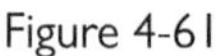 $\frac{AB}{QP}=\phi^2$.

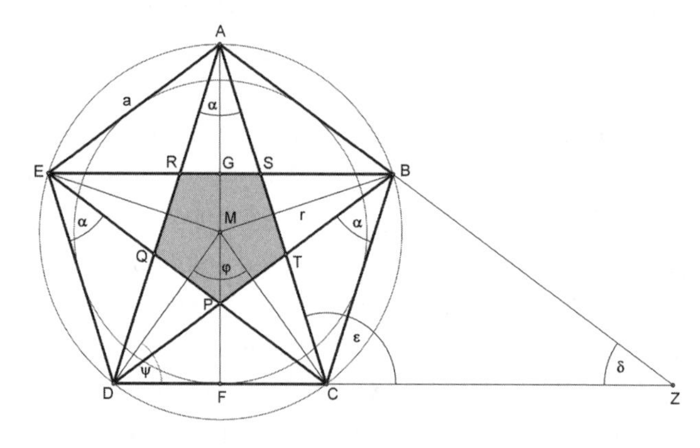


Figure 4-62

Since we have had so many appearances of the golden ratio in this configuration, we will summarize some of the main relationships that we just highlighted.

- The ratio of the diagonal of the (larger) pentagon to one of its sides is

$$\phi = \frac{1+\sqrt{5}}{2}.$$

- Diagonals of the pentagon divide each other into the golden ratio.
- The pentagon's diagonals are partitioned into the golden ratio at two points.
- The side of the pentagon (e.g., *AE*) and the external side of the pentagram (e.g., *AR*) are in the golden ratio.

From the diagram in figure 4-63, we can clearly see that the ratio of the sides of the two pentagons is also related to $\phi$, with the following ratio: $\phi^2 : 1$. Furthermore, from the diagram in figure 4-63, we can see in another way what we already stated, namely that the points of intersection of the diagonals of the pentagon partition a diagonal into segments that maintain the golden ratio.

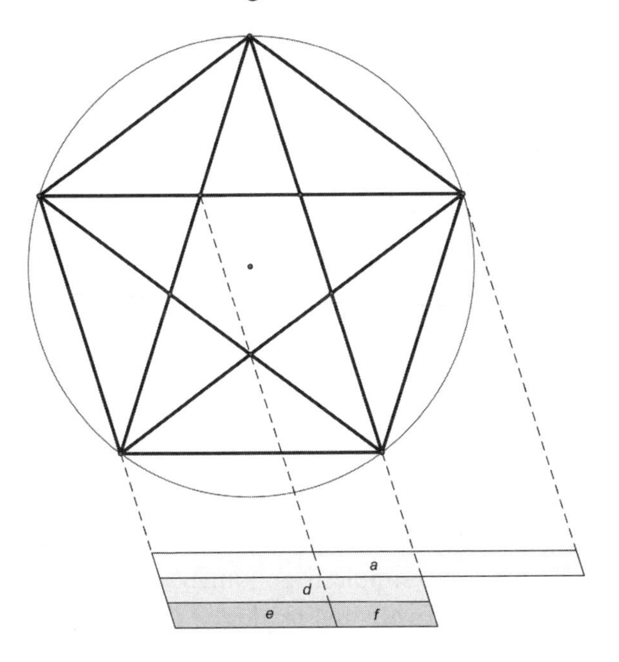

Figure 4-63

We can summarize this as follows:

$$d=\phi\cdot a=\phi^2\cdot e=\phi^3\cdot f$$
$$a=\phi\cdot e=\phi^2\cdot f$$
$$e=\phi\cdot f.$$

Because $d=a+e$, it follows that $\phi^3\cdot f=\phi^2\cdot f+\phi\cdot f$. This leads us to the now familiar relationship $\phi^3=\phi^2+\phi$. Notice the consistency of these relationships both in geometric contexts and in algebraic contexts. Just as we would expect in mathematics!

Having thoroughly investigated the relationships between the parts of the pentagon and the parts of the pentagram, as well as their interrelations, we now turn to the relationship between the pentagon and its circumscribed circle—more specifically to that circumscribed circle's radius, often called the *circumradius*.

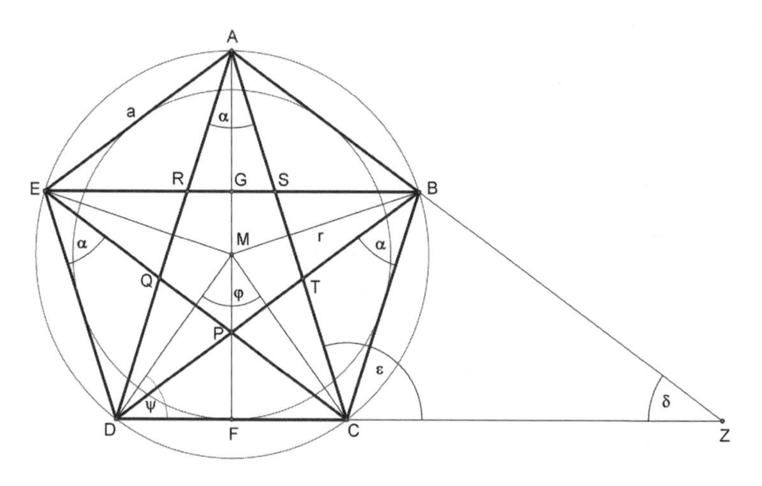


Figure 4-64

To find the radius of the circumscribed circle, $r$, in terms of the side of the pentagon, we will use some simple trigonometry (fig. 4-64). Since the radii about point $M$ make five equal angles, $\angle CMD=\varphi=72°$. So that half that angle $\frac{\varphi}{2}=36°$.

We now take

$$\sin\frac{\phi}{2}=\frac{CF}{CM}=\frac{\frac{a}{2}}{r},$$

which can be transformed to

$$r=\frac{a}{2}\cdot\frac{1}{\sin\frac{\phi}{2}}, \text{ or } r=\frac{a}{2}\cdot\frac{1}{\sin 36^\circ}.$$ [14]

We now can get the value of

$$r=\frac{a}{2}\cdot\frac{1}{\sin 36^\circ}=\frac{a}{2}\cdot\frac{2\phi}{\sqrt{\phi^2+1}}=a\cdot\frac{\phi}{\sqrt{\phi^2+1}}=a\cdot\sqrt{\frac{5+\sqrt{5}}{10}},$$ [15]

also $a=\frac{r}{2}\cdot\sqrt{\frac{50-10\sqrt{5}}{5}}=\frac{r}{2}\cdot\sqrt{10-2\sqrt{5}}=r\cdot\sqrt{\frac{5-\sqrt{5}}{2}}=r\cdot\frac{\sqrt{\phi^2+1}}{\phi}.$ [16]

Therefore, the ratio of the radius of the circumscribed circle to the side of the pentagon is

$$\frac{r}{a}=\frac{a\cdot\sqrt{\frac{5+\sqrt{5}}{10}}}{a}=\sqrt{\frac{5+\sqrt{5}}{10}}=\frac{\phi}{\sqrt{\phi^2+1}}\approx 0.8506508083.$$

The reciprocal of this is

$$\frac{a}{r}=\sqrt{\frac{10}{5+\sqrt{5}}}=\frac{\sqrt{\phi^2+1}}{\phi}\approx 1.175570504.$$

Again, we find the relationship between parts of the pentagon expressible in terms of the golden ratio. However, having shown the relationship of the circumscribed circle and the side of the pentagon, it is incumbent on us to now show the relationship of the inscribed circle with the pentagon's side. That is, we now seek a relationship between the side of the pentagon, $a$, and the inscribed circle's radius, $\rho$.

We begin with representing the height, $b=AF$, of $\Delta ACD$ as follows:

$$b=d\cdot\cos\frac{\alpha}{2}=a\cdot\phi\cdot\cos\frac{36^\circ}{2}=a\cdot\phi\cdot\sqrt{\frac{5+\sqrt{5}}{8}}$$

$$=\frac{a}{2}\cdot\sqrt{5+2\sqrt{5}}=\frac{a}{2}\cdot\phi\sqrt{\phi^2+1}.$$

With $b=AF=\frac{a}{2}\cdot\sqrt{5+2\sqrt{5}}=\frac{a}{2}\cdot\phi\cdot\sqrt{\phi^2+1}=a\cdot\sqrt{\frac{5+2\sqrt{5}}{4}}$,

and $r=AM=\frac{a}{10}\cdot\sqrt{50+10\sqrt{5}}=a\cdot\sqrt{\frac{5+\sqrt{5}}{10}}$,

we are able to get the inscribed circle's radius,

$$\rho=FM=AF-AM=b-r$$

$$\rho=b-r=a\cdot\sqrt{\frac{5+2\sqrt{5}}{4}}-a\cdot\sqrt{\frac{5+\sqrt{5}}{10}}=\frac{a}{\sqrt{20}}\cdot\sqrt{5+2\sqrt{5}}=a\cdot\sqrt{\frac{5+2\sqrt{5}}{20}}.$$

(It's not trivial to prove $\sqrt{5+2\sqrt{5}}=\sqrt{25+10\sqrt{5}}-\sqrt{10+2\sqrt{5}}$, so we provide this in the appendix.)

Therefore, we can express

$$\rho=a\cdot\sqrt{\frac{5+2\sqrt{5}}{20}}=a\cdot\frac{\phi^2}{2\sqrt{\phi^2+1}}.$$

Now to set up the desired ratios, we first have

$$\frac{\rho}{a}=\frac{a\cdot\sqrt{\frac{5+2\sqrt{5}}{20}}}{a}=\sqrt{\frac{5+2\sqrt{5}}{20}}=\frac{\phi^2}{2\sqrt{\phi^2+1}}\approx 0.6881909602.$$

Or the reciprocal value:

$$\frac{a}{\rho}=\frac{2\sqrt{\phi^2+1}}{\phi^2}=\frac{\sqrt{20}}{\sqrt{5+2\sqrt{5}}}=\frac{\sqrt{20}\sqrt{5-2\sqrt{5}}}{\sqrt{5}}=2\cdot\sqrt{5-2\sqrt{5}}$$
$$\approx 1.453085056.$$

In both cases, you see that, again, the ratio involves the golden ratio.

It now remains for us to inspect the ratio of the two radii—namely the radius of the circumscribed circle and the radius of the inscribed circle. Using the previously determined values of the radii, we have the following ratio:

$$\frac{r}{\rho}=\frac{\frac{\phi}{\sqrt{\phi^2+1}}}{\frac{\phi^2}{2\sqrt{\phi^2+1}}}=\frac{2}{\phi}=\sqrt{5}-1\approx 1.236067976,$$

or the reciprocal ratio:

$$\frac{\rho}{r}=\frac{\phi}{2}=\frac{\sqrt{5}+1}{4}\approx 0.8090169943.$$

Now that we have found just about all the linear comparisons involving the pentagon and the pentagram through the plethora of golden ratio appearances, we might see how the areas of the various previously mentioned parts compare.

We first compare the areas of the various triangles that we identified in figure 4-55, which are repeated here for convenience as figure 4-65.

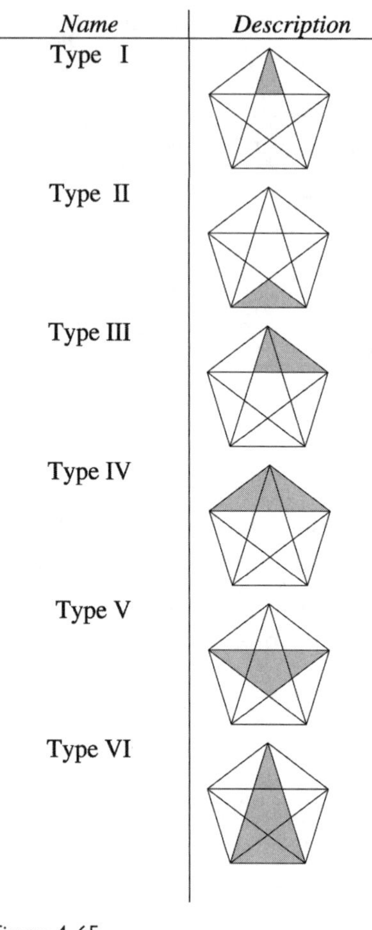

*Name*
*Description*
Type I
Type II
Type III
Type IV
Type V
Type VI


Figure 4-65

To make this comparison of areas more manageable, we provide a chart with the ratios of the areas of these triangles, which allows for direct comparisons. For example, to find the ratio of the area of a type III triangle to that of a type V triangle, we go to the intersection of the appropriate row and column to get

$$\frac{\sqrt{5}-1}{2}.$$

In the same way, we can find the ratio of the area of a type V triangle to that of a type III triangle, namely

$$\frac{\sqrt{5}+1}{2},$$

which, as you would have expected, is the reciprocal of the previously found value. You will also note in the chart of figure 4-66 that the areas of triangles of types IV and V are the same. This we can clearly see in figure 4-65.

| $\text{Area}_{\text{Type } i}:\text{Area}_{\text{Type } j}$ | Type I | Type II | Type III | Type IV | Type V | Type VI |
|---|---|---|---|---|---|---|
| Type I | 1 | $\frac{\sqrt{5}-1}{2}$ | $\frac{3-\sqrt{5}}{2}$ | $\sqrt{5}-2$ | $\sqrt{5}-2$ | $\frac{7-3\sqrt{5}}{2}$ |
| Type II | $\frac{\sqrt{5}+1}{2}$ | 1 | $\frac{\sqrt{5}-1}{2}$ | $\frac{3-\sqrt{5}}{2}$ | $\frac{3-\sqrt{5}}{2}$ | $\sqrt{5}-2$ |
| Type III | $\frac{\sqrt{5}+3}{2}$ | $\frac{\sqrt{5}+1}{2}$ | 1 | $\frac{\sqrt{5}-1}{2}$ | $\frac{\sqrt{5}-1}{2}$ | $\frac{3-\sqrt{5}}{2}$ |
| Type IV | $\sqrt{5}+2$ | $\frac{\sqrt{5}+3}{2}$ | $\frac{\sqrt{5}+1}{2}$ | 1 | 1 | $\frac{\sqrt{5}-1}{2}$ |
| Type V | $\sqrt{5}+2$ | $\frac{\sqrt{5}+3}{2}$ | $\frac{\sqrt{5}+1}{2}$ | 1 | 1 | $\frac{\sqrt{5}-1}{2}$ |
| Type VI | $\frac{3\sqrt{5}+7}{2}$ | $\sqrt{5}+2$ | $\frac{\sqrt{5}+3}{2}$ | $\frac{\sqrt{5}+1}{2}$ | $\frac{\sqrt{5}+1}{2}$ | 1 |

Figure 4-66

We shall now compare the area of the original pentagon to that of the pentagram. To accomplish this, we will consider the area of the pentagon as the sum of the areas of two type IV triangles and one type VI triangle. This should be intuitively clear from the symmetry in the diagrams shown in figure 4-65.

$$\text{Area}_{\text{Pentagon}} = 2 \cdot \text{Area}_{\Delta\text{Type IV}} + \text{Area}_{\Delta\text{Type VI}}$$

$$= 2a^2 \cdot \frac{\sqrt{2}\cdot\sqrt{5+\sqrt{5}}}{8} + a^2 \cdot \frac{\sqrt{5+2\sqrt{5}}}{4}$$

$$= \frac{5a^2}{4} \cdot \frac{\phi^2}{\sqrt{\phi^2+1}} \approx 1.720477400 \cdot a^2.$$

To get the area of the pentagram, we will remove five of the type II triangles from the area of the pentagon:

$$\text{Area}_{\text{Pentagram}} = \text{Area}_{\text{Pentagon}} - 5 \cdot \text{Area}_{\Delta\text{Type II}} = a^2 \cdot \frac{\sqrt{25-10\sqrt{5}}}{2}$$

$$= \frac{\sqrt{5}a^2}{2} \cdot \frac{\sqrt{\phi^2+1}}{\phi^2} \approx 0.8122992405 \cdot a^2.$$

We are now ready to establish the ratio of the areas of the pentagon and the pentagram.

$$\frac{\text{Area}_{\text{Pentagon}}}{\text{Area}_{\text{Pentagram}}} = \frac{\frac{\sqrt{25+10\sqrt{5}}}{4}}{\frac{\sqrt{25-10\sqrt{5}}}{2}} = \frac{\sqrt{5}+2}{2} = \phi + \frac{1}{2} \approx 2.118033988.$$

We established earlier that the ratio of the sides of the two pentagons is $\frac{a}{f} = \phi^2$. Since the ratio of the areas of two similar figures (in this case, pentagons) is the square of the ratio of the corresponding linear parts, it follows directly that

$$\frac{\text{Area}_{\text{Pentagon}}}{\text{Area}_{\text{Smaller Pentagon}}} = \phi^4.$$

In other words, the ratio of the two pentagons is $\phi^4 : 1$. This can be further justified as follows:

$$\frac{\text{Area}_{\text{Pentagon}}}{\text{Area}_{\text{Smaller Pentagon}}}=\frac{\frac{\sqrt{25+10\sqrt{5}}}{4}}{\frac{\sqrt{2}\cdot\sqrt{125-55\sqrt{5}}}{8}}=\frac{3\sqrt{5}+7}{2}=\phi^4\approx 6.854101966.$$

This now completes our comparison of the various parts of this inscribed regular pentagon—each time the comparison made relies on the omnipresent golden ratio.

## POLYGON CONSTRUCTIONS

As we lead up to the construction of a regular pentagon, we should briefly review the procedures for construction of other familiar regular polygons. (Bear in mind that when we speak of geometric constructions, the tools we consider using are the unmarked straightedge and compasses, and *not* such tools as a protractor.) To construct a regular polygon of three sides—better known as an *equilateral triangle*—we simply draw line segment $AB$, and with this length as radius, we construct two circles as shown in figure 4-67 and complete the triangle.

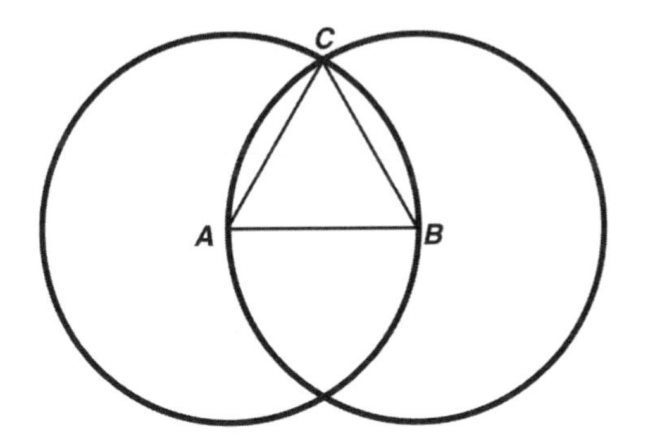


Figure 4-67

In figure 4-68 we show the construction of a regular hexagon, which simply involves repeating six times the process shown for the equilateral triangle. To make matters a bit simpler, we shall use the symbol ⊙ to indicate drawing a circle and the symbol ⊥ to indicate the drawing of a perpendicular line. The regular four-sided polygon—better known as a *square*—is also quite simple to construct. The construction can be done in a number of ways. One method is shown in figure 4-69. Here we simply begin with a line segment *AB* and then construct perpendiculars at each end and complete by drawing the circular arcs shown to locate the other two vertices of the square.

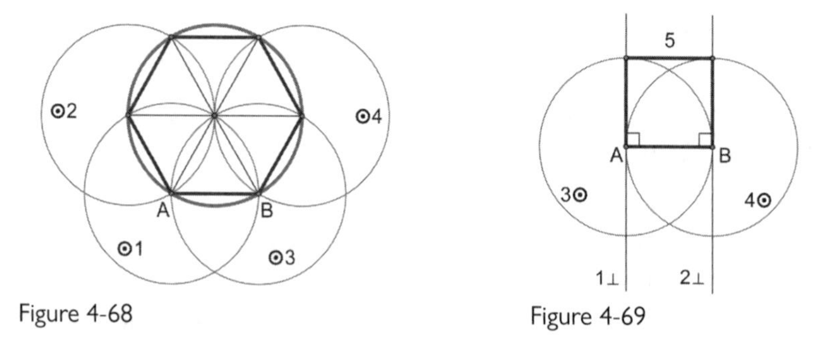


Figure 4-68

Figure 4-69

We now come to the construction of the regular pentagon. This is by no means simple—as some might describe the previous polygon constructions. The construction of the regular pentagon would be quite difficult if we were unfamiliar with the golden section. Moreover, we can say that the pentagon is constructible with unmarked straightedge and compasses, if we can use these tools to construct the golden section.

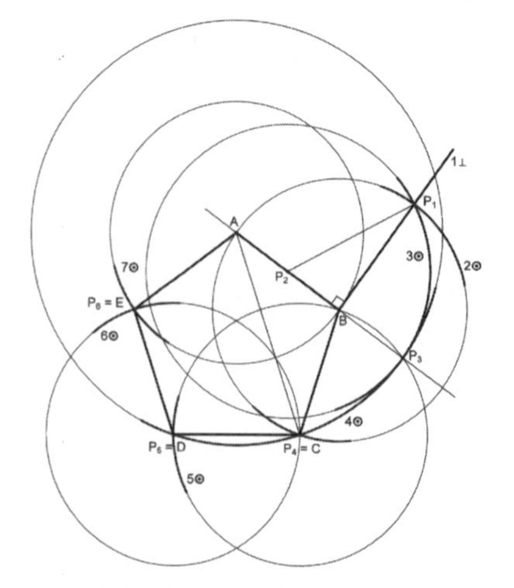


Figure 4-70

Unlike the previous three constructions, the one for a regular pentagon with an unmarked straightedge and compasses, figure 4-70 looks a bit overwhelming, but it can be rather easy to follow.

We begin with line segment $AB$, which will be the length of the side of the pentagon we are about to construct.

1⊥: At the point $B$, construct a perpendicular to the segment $AB$.

2⊙: Construct a circle at point $B$ with radius $AB$ and mark the point at which it cuts the perpendicular point $P_1$.

3⊙: Construct the midpoint $P_2$ of $AB$. Then with $P_2$ as center, construct a circle with radius $P_1P_2$ cutting $AB$ (extended) at $P_3$.

4⊙: Construct a circle with center at $A$ and with radius $AP_3$ cutting the circle that we drew in step 2 at point $P_4$ (we will call this point $C$). Then draw $BC$.

5⊙: Construct a circle with $P_4$ as center and with radius $BP_4$. Mark the point of intersection of this circle with the circle we drew in step 4 as point $P_5$ (which we will also call $D$). This circle will also contain $P_3$. Then draw $CD$.

6⊙: Construct a circle with center at $P_5$ with radius $P_4P_5$.

7⊙: Construct a circle with center at point $A$ and with radius $AB$ cutting the circle drawn in step 6 at point $P_6$ (we will call this point $E$). Then draw $DE$.

8: By drawing line segment $EA$, the pentagon is completed.

A brief explanation as to why this construction produces a regular pentagon is based on the construction we made of the golden section. In isosceles $\Delta ABC$, where we let $AC=d$ and $AB=a$ , the ratio of these two segments is $d : a=\phi$, which is the golden ratio. Using the earlier construction of the diagonal, we have $d=a\cdot\phi$ (see chap.1, p. 16). Once $\Delta ABC$ is established, by drawing circles with radius $a$ with centers at points $A$, $C$, and $D$, we get the remaining vertices of the pentagon.

By the way, as an alternative to the above construction, you will notice that during this procedure, when we arrived at $\Delta ABC$, we constructed an angle of 36°, namely $\angle BAC$. This angle would allow you to construct a regular decagon by locating the ten points on a circle with a central angle of 36°. Connecting these ten points consecutively determines a regular decagon.

We might also have arrived at a regular decagon from our original construction of the pentagon, by simply locating the center of the pentagon, constructing the circumscribed circle, and then bisecting each of the arcs determined by the pentagon's vertices to get the additional five vertices for the decagon.

## The Pentagon in Other Configurations

Having now constructed—and analyzed—the pentagon, we shall inspect some of the lovely relationships that can evolve from the pentagon. We begin by rotating the shaded regular pentagon in figure 4-71 by 72° at point *W*, and then we rotate the image of this pentagon again 72° at point *X*, and repeat this 72° rotation at points *Y* and *Z*, until we reach point *E*, at which point the original (shaded) pentagon will be in its original orientation.

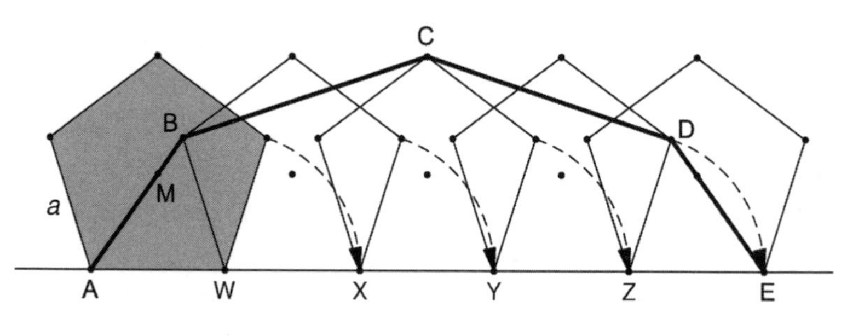


Figure 4-71

By joining the points *A*, *B*, *C*, *D*, and *E*, as shown in figure 4-71, we reveal a rather interesting configuration. We have an arch *ABCDE* where parts of it are in the golden ratio, as $\frac{BC}{AB}=\frac{CD}{DE}=\phi$. (So as not to distract from our admiration of the pentagon, the justification for this property and those that follow can be found in the appendix.) We might then call this a *golden arch*. Although it may appear to be counterintuitive, the area under the arch *ABCDE* is three times that of the original pentagon (shaded). Furthermore, an interesting aspect of this configuration is that the midpoint, $M$, of the original pentagon (shaded) lies on line segment $AB$. Furthermore, we have $\angle BAE = \angle AED = 54°$ and $\angle ABC = \angle BCD = \angle CDE = 144°$.

To further add to our admiration of the pentagon (with side of length $a$) consider the following: We have seen earlier that by drawing the pentagon's diagonals, we create another pentagon. This process can continue indefinitely with each successive pentagon a regular pentagon and therefore similar to the rest of the pentagons. To get from one pentagon's radius (i.e., the center of its circumscribed circle) to the next one we multiply by

$$k = \frac{3-\sqrt{5}}{2} = \frac{1}{\phi^2}.$$

Once again, our now-familiar golden section appears!

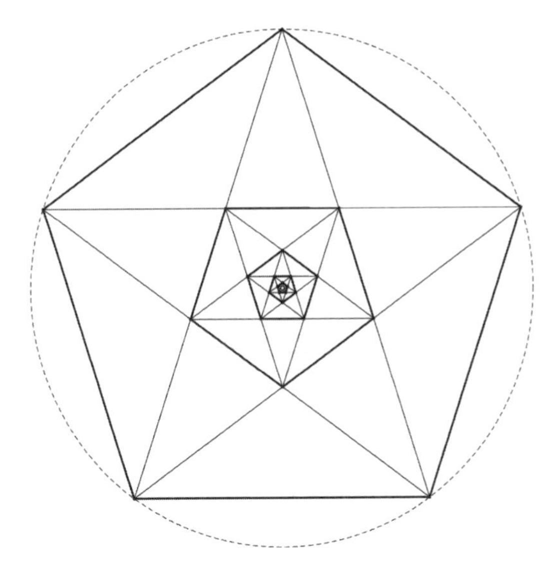

Figure 4-72

Using the multiplication factor $\frac{1}{\phi^2}$ between successive pentagons, we can generate the side lengths of the successive pentagons. Beginning with the outermost pentagon with side length $a$, we can multiply

$a \cdot \frac{1}{\phi^2}$ to get the side length of the next smaller pentagon, $a_1 = \frac{a}{\phi^2}$. The next smaller-size pentagon with side length $a_2$ is then found in the same way, by multiplying $a_1$ by $\frac{1}{\phi^2}$.

That is, $a_2 = \frac{a}{\phi^2} \cdot \frac{1}{\phi^2} = \frac{a}{\phi^4}$.

Continuing in this way, we have $a_3 = a_2 \cdot \frac{1}{\phi^2} = \frac{a}{\phi^4} \cdot \frac{1}{\phi^2} = \frac{a}{\phi^6}$.

Similarly, we then get $a_4 = a_3 \cdot \frac{1}{\phi^2} = \frac{a}{\phi^6} \cdot \frac{1}{\phi^2} = \frac{a}{\phi^8}$.

This can continue on indefinitely with the general term being

$$a_n = a_{n-1} \cdot \frac{1}{\phi^2} = \frac{a}{\phi^{2n}} \; (n = 1, 2, 3, \ldots, \text{ and } a_0 = a).$$

If we wish to generate successive pentagons getting larger (i.e., from inside outward), we would use the multiplication factor $\phi^2$ instead of $\frac{1}{\phi^2}$ (fig. 4-73).

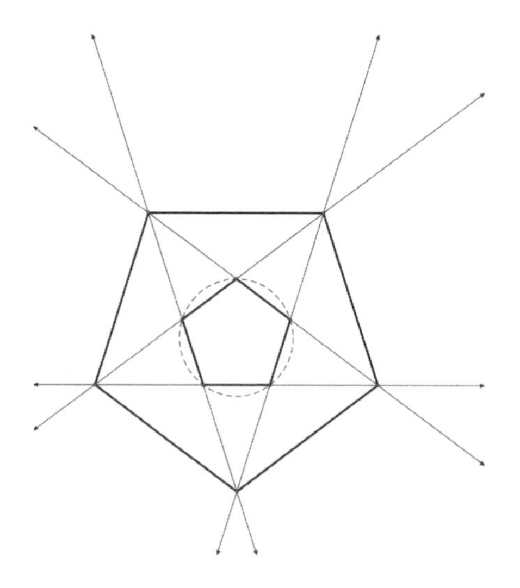

Figure 4-73

## THE GOLDEN ELLIPSE

An ellipse is recognized as an oval-looking shape. Yet its definition is a set of points the sum of whose distances to two given points is a constant. In figure 4-74 we have an ellipse where all points, $P$, on the ellipse are so situated that the sum of the distances to the two fixed points (these fixed points are called the *foci* of the ellipse), $F_1$ and $F_2$, is a constant. That is, for any point $P$ on the ellipse, $PF_1+PF_2$ is the same. We obtain the point $C$ on the ellipse (fig. 4-74) by getting the intersection with the perpendicular bisector of $F_1F_2$. This will allow us to identify the sum, $PF_1+PF_2$, which is $2 \cdot CF_1=2a$. If line $F_1F_2$ were extended to meet the ellipse at points $A$ and $B$, and if point $P$ were to assume the position of these points $A$ and $B$, then $AB=2a$.

We will designate $CD=2b$, and we have $CF_1=CF_2=a$ and $F_1F_2=2e$. Then, by the Pythagorean theorem,

$$e=\sqrt{a^2-b^2} \text{ (with } a \geq b).$$

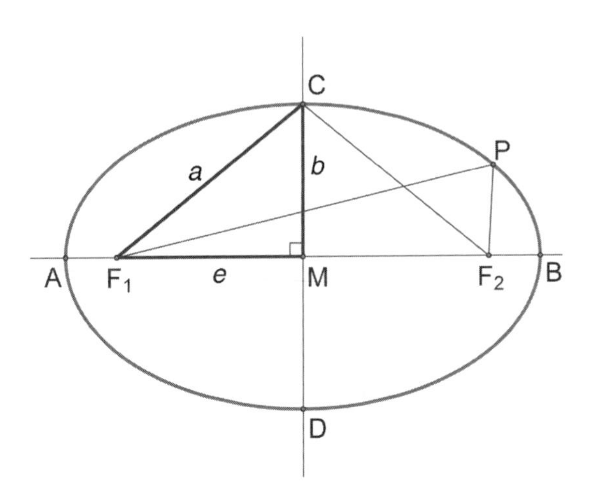


Figure 4-74

If the two foci, $F_1$ and $F_2$, were to coincide, then they would share point $M$ and the ellipse would be a circle. Using this definition of an ellipse, a possible construction of an ellipse would be to fix a piece of string at two foci and then trace out an ellipse by keeping the string taut. In figure 4-74, the string length would be $2a$.

The formula for the area of an ellipse is $\text{Area}_{\text{Ellipse}} = \pi ab$. If $a$ and $b$ are equal, we get a circle, and then the area is the familiar $\pi aa = \pi a^2$.

Consider now a circle whose diameter is the distance between the two foci, $F_1$ and $F_2$, of a given ellipse (whose semiaxes are $a$ and $b$) as is shown in figure 4-75. Clearly, this ellipse could take on many different shapes, while the circle's shape would remain constant. One of these ellipses would have the same area as the circle. We will be interested to see what the ratio $\frac{a}{b}$ will be when these shapes have equal areas. One might suspect from the theme of this book that the golden ratio will emerge. Well, let's see.

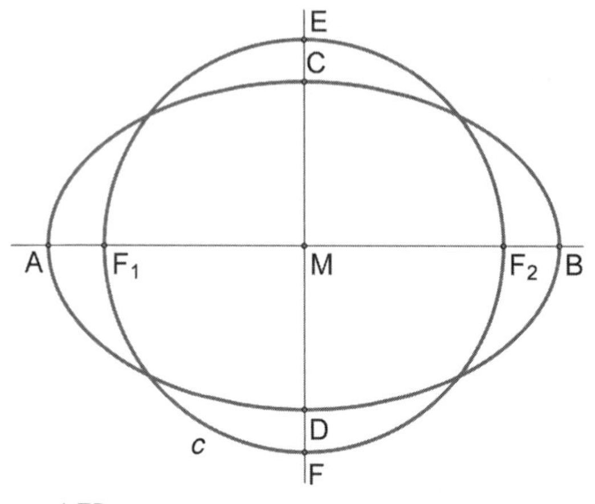


Figure 4-75

In figure 4-75, the circle with center $M$ and radius, $r = MF_1 = MF_2 = e$, has an area, $\text{Area}_{\text{Circle c}} = \pi e^2 = \pi(a^2 - b^2)$. For the ellipse, $\text{Area}_{\text{Ellipse}} = \pi ab$. When the two figures have equal areas we have $\pi(a^2 - b^2) = \pi ab$, or $a^2 - b^2 = ab$. We will now divide both sides of this equation by $b^2$ to get

$$\frac{a^2 - b^2}{b^2} = \frac{ab}{b^2}$$

$$\frac{a^2}{b^2} - 1 = \frac{a}{b}$$

$$\frac{a^2}{b^2} - \frac{a}{b} - 1 = 0.$$

By now, this should look familiar, especially if we replace $\frac{a}{b}$ with $x$ to get the now famous equation of the golden section, namely $x^2 - x - 1 = 0$, whereupon $x = \frac{a}{b} = \phi$. (Remember, the negative root of this equation, $-\frac{1}{\phi}$ is not relevant to us.) Once again, where it would be least expected, the golden section emerges as the ratio of the two semiaxes of the ellipse that has the same area as the circle with the diameter equal to the distance between the foci. We might well call this ellipse the *golden ellipse*.[17]

## THE GOLDEN CUBOID

In lay terms, a cuboid is essentially a box—that is, a rectangular solid with all faces as rectangles and mutually perpendicular. Let's consider one that has edge-lengths $a$, $b$, and $c$, a diagonal length of 2, and volume equal to 1 cubic unit. We seek to find the values of $a$, $b$, and $c$, so we could then get the ratio to one another. The ratio of the areas of the three faces (sides) of the cuboid will also sound familiar.

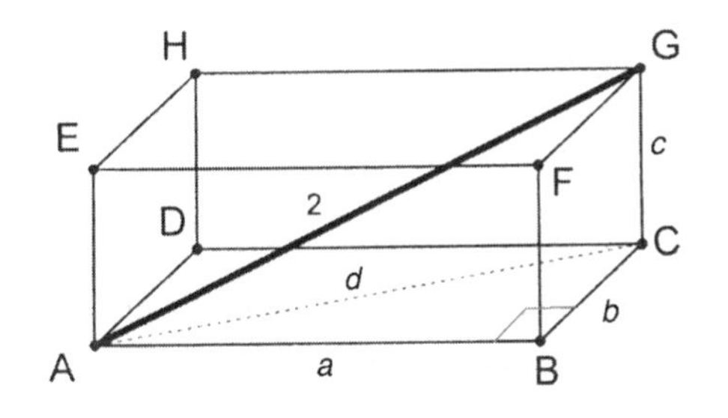


Figure 4-76

From the given information, the volume of the cuboid can be expressed as $abc = 1$. Applying the Pythagorean theorem twice:

For $\Delta ABC$, we get $d^2 = a^2 + b^2$, and for $\Delta ACG$, we get $d^2 + c^2 = 2^2 = 4$.

By substitution, we get $a^2 + b^2 + c^2 = 2^2 = 4$.

To make things a bit simpler, we shall let $b = 1$; then $ac = 1$. From the above equations we now have $a^2 + c^2 = 4 - b^2 = 4 - 1 = 3$. By substituting

$c=\frac{1}{a}$ in the previous equation, we arrive at $a^2+\frac{1}{a^2}=3$, or in simpler form: $a^4-3a^2+1=0$. Now if we let $a^2=x$, then we have the following equation: $x^2-3x+1=0$, whose roots are

$$\frac{3}{2}\pm\sqrt{\frac{9}{4}-1}=\frac{3}{2}\pm\frac{\sqrt{5}}{2},$$

or in more familiar form:

$$\frac{3+\sqrt{5}}{2}=\phi^2=\phi+1 \text{ and } \frac{3-\sqrt{5}}{2}=\frac{1}{\phi^2}=\phi^{-2}.$$

Were we to try to factor this equation, we would—remarkably—get the following factors:

$$x^2-3x+1=(x-x_1)\cdot(x-x_2)=0$$

$$(x-\phi^2)\cdot(x-\frac{1}{\phi^2})=0, \text{ then } x=\phi^2, \text{ or } x=\frac{1}{\phi^2}.$$

If we now substitute back into the previous equation for the values just arrived at, we get $a^2=\phi^2$ or $a^2=\frac{1}{\phi^2}$. Remember that $c=\frac{1}{a}$ and $a\geq c$; therefore, $a=\phi$ and $c=\frac{1}{\phi}$. This amazingly gives us the ratio of the three edges of the cuboid: $a : b : c=\phi : 1 : \frac{1}{\phi}$.

Now to our second concern about this cuboid: the ratio of the areas of the three faces of the cuboid. If we let the areas of the three faces be represented by $A_1$, $A_2$, and $A_3$, and with $b=1$,

$$A_1=a\cdot b=\phi\cdot 1=\phi,$$

$$A_2=a\cdot c=\phi\cdot\frac{1}{\phi}=1, \text{ and}$$

$$A_3=b\cdot c=1\cdot\frac{1}{\phi}=\frac{1}{\phi}.$$

Once again we have the golden section appearing when you might not have expected it, as follows: $A_1 : A_2 : A_3=\phi : 1 : \frac{1}{\phi}$, which could also be expressed as $\phi^2 : \phi : 1$. This would seem to justify calling this cuboid a *golden cuboid.*

To further enhance its features, if we can take a golden cuboid and cut off two interior rectangular solids (cuboids) with a pair of opposite square faces as shown in figure 4-77 (with the squares shaded), then the remaining solid will be a golden cuboid—quite astonishing! The golden cuboid has sides in the following ratio: $1 : \frac{1}{\phi} : \frac{1}{\phi^2} = \phi : 1 : \frac{1}{\phi} = \phi^2 : \phi : 1$.

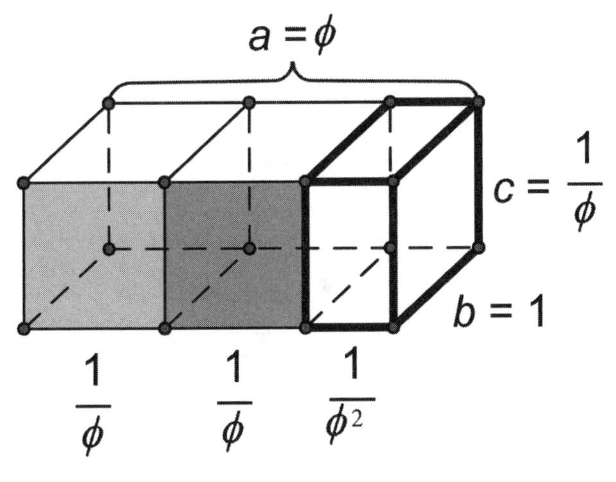

Figure 4-77

If this were not enough, the golden cuboid gives us more cause to rejoice when it comes to promoting the golden section. In figure 4-78, we find four types of rectangular solids marked A, B, C, and D.

Type A has an edges ratio of

$$a : b : c = 1 : \frac{1}{\phi} : \frac{1}{\phi^2} = \phi : 1 : \frac{1}{\phi} = \phi^2 : \phi : 1, \text{ or}$$
$$(\phi + 1) : \phi : 1 \text{ (since } \phi + 1 = \phi^2\text{)}.$$

Type B has an edges ratio of $a : b : c = 1 : \phi : \phi$.

Type C has an edges ratio of $a : b : c = \phi : \phi : \phi^2 = 1 : 1 : \phi$.

Type D is a cube with sides of length $\phi$.

If a cuboid (type B) of edge lengths $\phi$, $\phi$, and 1 has a cube (type D) of edge length $\phi$ attached to it as in figure 4-78, we get a cuboid (type C) with dimensions $\phi$, $\phi$, $\phi+1(=\phi, \phi, \phi^2)$. This new cuboid can then be enlarged by tagging on a cuboid (type A) of dimensions $\phi^2, \phi, 1$. This gives us a large cuboid of dimensions $\phi+1, \phi+1, \phi(=\phi^2, \phi^2, \phi)$, which (by removing the common factor of $\phi$) is consequently similar to the type B cuboid with edge dimensions of $\phi$, $\phi$, and 1. Essentially, we break up this cuboid or build it up with the help of cuboids of edges in ratios of the golden section.

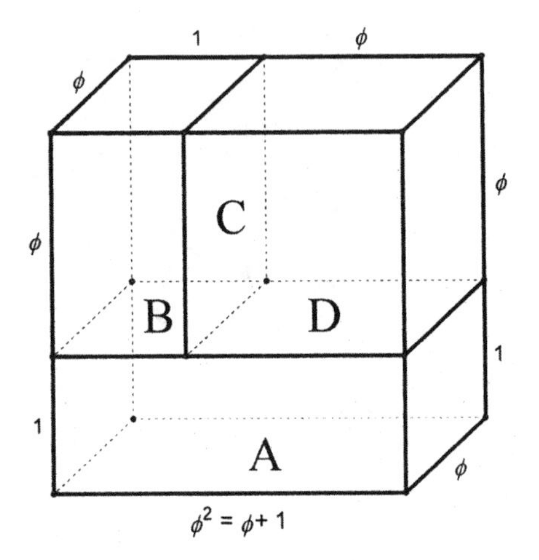


Figure 4-78

The volumes of these cuboids also reveal for us applications of the golden section. Let the volumes of the respective cuboids in figure 4-78 be represented by $V_A$, $V_B$, $V_C$, and $V_D$.

Then

$V_A=\phi^2\cdot\phi\cdot 1=\phi^3$ or $V_A=\phi^2\cdot\phi\cdot 1=(\phi+1)\,\phi=\phi^2+\phi=\phi+1+\phi=2\phi+1$,
$V_D=\phi^2$,
$V_B=\phi\cdot 1\cdot\phi=\phi^2=\phi+1$,
$V_C=V_B+V_D=\phi+1+\phi^2=3\phi+2$, and
$V_{\text{Complete Cuboid}}=V_A+V_C=5\phi+3$.

This enables us to set up the following ratios (obtained by using the numerical value $\phi=\frac{\sqrt{5}+1}{2}$):

$$\frac{V_A}{V_B} = \phi,\ \frac{V_C}{V_A} = \phi,\ \frac{V_C}{V_B} = \phi^2,\ \frac{V_{\text{Complete Cuboid}}}{V_A} = \phi^2,$$

$$\frac{V_{\text{Complete Cuboid}}}{V_B} = \phi^3, \text{ and } \frac{V_{\text{Complete Cuboid}}}{V_C} = \phi.$$

Were this not enough, we can also observe that in an extension of each of the faces of a cube (type D), there is at least one golden rectangle.

## THE GOLDEN POLYHEDRA

A polyhedron is a solid figure (or its surface) consisting of polygon faces—the polyhedron's edges are formed by the intersection of two polygon sides, and the polyhedron's vertices are formed by the intersection of three or more edges. Polyhedra are named by the number of faces they have. There are only five *regular* polyhedra—that is, those that have congruent polygons faces, congruent vertices, and congruent edges (see fig. 4-79). These are typically called *Platonic solids*, named after the eminent philosopher Plato (ca. 427–ca. 347 BCE), who first popularized them in his work *Timaeus*, where these solids represented the elements of the universe.

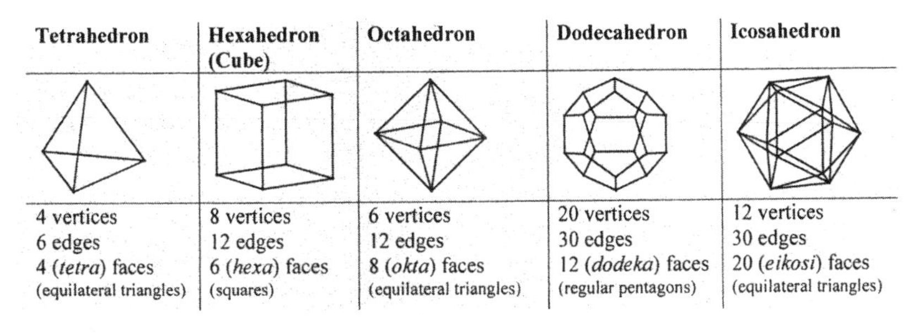

| **Tetrahedron** | **Hexahedron (Cube)** | **Octahedron** | **Dodecahedron** | **Icosahedron** |
|---|---|---|---|---|
| 4 vertices<br>6 edges<br>4 (*tetra*) faces<br>(equilateral triangles) | 8 vertices<br>12 edges<br>6 (*hexa*) faces<br>(squares) | 6 vertices<br>12 edges<br>8 (*okta*) faces<br>(equilateral triangles) | 20 vertices<br>30 edges<br>12 (*dodeka*) faces<br>(regular pentagons) | 12 vertices<br>30 edges<br>20 (*eikosi*) faces<br>(equilateral triangles) |

Figure 4-79

Actually, much credit should be given to Euclid (ca. 300 BCE), who in his *Elements* (book 13, 13–17) describes the construction of these five polyhedra using only a straightedge and compasses, and then, as a crowning glory, proves that these are the only such regular polyhedra.

The Platonic solids have many interesting properties that we will investigate. For example, each of the five Platonic solids can be inscribed in a sphere and can have a sphere inscribed in them. Then there is the famous formula discovered by the Swiss mathematician Leonhard Euler (1707–1783) that connects the number of vertices ($v$), edges ($e$), and faces ($f$) of any convex polyhedron: $v+f=e+2$.[18] Curiously, the octahedron is the only one of the Platonic solids that can be colored as a checkerboard—with no same colors sharing a common edge.

Yet as expected, our interest in the Platonic solids is with those that demonstrate the golden section. We will therefore concentrate on three of these: the octahedron, the dodecahedron, and the icosahedron. These are shown from various viewpoints in figures 4-80, 4-81, and 4-82.

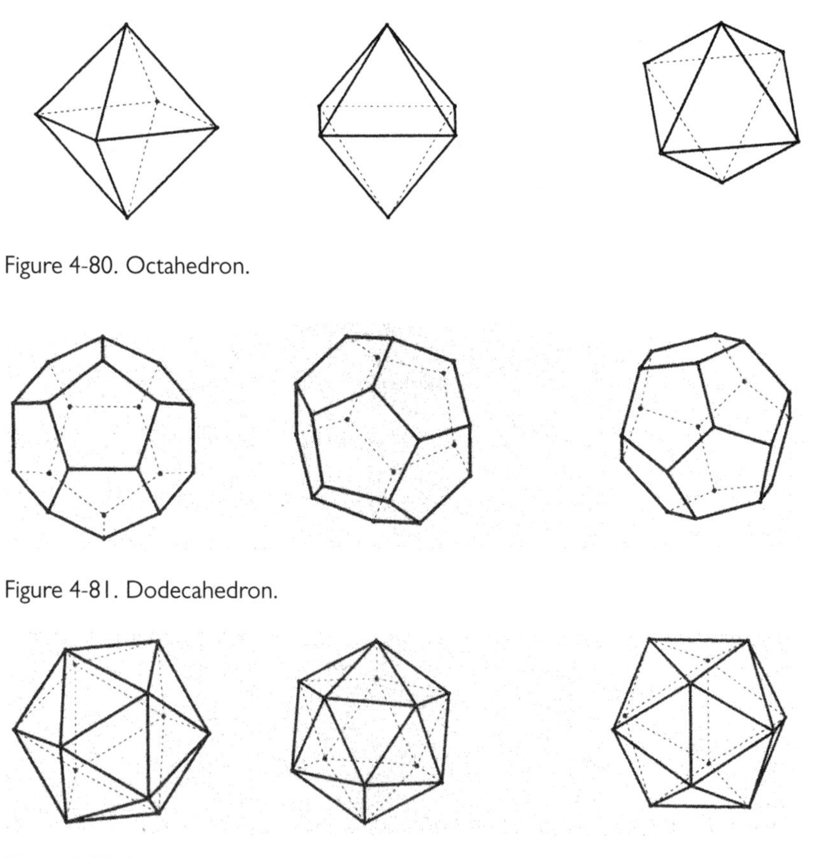

Figure 4-80. Octahedron.

Figure 4-81. Dodecahedron.

Figure 4-82. Icosahedron.

We now consider a rather surprising characteristic for the icosahedron, namely that the twelve vertices can be connected to form three congruent golden rectangles that in pairs also happen to be perpendicular to one another. (See figs. 4-83 and 4-84.) This was first discovered by Luca Pacioli (ca. 1446–1517) in his book *De Divina Proportione*.

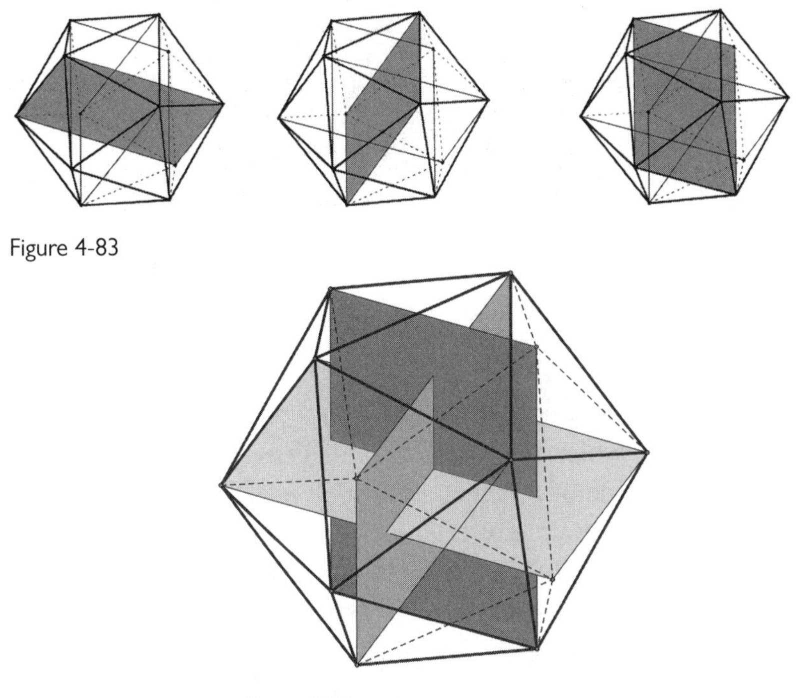

Figure 4-83

Figure 4-84

In figure 4-83, we show the three congruent and perpendicular golden rectangles separately, and then in figure 4-84 we show them together in one icosahedron. To understand this, consider the five equilateral triangles that share vertex $P$, shown in figure 4-85. They form a pyramid with the regular pentagonal base $ABCDE$.[19]

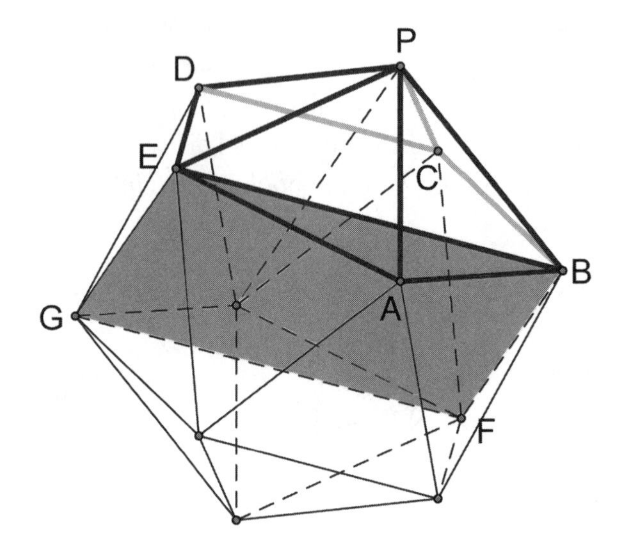


Figure 4-85

We shall accept without proof that *BEGF* is a rectangle with opposite sides *BF* and *EG*, which are also edges of the icosahedron (see fig. 4-85). The longer side of this rectangle (*BE*) is also a diagonal of the pentagon *ABCDE*. We already established, in our discussion of pentagons and pentagrams, that in a regular pentagon the ratio of the diagonal to a side is the golden ratio, $\frac{BE}{AE}=\phi$. Therefore, $\frac{BE}{BF}=\frac{BE}{AE}=\phi$, making *BEGF* a golden rectangle. Analogously, the other two rectangles shown in figure 4-84 are also golden rectangles.

To get a clearer picture of what this actually looks like, we can construct three pieces of cardboard in the shape of three congruent golden rectangles. We then make a slit down the middle and parallel to the longer side of each piece of cardboard. The slit should be long enough to allow for the width of another of the pieces to fit. (You may have to extend one slit to the side of the rectangle.) We then place the rectangles so that each one goes through the slit of another, as shown in figure 4-86. By carefully connecting the vertices of this structure with string, we are able to construct the icosahedron. (See fig. 4-84.) In a regular icosahedron it is possible to erect five such individual structures.

Since the pyramids formed at each vertex of the icosahedron have a base in the shape of a pentagon—for example, *ABCDE* in figure 4-85—we notice the close connection that the golden ratio has with the icosahedron.

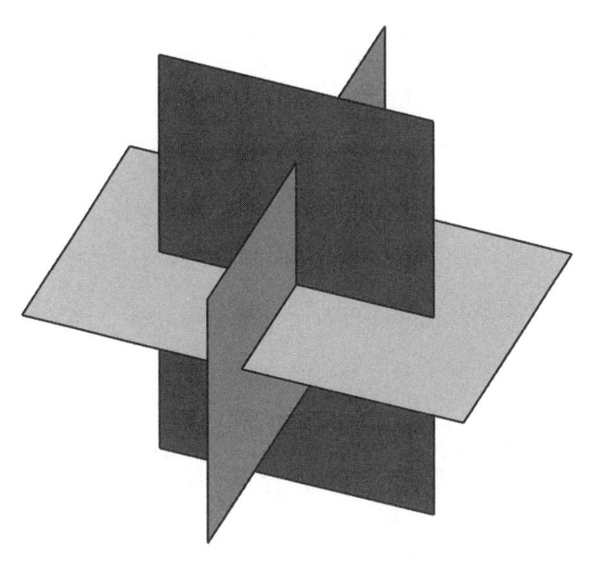

Figure 4-86

In the interior of the icosahedron, there are numerous regular pentagons, and at each vertex five equilateral triangles meet. This should facilitate your finding other parts of this figure that show the relationship of the golden section.

You will recall that the octahedron has twelve edges and the icosahedron has twelve vertices. This allows us to "encase" a regular icosahedron in a regular octahedron by having each edge of the octahedron contain one vertex of the icosahedron. (See fig. 4-87.) Moreover, the edges of the octahedron are partitioned by the vertices of the icosahedron in the golden ratio.

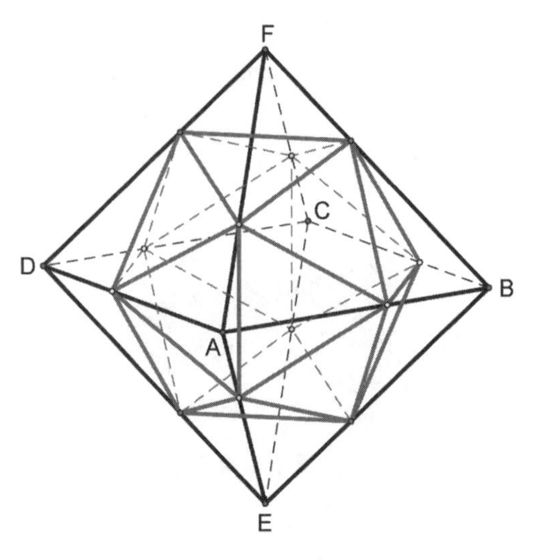


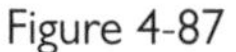
Figure 4-87

To justify this unexpected relationship, we must first notice that the six vertices of the octahedron, *A*, *B*, *C*, *D*, *E*, and *F*, are also the vertices of three mutually perpendicular squares *ABCD*, *AECF*, and *BEDF*. Furthermore, the sides of the squares are the edges of the octahedron. This can be seen in figure 4-88.

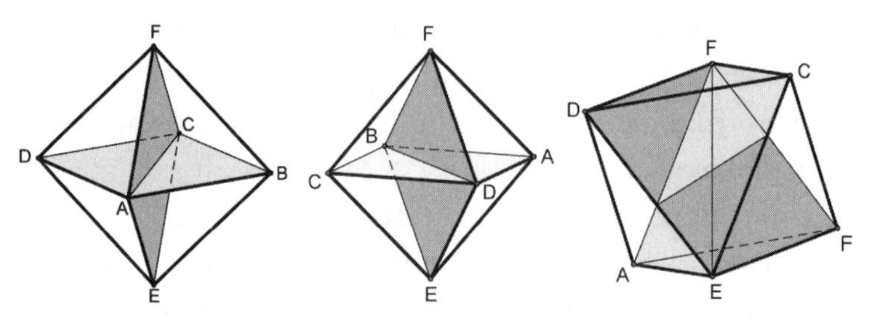


Figure 4-88

If we now select the points on the twelve edges of the octahedron that cut each side into the golden ratio as shown in figures 4-89 and 4-90, we will have determined the vertices of an icosahedron. This then produces the perpendicular golden rectangles that we identified above (fig. 4-84)—and, furthermore, justifies their perpendicularity.

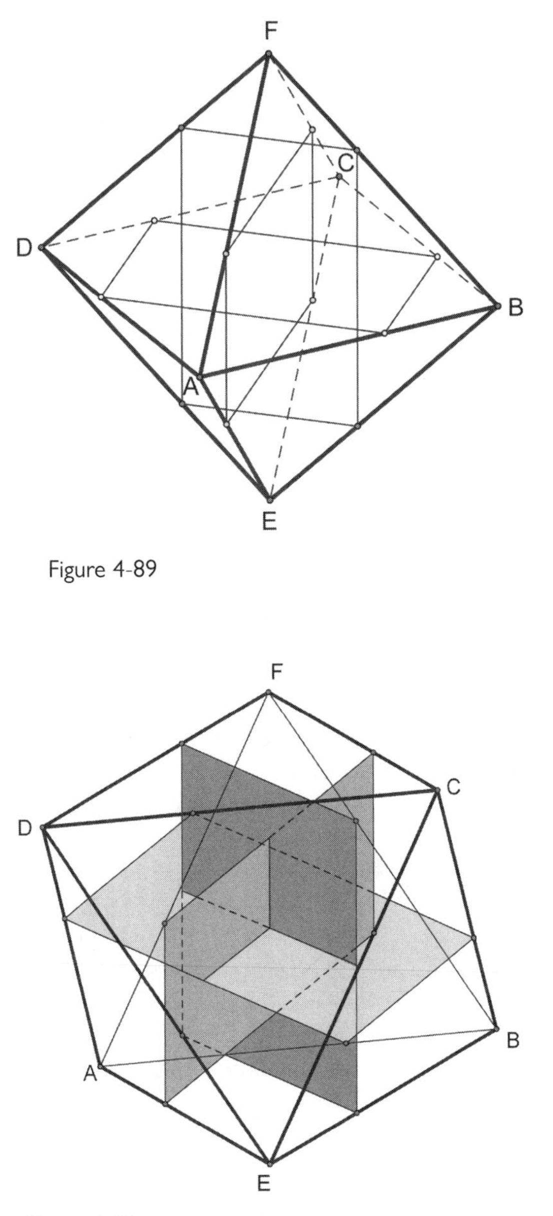


Figure 4-89

Figure 4-90

Perhaps the icosahedrons shown in figure 4-91 provide a clearer picture of each of the golden rectangles and how they fit inside the icosahedron.

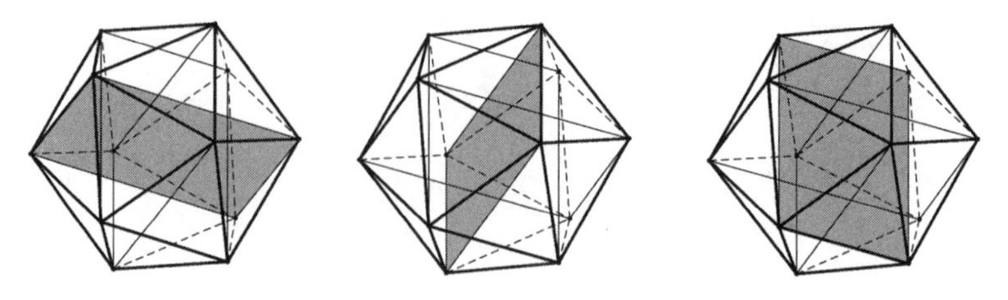

Figure 4-91

Thus we now have shown some of the ways the golden ratio manifests itself in the Platonic solids.

## Dual Polyhedra

Recall the five Platonic solids that we have in figure 4-92 and notice how these solids might relate to each other. Notice how the number of vertices of the hexahedron (cube) is the same as the number of faces of the octahedron and vise versa. The same relationship holds for the dodecahedron and the icosahedron. These polyhedra are called *duals* of one another. Interestingly, the tetrahedron is a *self-dual*! This relationship leads to a lovely geometric phenomenon.

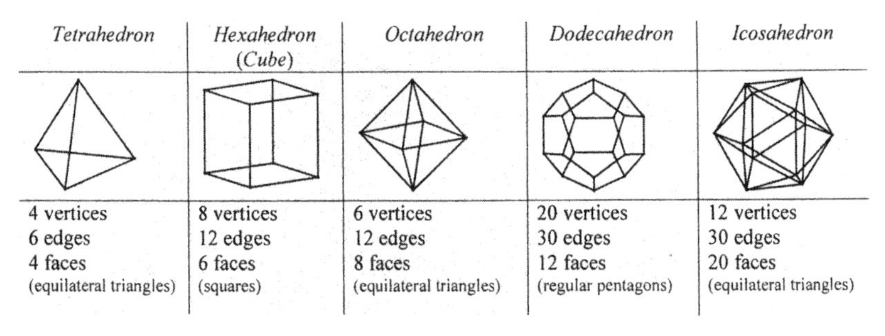

| *Tetrahedron* | *Hexahedron (Cube)* | *Octahedron* | *Dodecahedron* | *Icosahedron* |
|---|---|---|---|---|
| 4 vertices<br>6 edges<br>4 faces<br>(equilateral triangles) | 8 vertices<br>12 edges<br>6 faces<br>(squares) | 6 vertices<br>12 edges<br>8 faces<br>(equilateral triangles) | 20 vertices<br>30 edges<br>12 faces<br>(regular pentagons) | 12 vertices<br>30 edges<br>20 faces<br>(equilateral triangles) |

Figure 4-92

Using the center of each of the polygonal faces, we can demonstrate geometrically the dual relationship these polyhedra have since they can be inscribed or circumscribed about each other as shown for the cube and the octahedron in figure 4-93.

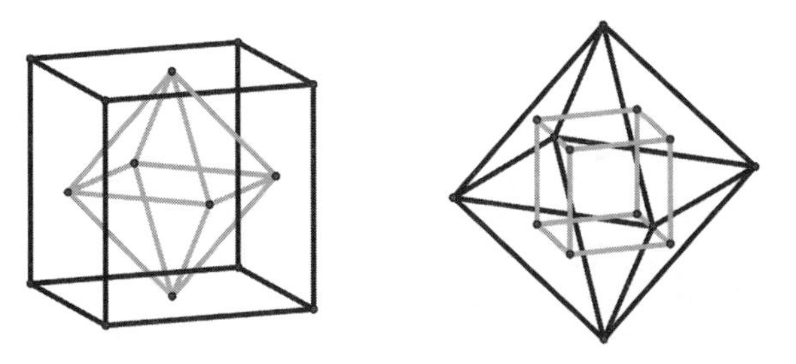

Figure 4-93

The same can be shown (fig. 4-94) for the dodecahedron and the icosahedron.

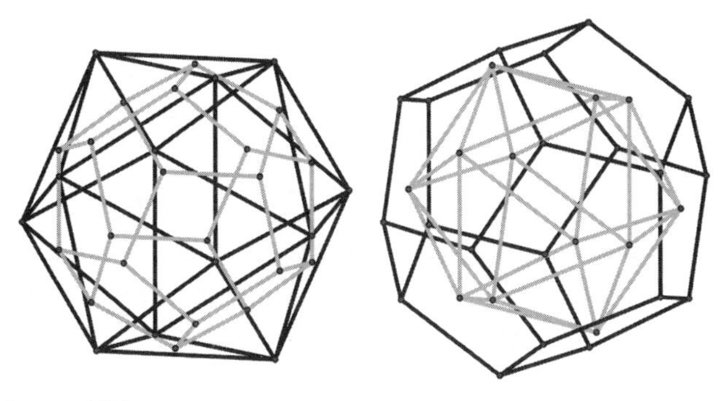

Figure 4-94

Yet here there are lots of appearances of the golden section. For example, the centers of the pentagonal faces of the icosahedron are the vertices of three (congruent) mutually perpendicular golden rectangles inscribed in them, as shown separately in figure 4-95 and combined in figure 4-96.

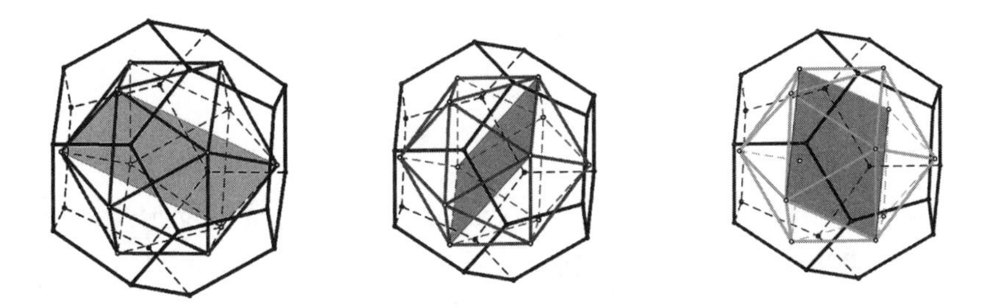

Figure 4-95

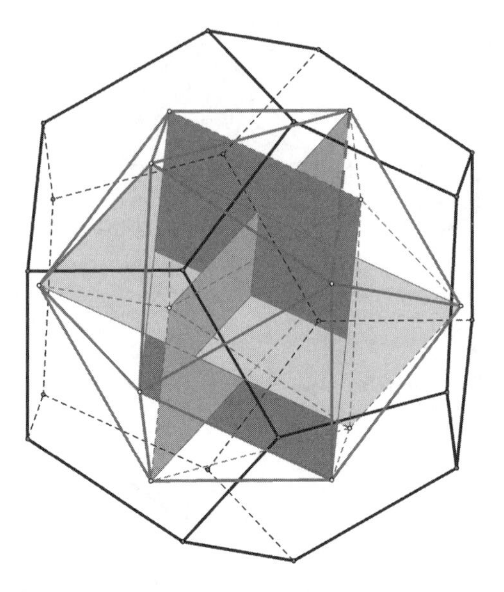

Figure 4-96

We then come to the tetrahedron, which is its self-dual and therefore can be inscribed (or circumscribed) in another tetrahedron as before, by simply joining the centers of the faces. (See fig. 4-97.)

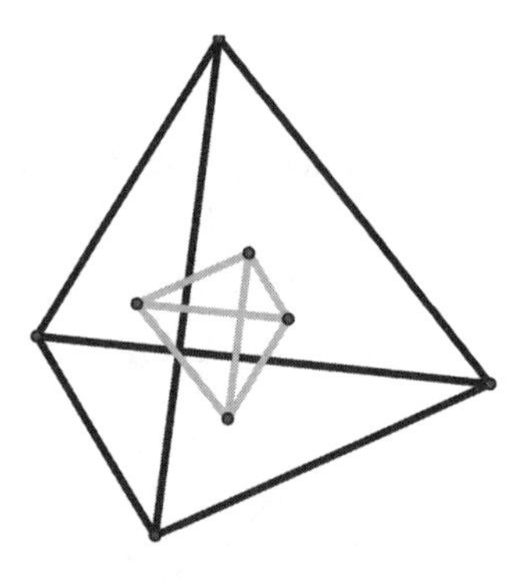

Figure 4-97

Now we shall return to the icosahedron and consider its edges in pairs—that is, three pairs opposite each other and parallel, which are shown in figure 4-98 as they form three golden rectangles.

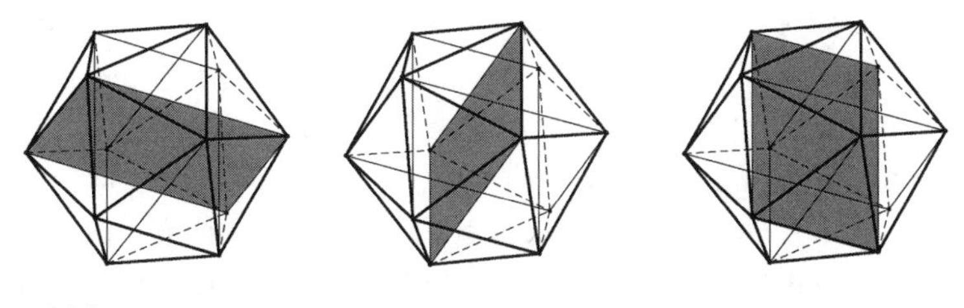

Figure 4-98

The remaining vertices of the icosahedron can be connected to form an inscribed cube. (See fig. 4-99.) The edges of this cube are at the same time the diagonals of the pentagonal faces, where the ratio of the diagonals to the faces is the golden ratio. Therefore, the ratio of the edge of the inscribed cube to the edge of the circumscribed dodecahedron is the golden ratio as well. In figure 4-100, $d : a=\phi$. One might then call this inscribed cube a *golden cube*, whose volume is $V=d^3=\phi^3a^3$.

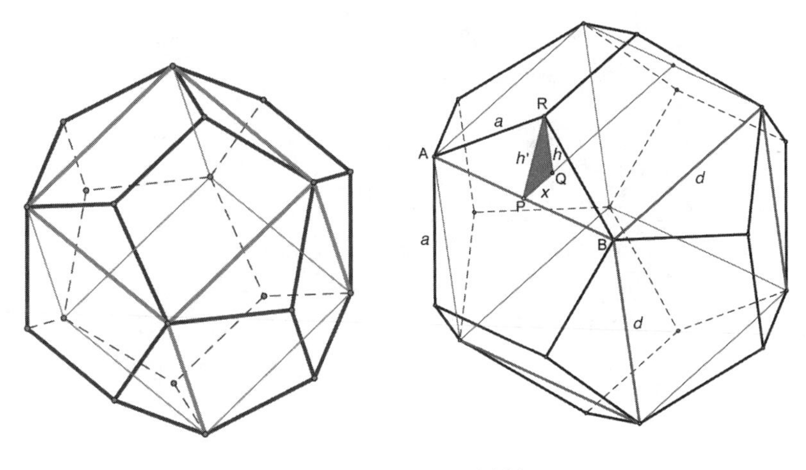


Figure 4-99

Figure 4-100

Further inspection of the dodecahedron (fig. 4-100) shows that each face of the cube is covered by a rooflike cap, the height of which is $h=\frac{1}{2\phi}$ and the length $a=\frac{1}{\phi}$, if the length of the edge of the inscribed cube is $d=1$. (The proof of this unusual relationship to the golden section is provided in the appendix.)

As mentioned earlier, Luca Pacioli, the fifteenth-century mathematician who befriended Leonardo da Vinci (1452–1519) and often pursued mathematical topics with him, wrote about the golden section in his famous book *De Divina Proportione* (written in 1496–1498, published in 1509). He had a dodecahedron in the now-famous painting[20] of him (fig. 4-101), which demonstrated a theorem by Euclid. He knew about the golden section's relationship to the dodecahedron. In the painting, you will notice a number of geometric objects and perhaps also identify the golden section in various places. This painting could well be the first sighting of the rhombicuboctahedron, which is suspended by a string and appears to be half filled with water. Some believe that this water line partitions the edges into the golden section. Where else might you suspect the appearance of the golden section?

Figure 4-101

Another surprising relationship between polyhedra is exhibited with the icosahedron that could be inscribed in a cube, where the length of the edge of the inscribed icosahedron is equal to the longer section of the edge of the cube that would be partitioned into the golden section. Only six of the twenty edges are placed symmetrically on the faces of the cube—as shown in figure 4-102.

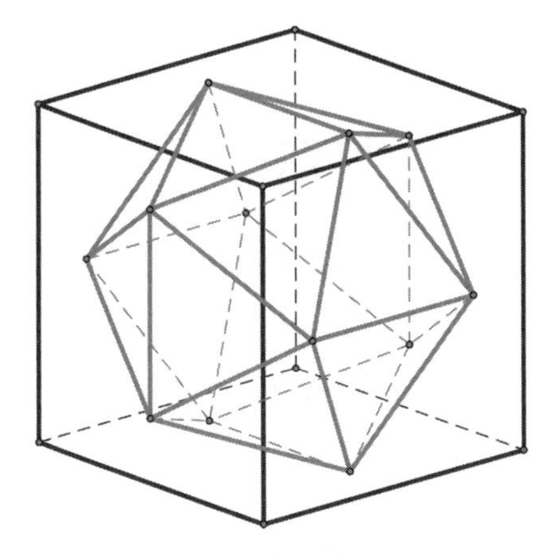

Figure 4-102

As with the regular pentagon, the dodecahedron is exhibited throughout the golden section. For starters, since the pentagon is thoroughly connected with the golden section, the dodecahedron—with its pentagonal faces—is also imbued with the golden section.

Here are a few of such golden section–related relationships involving the dodecahedron:

With edge length $a$, we have the following.

Radius, $R$, of the circumscribed sphere (circumradius):

$$R=\frac{a}{4}(\sqrt{15}+\sqrt{3})=\frac{\sqrt{3}}{2}\cdot\phi\cdot a.$$

Radius, $r$, of the inscribed sphere (inradius):

$$r=\frac{a}{20}\sqrt{250+110\sqrt{5}}=\frac{\phi^2}{2\sqrt{3-\phi}}\cdot a=\frac{\sqrt{\phi^5}}{2\cdot\sqrt[4]{5}}\cdot a.$$

Surface area, $S$:

$$S=3a^2\sqrt{25+10\sqrt{5}}=\frac{15\phi}{\sqrt{3-\phi}}\cdot a^2=\frac{15\phi^2}{\sqrt{\phi^2+1}}\cdot a^2.$$

Volume, *V*:

$$V=\frac{a^3}{4}(15+7\sqrt{5})=\frac{5\phi^3}{2(3-\phi)}\cdot a^3=\frac{5\phi^5}{2(\phi^2+1)}\cdot a^3.$$

Each polyhedral angle[21] of the dodecahedron has

$$\cos\theta=-\frac{\sqrt{5}}{5}\text{, also }\theta=\text{arc cos }(-\frac{\sqrt{5}}{5})\approx 116.57° \ (\approx 116°\ 34').$$

The angle, $\eta$, formed by two faces sharing a common edge—known as a *dihedral angle*—is given by the following:

$$\cos\eta=-\sqrt{\frac{5-\sqrt{5}}{10}}\text{, also }\eta=\text{arc cos}\left(-\sqrt{\frac{5-\sqrt{5}}{10}}\right)\approx 121.72° \ (\approx 121°\ 43').$$

It would appear that any aspect of the dodecahedron seems to involve the golden section. There are more appearances. We shall pursue some and leave some for the reader.

This time we shall consider another aspect of the dodecahedron. Imagine the dodecahedron placed on a table (fig. 4-103) and consider its height. We will let the dodecahedron's height be $h_3=1$, and at its base, we will call the height $h_0=0$. The two vertices (as we look at the solid laterally) we will call $h_1$ and $h_2$. The distance between the two planes (the pentagons at the top and bottom, each with side length $a$) is the diameter, $s$, of the inscribed sphere. Recall that the diagonal of the pentagonal faces is $d=a\cdot\phi$. We will use some of our earlier calculations, from when we showed that the parts of the pentagon were related to the golden section, $\phi$. To make this easier to follow, we provide in figure 4-104 a different viewpoint of the dodecahedron.

## Dodecahedron

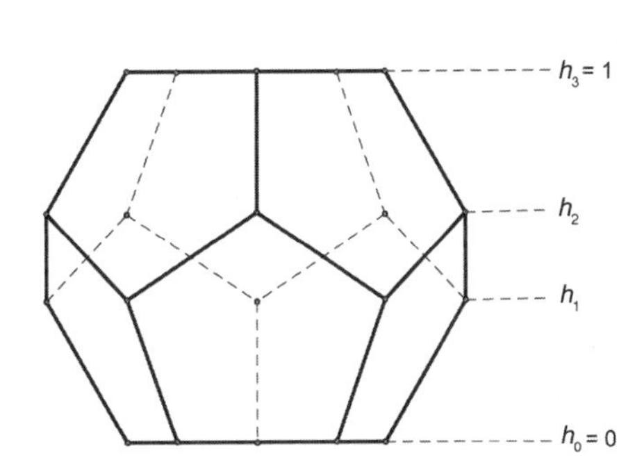


Figure 4-103

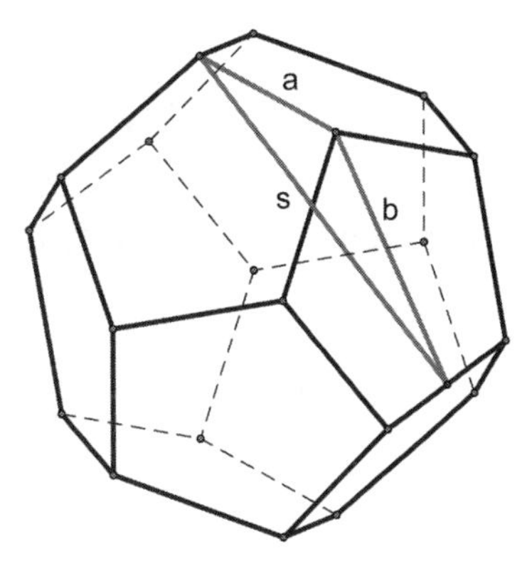


Figure 4-104

## Front View of Pentagonal Face

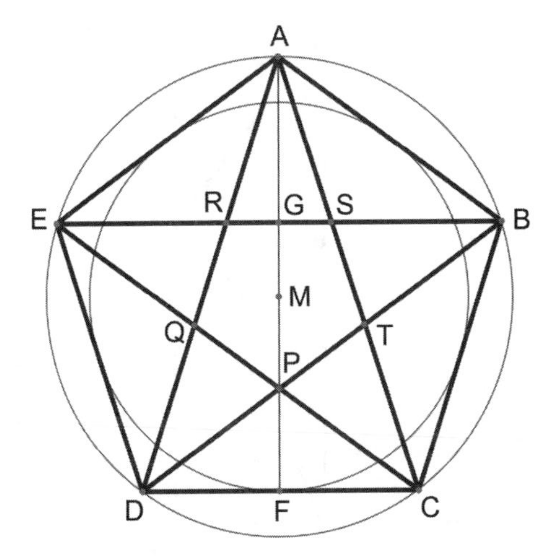


Figure 4-105

In figure 4-105, we show one of the pentagonal faces where we have the following:

$$FG = AF - AG = b - c = \frac{a}{2} \cdot \phi\sqrt{\phi^2 + 1} - \frac{a}{2} \cdot \frac{\sqrt{\phi^2 + 1}}{\phi} = \frac{a}{2} \cdot \sqrt{\phi^2 + 1} = \frac{a}{2} \cdot \sqrt{\frac{5 + \sqrt{5}}{2}}$$

(see p. 142 for *a*, *b*, and *c* designations).

This then allows us to set up the ratio

$$FG : AG = \frac{a}{2} \cdot \sqrt{\phi^2 + 1} : \frac{a}{2} \cdot \frac{\sqrt{\phi^2 + 1}}{\phi},$$

which then can be simplified to

$$\frac{FG}{AG} = \frac{\sqrt{\phi^2 + 1}}{\frac{\sqrt{\phi^2 + 1}}{\phi}} = \phi,$$

which we actually already discovered earlier.

**Side View of Dodecahedron**

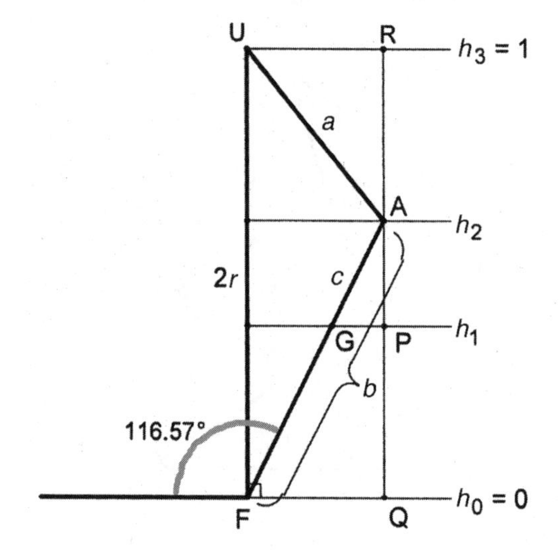


Figure 4-106

In figure 4-106, we see the side view of the dodecahedron, in which we will focus on $\Delta AFQ$, where $FQ$ and $GP$ are parallel and $AQ$ is perpendicular to the base $FQ$.

As a result of the symmetry in the pentagon, where $CT = AS$ (fig. 4-105), we also have $PQ = AR$ (fig. 4-106). Then $AP = 1 - 2PQ$.

We can then obtain the height of the dodecahedron (which is also the diameter of the inscribed sphere, $s$):

$$1 = h_3 = s = 2r = \frac{a}{10}\sqrt{250+110\sqrt{5}};$$

therefore, $a = \dfrac{\sqrt{2}\sqrt{25-11\sqrt{5}}}{2} \approx 0.4490279765.$

Since in $\Delta AFQ$ we have $GP \parallel FQ$, we can state the following:

$$\frac{PQ}{FG} = \frac{AP}{AG} = \frac{1-2PQ}{AG},$$

which give us

$$PQ = \frac{FG}{AG}(1-2PQ).$$

Since

$$\frac{FG}{AG} = \phi, \text{ we get } PQ = \phi(1-2PQ) = \phi - 2\phi PQ.$$

Then $PQ = \dfrac{\phi}{1+2\phi} = \dfrac{\phi}{1+\phi+\phi} = \dfrac{\phi}{\phi^2+\phi} = \dfrac{\phi}{\phi(\phi+1)} = \dfrac{1}{\phi+1} = \dfrac{1}{\phi^2}.$

From the symmetry of the dodecagon, we can conclude that $PQ = AR$. We can then represent $AQ = 1 - AR = 1 - PQ = 1 - \frac{1}{\phi^2} = \frac{1}{\phi}$.

Again, to our expected satisfaction, the heights of various points of a dodecahedron are measured in terms of the golden section—a measure that does not seem to escape us as we investigate the dodecahedron.

We can take this a step further by noting that the distance from the midpoint of an edge to the center of the dodecahedron and the length of half a diagonal of one of the pentagonal faces is in the ratio of $\phi : 1$.

This also happens to be the ratio of the radius of the inscribed sphere of a dodecahedron to the radius of the inscribed circle of one of the pentagonal faces.

## THE GOLDEN PYRAMID

Suppose we now take a regular pentagon and use it as a base for a pyramid with sides composed of five congruent golden triangles (fig. 4-107).

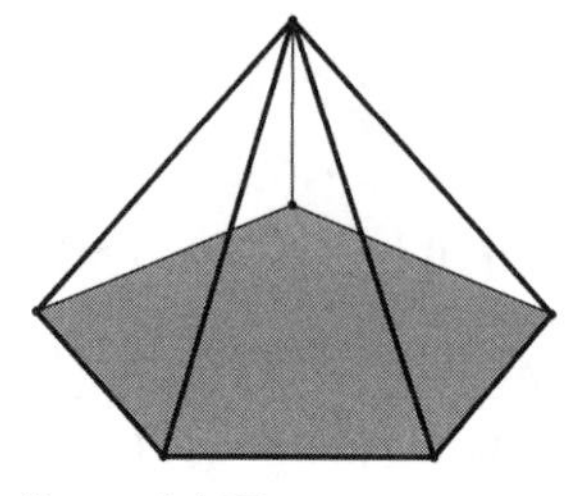

Figure 4-107

This right pyramid can be considered a *golden pyramid* since it is thoroughly representative of the golden section, as can be expected, given the parts that were used to construct it are imbued with this now-famous ratio. We might also have looked at the construction of the golden pyramid by taking a pentagon and extending its sides to form a pentagram as in figure 4-108. Then, by folding up the triangular portions of the pentagram, we would arrive at the same golden pyramid as above.

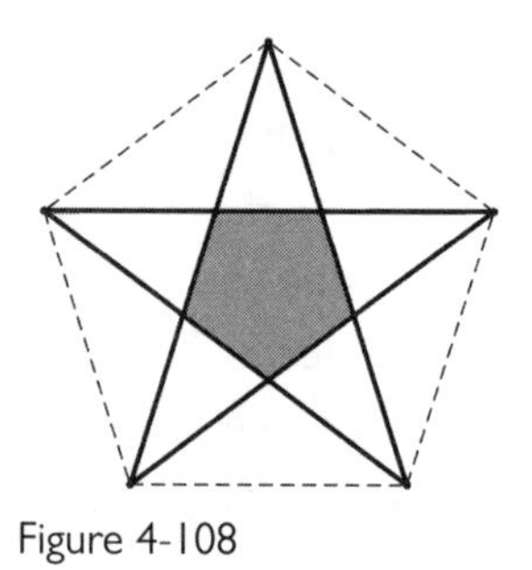

Figure 4-108

As you would expect by now, the golden section can be found throughout the golden pyramid. In our search for the golden section, we do not have to inspect the sides or the base, since those have been considered earlier, yet it would be interesting to inspect the altitude, $CM$, of the pyramid (fig. 4-109).

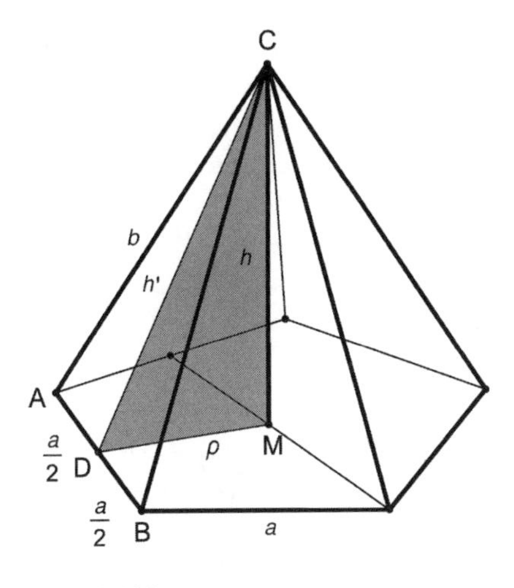


Figure 4-109

We know that $AC = \phi AB$, or in terms of the lengths, $b = \phi a$. When we apply the Pythagorean theorem to right triangle $ACD$, we get

$$h' = CD = \sqrt{AC^2 - AD^2} = \sqrt{b^2 - \frac{a^2}{4}}.$$

However, since $b = \phi a$, and $\phi = \frac{\sqrt{5}+1}{2}$, we then have

$$CD = \sqrt{a^2\phi^2 - \frac{a^2}{4}} = \frac{a}{2}\sqrt{4\phi^2 - 1} = \frac{a}{2}\phi\sqrt{\phi^2 + 1} = \frac{a}{2}\sqrt{5 + 2\sqrt{5}}.$$

The radius of the inscribed circle of the pentagon is

$$\rho = DM = \frac{a}{2} \cdot \frac{\phi^2}{\sqrt{\phi^2 + 1}} = a \cdot \sqrt{\frac{5 + 2\sqrt{5}}{20}} = \frac{a}{2} \cdot \sqrt{\frac{5 + 2\sqrt{5}}{5}}.$$

Then, applying the Pythagorean theorem to right $\Delta CDM$, we are able to get the height of the golden pyramid:

$$h = CM = \sqrt{CD^2 - DM^2} = \sqrt{h'^2 - \rho^2} = \sqrt{\frac{a^2 4\phi^2 - 1}{4} - \frac{a^2\phi^4}{4(\phi^2+1)}}$$

$$= \frac{\phi^2}{\sqrt{\phi^2+1}} \cdot a = \sqrt{\frac{5+2\sqrt{5}}{5}} \cdot a.$$

The ratio of

$$\frac{CM}{DM} = \frac{h}{\rho} = \frac{\frac{\phi^2}{\sqrt{\phi^2+1}} \cdot a}{\frac{\phi^2}{2\sqrt{\phi^2+1}} \cdot a} = 2,$$

which tells us that the height of the golden pyramid, is twice the length of the radius of the inscribed circle of the pentagon base.

Calculating the total surface area of the golden pyramid requires us to first get the area of the pentagonal base:

$$\frac{5\phi^2}{4\sqrt{\phi^2+1}} \cdot a^2 = \frac{\sqrt{25+10\sqrt{5}}}{4} \cdot a^2$$

and then the area of the five golden triangles,

$$5a^2\left(\frac{\phi\sqrt{\phi^2+1}}{4}\right) = 5a^2\left(\frac{\sqrt{5+2\sqrt{5}}}{4}\right).$$

Thus the total surface area of the golden pyramid is the sum of these:

$$\frac{5a^2\phi(\phi^2+\phi+1)}{4\sqrt{\phi^2+1}} = \frac{5a^2\phi(\phi^2+\phi^2)}{4\sqrt{\phi^2+1}} = \frac{5a^2 \cdot 2\phi^3}{4\sqrt{\phi^2+1}} = \frac{5a^2\phi^3}{2\sqrt{\phi^2+1}}$$

$$= a^2\sqrt{\frac{125+55\sqrt{5}}{8}} = \frac{a^2}{2}\sqrt{\frac{125+55\sqrt{5}}{2}}.$$

To complete our calculation of the dimensions of the golden pyramid, we need to calculate the volume.[22]

The volume of the golden pyramid is

$$\frac{1}{3}\cdot \text{Area}_{\text{Pentagon Base}}\cdot h=\frac{1}{3}\cdot \frac{5\phi^2}{4\sqrt{\phi^2+1}}\cdot a^2\cdot \frac{\phi^2}{\sqrt{\phi^2+1}}\cdot a=\frac{5\phi^4}{12(\phi^2+1)}\cdot a^3$$

$$=\frac{\phi^2(\phi^2+1)}{12}\cdot a^3=\frac{5+2\sqrt{5}}{12}\cdot a^3.$$

As expected, all dimensions and measurements involve the golden section!

If we were to take twelve congruent golden pyramids and place them on the faces of a dodecahedron whose faces are congruent to the base of these pyramids, we would get a *small stellated dodecahedron* (see fig. 4-110). We could also arrive at this attractive figure by extending all the edges of a regular dodecahedron until they meet other extended edges. The dodecahedron has twelve vertices, thirty edges, and twelve faces.

Figure 4-110

Perhaps the earliest appearance of the small stellated dodecahedron was in about 1430 as a mosaic by Paolo Uccello on the floor of San Marco cathedral in Venice, Italy. It was "rediscovered" by Johannes Kepler (1571–1630), who used the term *urchin* in his work *Harmonice Mundi* in 1619, and again by the French mathematician Louis Poinsot (1777–1859) in 1809. This solid has sixty faces, thirty-two vertices, and ninety edges. In addition to the famous five Platonic solids—all of which are convex—there are four other symmetric solids—although not convex—named after their discoverers: *Kepler-Poinsot solids*.[23]

## The "Golden State" of the USA

Figure 4-111

As an amusing side note, although we all refer to California, the most populous state of the United States and the thirty-first to enter the union, on September 9, 1850, as the *Golden State* (fig. 4-111), perhaps because of the famous gold rush that began on January 24, 1848, Professor Monte Zerger has shown that there may be more reason to designate Illinois as the "golden state."[24] His rationale stems from the following. The telephone area code in southern Illinois is 618. Remember, that $\frac{1}{\phi} = .618\ldots$. Furthermore, the area code $309 = \frac{618}{2}$ is also in Illinois. The zip code 618 is also to be found in the state of Illinois. Now for the geometric or geographic justification for his claim. If we partition the map of the contiguous forty-eight states with the golden section both horizontally and vertically, we find the point of intersection of these two lines near Decatur, Illinois (see fig. 4-112). Naturally, had we drawn the horizontal and vertical lines at other sides of the map, we could have just as easily designated the states of Colorado, Mississippi, or Texas, as golden section intersection points. However, the area codes and the zip codes steer us to Illinois.

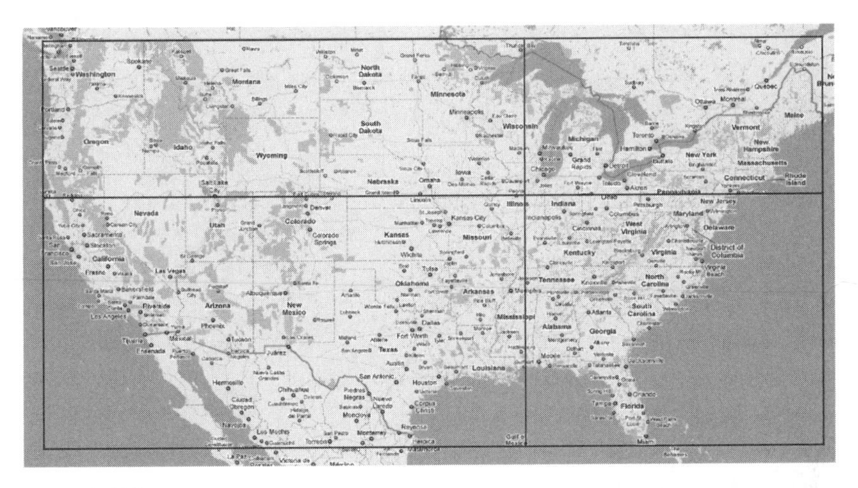


Figure 4-112

So as you can see, we can also find evidence of the golden section when we bring into question some time-honored uses of terms. Despite all this evidence and the many other connections one can concoct to show that Illinois has the Fibonacci numbers embedded in various parts, the Golden State still remains California. After all, we can't really change the name of the Golden State Warriors basketball team or move them to Chicago, Illinois, the home of the Chicago Bulls basketball team.

Chapter 5

# Unexpected Appearances of the Golden Ratio

Up to now, we have investigated the golden section geometrically, algebraically, and numerically. We shall now embark on an unusual adventure—exploring the many curious ways in which the golden section appears where you might least expect it. Just as the value of $\pi$, which emanates from its relationship to the circle, can be found in a host of other contexts, so too can the golden section, $\phi$, be found—in as many interesting and unanticipated places. This potpourri of sightings of the golden section will vary greatly, which we hope will add to the never-ending fascination that this ubiquitous number has provided us over the millennia. We see these as mathematical curiosities and have named them accordingly.

## CURIOSITY 1

In an equilateral triangle, $\Delta ABC$, each side of length $s$ is partitioned (with the same orientation) into the segments $a$ and $b$, which are in the golden ratio (fig. 5-1). The result is that an inscribed equilateral triangle, $\Delta DEF$, is created with side length $c$. Although this figure has the golden section built into the construction, it is amazing how the golden ratio emerges in a multitude of aspects of this figure.

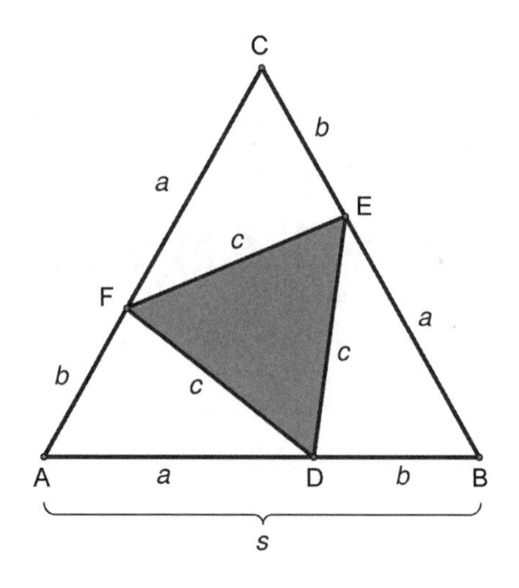


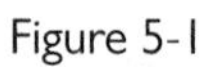
Figure 5-1

Here are some of the appearances of $\phi$ in figure 5-1:

1. $c = \frac{s}{\phi}\sqrt{1 + \frac{1}{\phi^2} - \frac{1}{\phi}}$.

2. $\text{Area}_{\Delta DEF} = \frac{\sqrt{3}(\phi^2 - \phi + 1)}{4\phi^4} \cdot s^2$.

3. The ratio of the areas of the two equilateral triangles is

$$\frac{\text{Area}_{\Delta ABC}}{\text{Area}_{\Delta DEF}} = \frac{\phi^2}{1 + \frac{1}{\phi^2} - \frac{1}{\phi}}.$$

4. The area of each of the three congruent triangles, $\Delta ADF$, $\Delta BDE$, and $\Delta CEF$, is

$$\text{Area}_{\Delta ADF} = s^2 \frac{\sqrt{3}}{4\phi^3}.$$

5. The ratio of the areas of the original equilateral triangle to one of the three congruent triangles is

$$\frac{\text{Area}_{\Delta ABC}}{\text{Area}_{\Delta ADF}} = \frac{2}{\phi} + 3.$$

6. The ratio of the area of the smaller equilateral triangle to one of the three congruent triangles is

$$\frac{\text{Area}_{\Delta DEF}}{\text{Area}_{\Delta ADF}} = \frac{2}{\phi}.$$

(The justifications for these are not complicated, but are relegated to the appendix so as not to break the flow of the presentation of curiosities.)

## CURIOSITY 2

We begin with a triangle, $\Delta ABC$ (fig. 5-2), with sides $BC=1$, $AC=x$, and $AB=x^2$. If $x<1$, then $BC$ is its largest side and $AB$ is its shortest side. Using the triangle inequality,[1] we get $x^2+x>1$. By adding $\frac{1}{4}$ to each side of the inequality:

$$x^2 + x + \frac{1}{4} > \frac{5}{4}, \text{ or } \left(x + \frac{1}{2}\right)^2 > \frac{5}{4}.$$

As we must be working with positive numbers, we get

$$x + \frac{1}{2} > \frac{\sqrt{5}}{2}, \text{ or } x > \frac{\sqrt{5}-1}{2} = \frac{1}{\phi}.$$

That is, the side of length $x$ must be such that $\frac{1}{\phi}<x<1$. Here the golden ratio takes the role of a limiting length.

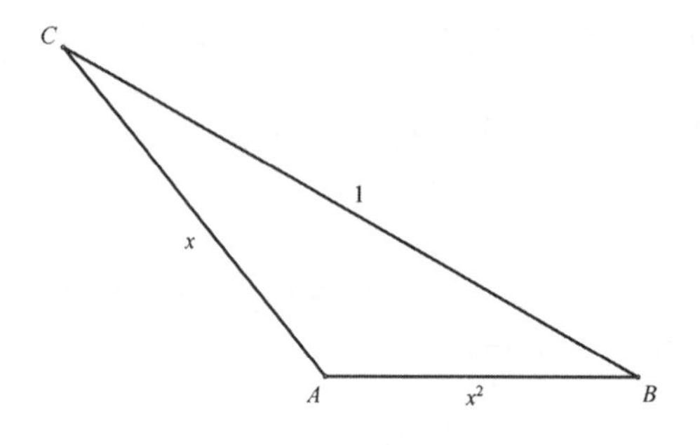


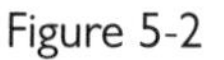
Figure 5-2

We now consider a circle *c* through point *B* and tangent to *AC* at point *A*, intersecting *BC* at *D*. (See fig. 5-3.)

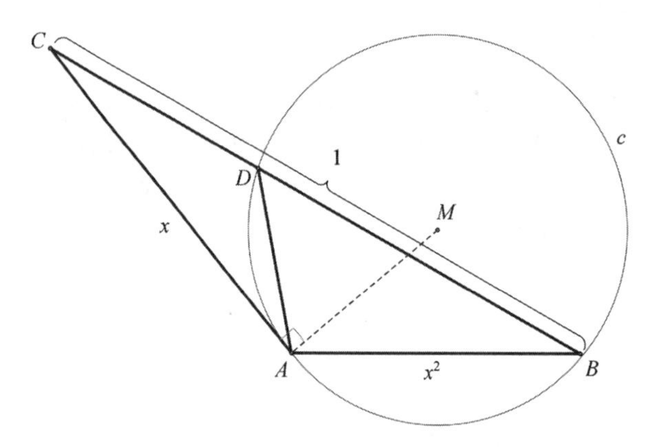


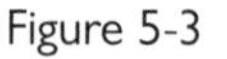
Figure 5-3

Because $\angle ABC = \frac{1}{2}\overset{\frown}{AD} = \angle DAC$, and $\angle ACB = \angle ACD$, we have $\Delta ABC \sim \Delta ACD$.

From this similarity, it then follows that

$\frac{AD}{AB} = \frac{AC}{BC}$, or $\frac{AD}{x^2} = \frac{x}{1}$, which leads to $AD = x^3$, and

$\frac{CD}{AC} = \frac{AC}{BC}$, or $\frac{CD}{x} = \frac{x}{1}$, which leads to $CD = x^2$.

From this we have the possibility of constructing a set of triangles, the sides having lengths $x^n$, $x^{n+1}$, $x^{n+2}$, where $n = 0, 1, 2, 3, \ldots$. A rather nice pattern!

## CURIOSITY 3

Here we will create a situation where the golden section will continuously reappear. In figure 5-4, the point $S$ divides the segment $AB$ into the golden section. From this, we can generate many more golden sections, as you will see in the following steps.

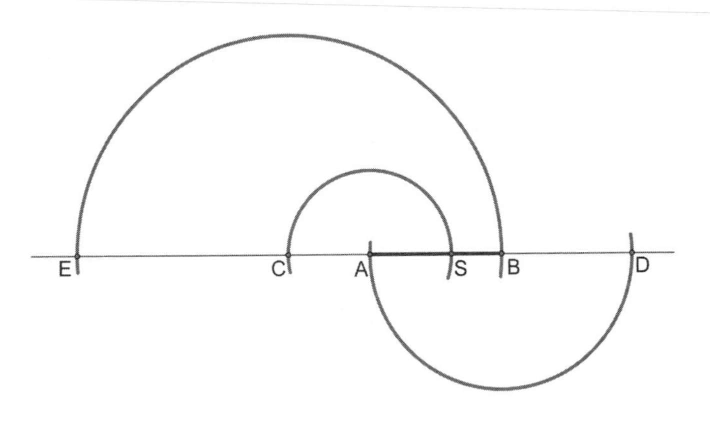


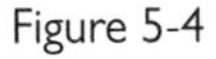
Figure 5-4

1. The circle with center at $A$ with radius $AS$ cuts the line $AB$ at a second point $C$. We then have $\frac{AB}{AC} = \phi$ and $\frac{BC}{AB} = \phi$.

2. The circle with $B$ as its center and radius $AB$ intersects the line $AB$ at a second point $D$. We then have $\frac{BC}{BD} = \phi$ and $\frac{CD}{BC} = \phi$.

What might you guess are the following ratios? $\frac{CD}{CE}$ and $\frac{DE}{CD}$. Yes, they are the golden ratio, $\phi$!

We can justify these appearances of the golden ratio as follows:

Because $\frac{AS}{BS} = \frac{AB}{AS} = \phi$ and $AC = AS$, it follows that $\frac{AB}{AC} = \frac{AB}{AS} = \phi$, and as well $\frac{BC}{AB} = \frac{AC + AB}{AB} = \frac{AC}{AB} + 1 = \frac{1}{\phi} + 1 = \phi$.

Analogously, we can show the rest of the ratios as equal to $\frac{BC}{AB}=\ldots=\phi$. You might want to continue this process to see the pattern that will evolve.

## CURIOSITY 4

Referring to figure 5-5,[2] we begin with line segment $AB$, and at $B$ we construct a perpendicular segment, $BC$, half the length of $AB$. So, if we let $AB=1$, then $BC=\frac{1}{2}$. We then construct a circle $c_1$ with its center at $A$ and radius $AC=r_1$ to intersect the line $AB$ at point $D$. At $D$, another perpendicular is erected the same length as $BC$. Therefore, $DE=\frac{1}{2}$. Finally, we construct a circle $c_2$ with its center at $D$ and with radius $DE=r_2$, cutting $AB$ at points $P$ and $Q$. Completely unexpectedly, it turns out that points $P$ and $Q$ enable us to partition the segments $AB$ and $AQ$ into the golden section.

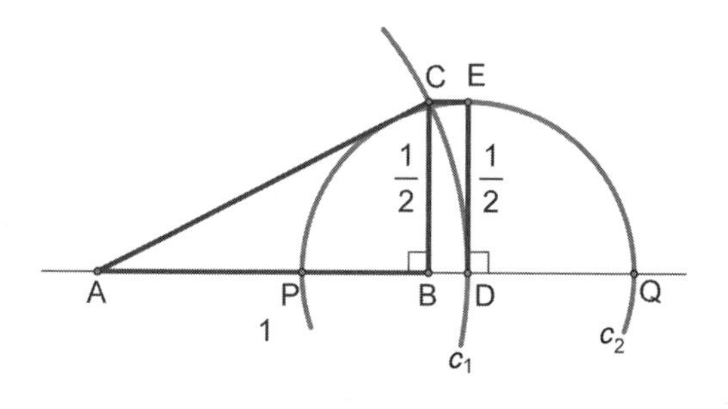


Figure 5-5

To see why this works—that is, to justify this oddity—we begin by applying the Pythagorean theorem to triangle $ABC$, to find

$r_1=AC=\frac{\sqrt{5}}{2}$, which tells us the other radius of the circle, $AD=\frac{\sqrt{5}}{2}$.

Since $AP=AD-DP=AC-DE=\frac{\sqrt{5}}{2}-\frac{1}{2}=\frac{\sqrt{5}-1}{2}=\frac{1}{\phi}$, and

$BP=DP-BD=DE-(AD-AB)=\frac{1}{2}-\left(\frac{\sqrt{5}}{2}-1\right)=\frac{3-\sqrt{5}}{2}=\frac{1}{\phi^2}$, we get, on the one hand,

$$\frac{AP}{BP} = \frac{\frac{1}{\phi}}{\frac{1}{\phi^2}} = \phi,$$

and, on the other hand,

$$\frac{AQ}{AB} = \frac{AD + DQ}{AB} = \frac{\sqrt{5} + 1}{2} = \phi,$$

as well as

$$\frac{AB}{BQ} = \frac{AB}{BD + DQ} = \frac{AB}{(AD - AB) + DQ} = \frac{1}{\frac{\sqrt{5}}{2} - 1 + \frac{1}{2}} = \frac{1}{\frac{1}{\phi}} = \phi.$$

Unexpected appearances of the golden section such as these make it so intriguing. Sometimes when you do not actually expect the golden ratio to appear, it just does.

## CURIOSITY 5

From our previous exploration of the golden section in polygons, we find it particularly ubiquitous in the pentagon, and consequently in the pentagram. It is now only fitting that we investigate the golden section's appearance in the regular hexagram: a six-pointed star in which each of the "points" is formed by an equilateral triangle and in which the center is a regular hexagon. We begin our search for the golden ratio in figure 5-6, with a regular hexagram. We will construct a circle with center $D$ and radius $DC$ to intersect the extension of line segment $AB$ at point $S$. Curiously enough—among other sightings of the golden section—we find that point $B$ partitions line segment $AS$ in the golden ratio. Let us see why this actually is true.

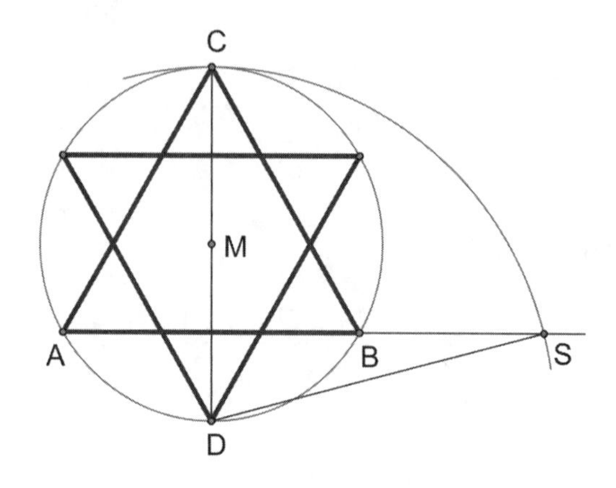


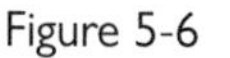
Figure 5-6

In figure 5-7, we will let $AB = a$ and, by symmetry, $AP = PB = \frac{a}{2}$. We also have $AG = GH = BH = \frac{a}{3}$, and $GP = PH = \frac{a}{6}$.

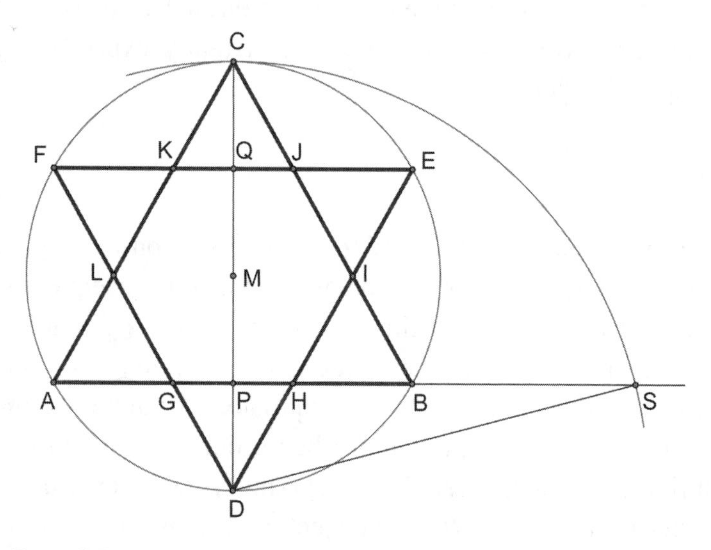


Figure 5-7

Applying the Pythagorean theorem to triangle $\Delta BCP$, we get $CP = DQ = \frac{a}{2}\sqrt{3}$.

When we apply the Pythagorean theorem to $\Delta DHP$, we get $DP = CQ = \frac{a}{6}\sqrt{3}$.

We have $CD = CP + DP = \frac{a}{2}\sqrt{3} + \frac{a}{6}\sqrt{3} = \frac{2a}{3}\sqrt{3}$.

The circle we constructed with center $D$ and radius $CD$ (which is also the diameter of the circumscribed circle of the hexagram) intersects the extension of line segment $AB$ at point $S$. With this point, $S$, we determine $\Delta DPS$, to which we shall again apply the Pythagorean theorem, to obtain

$$PS^2 = DS^2 - DP^2 = CD^2 - DP^2 = \left(\frac{2a}{3}\sqrt{3}\right)^2 - \left(\frac{a}{6}\sqrt{3}\right)^2 = \frac{5a^2}{4},$$

which then has $PS = \frac{a}{2}\sqrt{5}$. Now that we have arrived at an expression involving $\sqrt{5}$, we begin to anticipate that the golden ratio is soon to appear.

To establish the golden ratio, we begin with

$$BS = PS - BP = \frac{a}{2}\sqrt{5} - \frac{a}{2} = \frac{a\left(\sqrt{5}-1\right)}{2}.$$

We are now ready to inspect the crucial ratios.

$$\frac{AB}{BS} = \frac{a}{\frac{a\left(\sqrt{5}-1\right)}{2}} = \frac{2}{\sqrt{5}-1} = \frac{\sqrt{5}+1}{2} = \phi$$

and

$$\frac{AS}{AB} = \frac{AP+PS}{AB} = \frac{\frac{a}{2}+\frac{a\sqrt{5}}{2}}{a} = \frac{\sqrt{5}+1}{2} = \phi.$$

Thus, we have again found the golden ratio, this time embedded in the hexagram—not a very well-known place to find it!

## CURIOSITY 6

The floral designs in figures 5-8 and 5-9 look rather attractive and can be seen in many contexts—toys, puzzles, and so on. Beyond its optical beauty, there is also the subtle beauty owing to its reliance on the golden section.

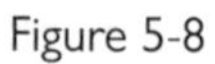

Figure 5-8

Figure 5-9

In figure 5-10, we can see the basis for its construction—a combination of circles, each of which is centered on one of the six equidistant points on a given circle.

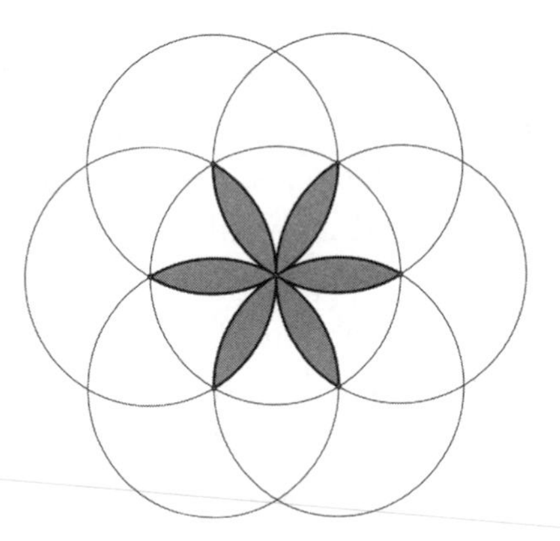

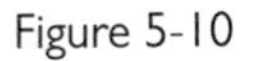
Figure 5-10

In figure 5-11, we have $AM=BM=CM=DM=EM=FM=r$, as well as $AB=BC=CD=DE=EF=AF=r$ (since *ABCDEF* is a regular hexagon). We then have the golden ratio in that $\frac{MS}{DS}=\frac{DM}{MS}=\phi$. Let's see why this is true.

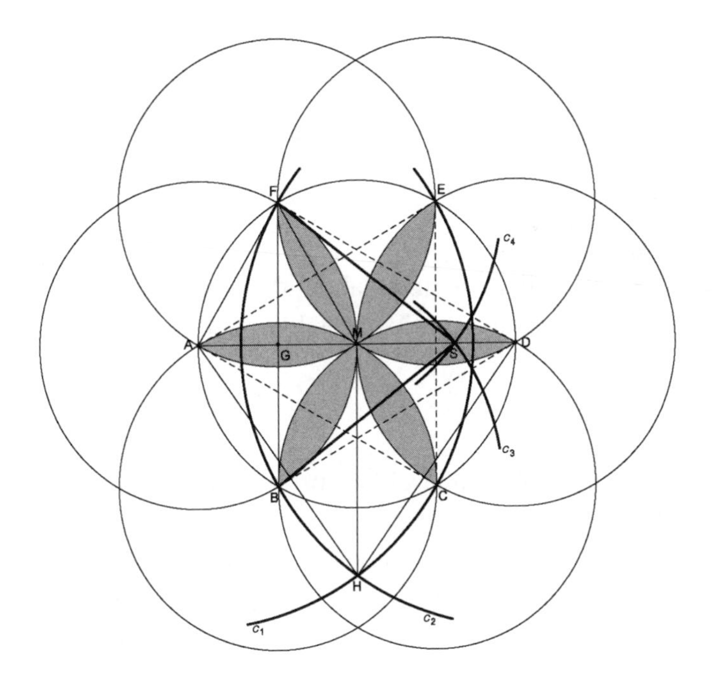

Figure 5-11

Building from our investigation of Curiosity 5, we have found that in figure 5-11, triangle $AFM$ is equilateral and $AM$ is the perpendicular bisector of $BF$. We then get $AG = GM = \frac{r}{2}$, and $GF = \frac{r}{2}\sqrt{3}$, where $r$ is the radius of one of the congruent circles. Also $BF = BG + GF = 2GF = r\sqrt{3}$.

Circle $c_1$ has its center at $A$ and radius $r_1 = r\sqrt{3}$, and circle $c_2$ has its center at $D$ and radius $r_2 = r_1 = r\sqrt{3}$.

Circles $c_1$ and $c_2$ intersect at point $H$ (and another point not shown) with $AH = DH (= BF) = r\sqrt{3}$ and $HM = r\sqrt{2}$.

Circle $c_3$ has its center at $B$ and radius $r_3 = HM = r\sqrt{2}$. Circle $c_4$ has its center at $F$ and intersects $AD$ at $S$.

We are now ready to show that $\frac{MS}{DS} = \frac{DM}{MS} = \phi$, the golden ratio. We first apply the Pythagorean theorem to triangle $BGS$ to get $GS^2 = BS^2 - BG^2 = (r\sqrt{2})^2 - (\frac{r}{2}\sqrt{3})^2 = \frac{5}{4}r^2$.

Therefore, $GS = \frac{r}{2}\sqrt{5}$. Then $MS = GS - GM = \frac{r}{2}\sqrt{5} - \frac{r}{2} = \frac{r}{2} \cdot (\sqrt{5} - 1)$.

With these values (in terms of $r$) we can find the desired ratios:

$$\frac{MS}{DS} = \frac{MS}{DM - MS} = \frac{\frac{r}{2}\left(\sqrt{5}-1\right)}{r - \frac{r}{2}\left(\sqrt{5}-1\right)} = \frac{\sqrt{5}-1}{3-\sqrt{5}} = \frac{\sqrt{5}+1}{2} = \phi$$

and

$$\frac{DM}{MS} = \frac{r}{\frac{r}{2}\left(\sqrt{5}-1\right)} = \frac{2}{\sqrt{5}-1} = \frac{\sqrt{5}+1}{2} = \phi \cdot$$

This justifies our earlier statement about the beautiful design (figs. 5-8 and 5-9) having the golden ratio embedded within it.

## CURIOSITY 7

An analogous situation to that of Curiosity 6 can be made for a similar design, but based on a regular pentagon rather than the hexagon used earlier. Here, too, the golden section will appear embedded in the design. As we inspect the design in figure 5-12, we find that we have five congruent circles centered at the five vertices of a regular pentagon, each containing the center of the pentagon. Thus, the design is similar to the previous one.

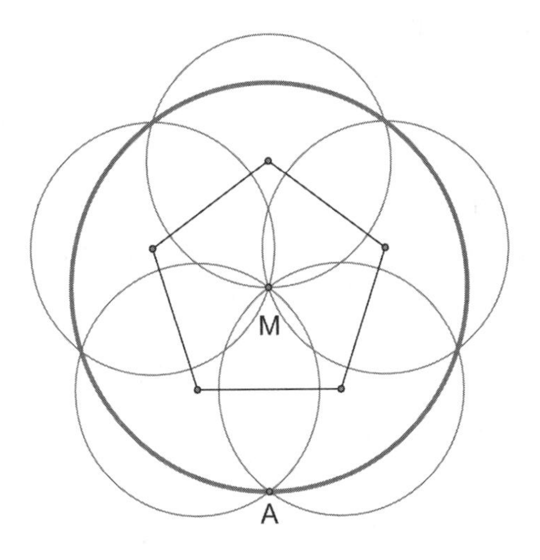


Figure 5-12

For the sake of convenience, we will let the radii of these circles have length $r=1$. In figure 5-13, we have a detailed enlargement of the diagram in figure 5-12, and we will call the centers of two of the intersecting circles $M_1$ and $M_2$. They intersect at points $A$ and $M$. We will now seek to find the length of $AM$.

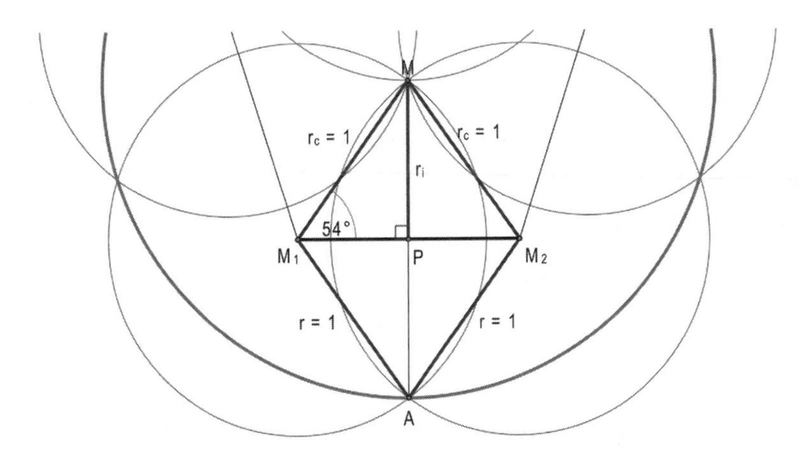


Figure 5-13

In chapter 4, we investigated the regular pentagon with respect to its involvement with the golden ratio. In figure 5-13, we present a close-up of a portion of figure 5-12 so that we can properly focus in on quadrilateral $MM_1AM_2$,which is a rhombus since $MM_1 = AM_1 = AM_2 = MM_2 = r = 1$. This now tells us that the radius of the circumscribed circle of the pentagon $r_c = MM_1 = MM_2 = 1$. The inscribed circle of the regular pentagon, which is tangent to the pentagon's side $M_1M_2$ at its midpoint $P$, is $MP = r_i$. These pentagon-related circles can be seen in figure 5-14.

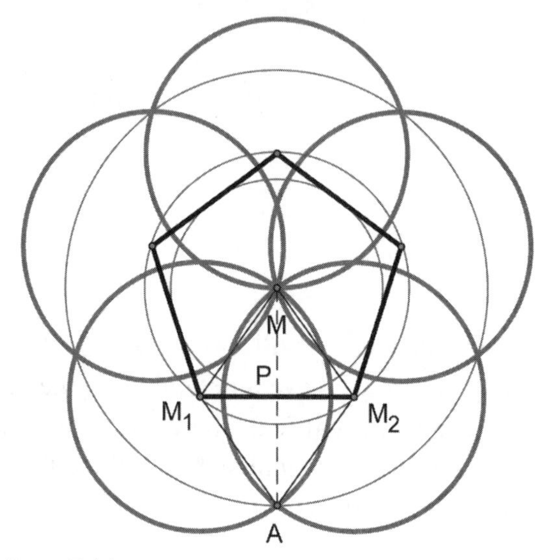


Figure 5-14

Once again looking at figure 5-13, and recalling our findings from chapter 4, we can better define the right $\Delta MM_1P$, by establishing that $\angle MM_1P = 54°$ and $\angle M_1MP = 36°$.

Then we see that

$$\sin \angle MM_1P = \frac{MP}{MM_1} = \frac{r_i}{r_u} = \frac{r_i}{1}$$

and

$$\sin 54° = \frac{\sqrt{5}+1}{4} = \frac{1}{2} \cdot \frac{\sqrt{5}+1}{2} = \frac{\phi}{2}, \text{ (see p. 125).}$$

So we can finally determine that $r_i = \frac{\phi}{2}$.

Because the diagonals of a rhombus are perpendicular bisectors of each other, $AM = 2 \cdot MP = 2r_i = 2 \cdot \frac{\phi}{2} = \phi$. Therefore, we have found that the length of one of the petals of the floral design is equal to the golden ratio.

## CURIOSITY 8

Again, we encounter here an unanticipated emergence of the golden section. In figure 5-15, we have two circles, $c_1$ and $c_2$, which are tangent at point $B$, have their centers at $M_1$ and $M_2$, and have radii $r_1 = AM_1 = BM_1$ and $r_2 = BM_2 = CM_2$. If the smaller circle, $c_2$, is so constructed that the point $C$ is the center of gravity of the shaded region (we call this shaded region a *lune*), then any chord of the larger circle, $c_1$, from point $B$ will be partitioned by the smaller circle, $c_2$, into the golden section. This means that point $C$ partitions $AB$ into the golden section, as well as any other chord of the larger circle from point $B$, such as, say, $DB$ (shown in fig. 5-16), where the point $E$, at which the smaller circle intersects $DB$, will also determine the golden section of that line segment, $DB$.

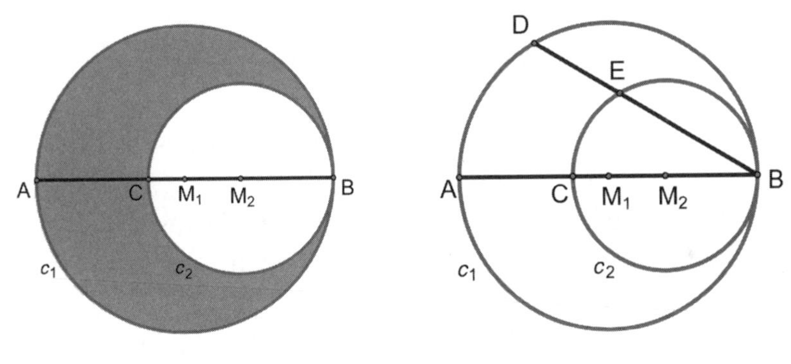


Figure 5-15

Figure 5-16

Let's use figure 5-16, where circle $c_2$ is so constructed that point $C$ is the center of gravity of the lune. The curiosity that we want to establish here is that circle $c_2$ is so constructed that the ratio of the radii $\frac{r_1}{r_2} = \phi$. This will then allow us to also show a further, and perhaps even more amazing, curiosity that the point $E$ (where circle $c_2$ intersects $BD$) partitions the chord $BD$ into the golden section.

To justify this curiosity, we begin by establishing the areas of the three figures—the two circles and the lune—in figure 5-16:

$\text{Area}_{c_1} = \pi r_1^2$, $\text{Area}_{c_2} = \pi r_2^2$, and $\text{Area}_{Lune} = \pi(r_1^2 - r_2^2)$.

We also need to express lengths in terms of the radii so we have $CM_1 = BC - BM_1 = 2r_2 - r_1$ and $M_1M_2 = BM_1 - BM_2 = r_1 - r_2$.

We now consider a balance scale with the fulcrum at point $M_1$. With $C$ the center of gravity of the lune and $M_2$ the center of gravity of the circle $c_2$, and knowing that they will balance the scale proportional to their weight, we get the following:

$$CM_1 \cdot \text{Area}_{Lune} = M_1M_2 \cdot \text{Area}_{c_2}, \text{ or } (2r_2 - r_1)\text{Area}_{Lune} = (r_1 - r_2)\text{Area}_{c_2}.$$

Therefore, $(2r_2 - r_1) \cdot \pi(r_1^2 - r_2^2) = (r_1 - r_2) \cdot \pi r_2^2$.

Dividing both sides of the equation by $\pi$ gives us

$$(2r_2 - r_1) \cdot (r_1 + r_2) \cdot (r_1 - r_2) = (r_1 - r_2) \cdot r_2^2.$$

Then, dividing both sides of the equation by $r_1 - r_2$ we get

$$(2r_2 - r_1) \cdot (r_1 + r_2) = r_2^2, \text{ or in other form: } r_2^2 + r_1r_2 - r_1^2 = 0.$$

Now dividing by $r_1^2$ and at the same time replacing $\frac{r_2}{r_1}$ with $x$, we get our now-familiar equation that will give us the golden ratio: $x^2 + x - 1 = 0$, where the (positive) root is $\frac{1}{\phi}$.

Thus $x = \frac{r_2}{r_1} = \frac{1}{\phi}$, or $r_1 = \phi \cdot r_2$, which is one of the relationships we wanted to demonstrate.

Not only are the radii in the golden ratio, but also the line segment $AB$ can now be shown to be partitioned into the golden ratio by point $C$. This we can show as follows:

$$\frac{BC}{AC} = \frac{BC}{AB - BC} = \frac{2r_2}{2r_1 - 2r_2} = \frac{r_2}{r_1 - r_2} = \frac{r_2}{\phi r_2 - r_2}$$

$$= \frac{r_2}{r_2(\phi - 1)} = \frac{1}{\phi - 1} = \frac{1}{\frac{\sqrt{5}+1}{2} - \frac{2}{2}} = \frac{2}{\sqrt{5} - 1} = \frac{\sqrt{5}+1}{2} = \phi.$$

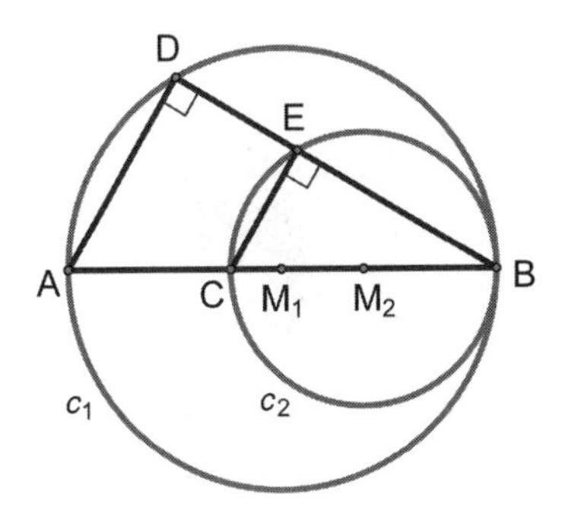


Figure 5-17

To complete our original claim about the smaller circle partitioning *any* chord of the larger circle that contains the common point of tangency into the golden ratio, we will consider any randomly selected chord, $DB$, of the larger circle, $c_1$, to see if it is then partitioned into the golden ratio by the smaller circle, $c_2$. In figure 5-17, $E$ is the point of intersection of the circle $c_2$ with chord $DB$. Since the angles at $D$ and $E$ are each inscribed in a semicircle, they are right angles. The right triangles $ABD$ and $CBE$ are similar and $AD \parallel CE$. Therefore, $\frac{BE}{DE} = \frac{BC}{AC} = \phi$.

When we reflect on this curiosity a bit, we notice how surprising it is that once again $\phi$ comes up when least expected.

## CURIOSITY 9

We now embark on a rather different configuration and search for the golden ratio embedded within it. We shall begin by taking a square with side length 2 and along two opposite sides construct two congruent semicircles each with radius $\frac{1}{2}$, as shown in figure 5-18. We will show that—surprisingly—the radius of the circle constructed tangent to the four semicircles is the reciprocal of the golden ratio.

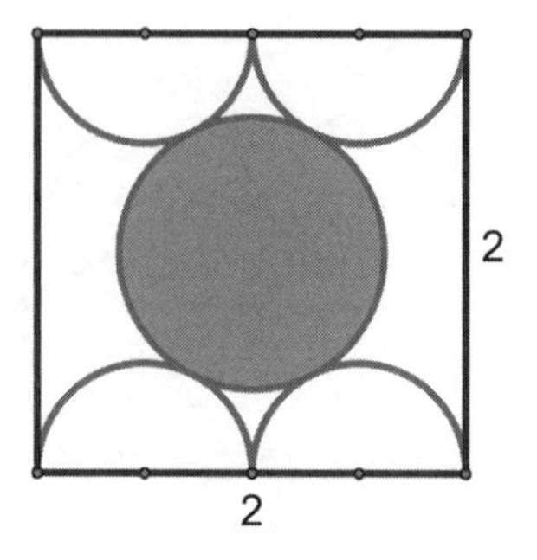


Figure 5-18

In figure 5-19, we repeat the above configuration, where $AB=BC=CD=AD=2$, yet this time with some auxiliary line segments. Clearly, the radii of the semicircles are $AG=EG=GP=\frac{1}{2}$.

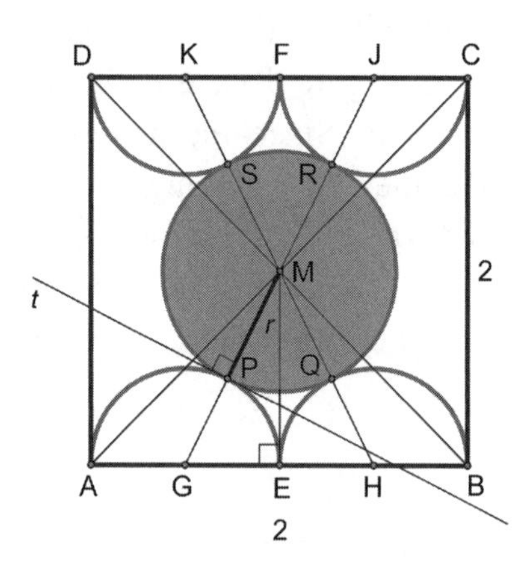


Figure 5-19

Applying the Pythagorean theorem to the right $\Delta EGM$, where $EM=1$, $EG=\frac{1}{2}$, and $GM=GP+MP=\frac{1}{2}+r$, we get

$$\left(\frac{1}{2}+r\right)^2 = 1^2 + \left(\frac{1}{2}\right)^2,$$

which results in $\frac{1}{4}+r+r^2=1+\frac{1}{4}$, and then gives us $r^2+r-1=0$. Once again, we find ourselves with the equation that will yield the golden section, $r=\frac{1}{\phi}$. (By now, we are accustomed to ignoring the root $r=-\phi$, since it is negative.) Thus, the radius of the center circle is

$$r=\frac{1}{\phi}=\frac{\sqrt{5}-1}{2}.$$

It is curious how, once again, unexpectedly the golden ratio appears.

## CURIOSITY 10

In this curiosity, we will actually begin with a constructed golden section and have it "automatically" assist us in constructing the golden section on other line segments. Our "tool" for doing this golden section construction is the *arbelos*,[3] a figure obtained by drawing three semicircles, two along the diameter of the third. The two smaller semicircles can be of any size, as long as the sum of their diameters equals the entire diameter of the third semicircle. Thus, as shown in figure 5-20, $AS+SB=AB$. This figure was known to Archimedes (287–212 BCE), who studied it extensively. The arbelos draws its name from the Greek, meaning *shoemaker's knife*, since it resembles the tool used by ancient shoemakers.

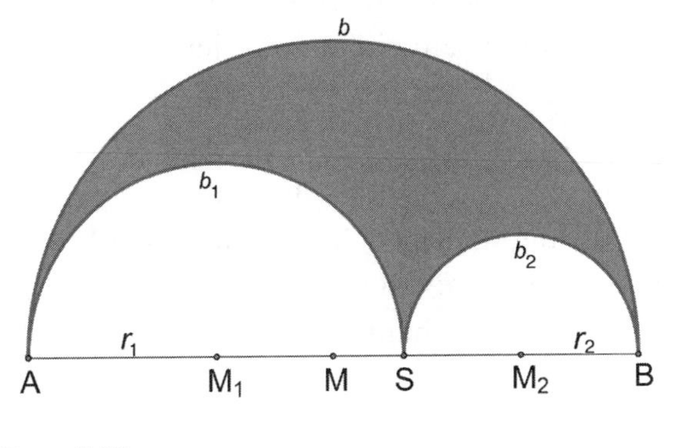


Figure 5-20

What distinguishes our arbelos from all the others is that we have constructed ours with one extra stipulation: The diameters of the two smaller circles are in the golden ratio—that is, $\frac{AS}{SB}=\phi$. (See fig. 5-20.)

In figure 5-21, we have $r_1 = AM_1 = SM_1$ and $r_2 = BM_2 = SM_2$. The radius, $r$, of the large semicircle is $r = r_1 + r_2 = AM = BM$. The usefulness of this special golden ratio arbelos is that, for any point $C$ on the larger semicircle, the two chords drawn to the semicircle's diametrical endpoints are related by the golden ratio. We might therefore want to call this the *golden arbelos*.

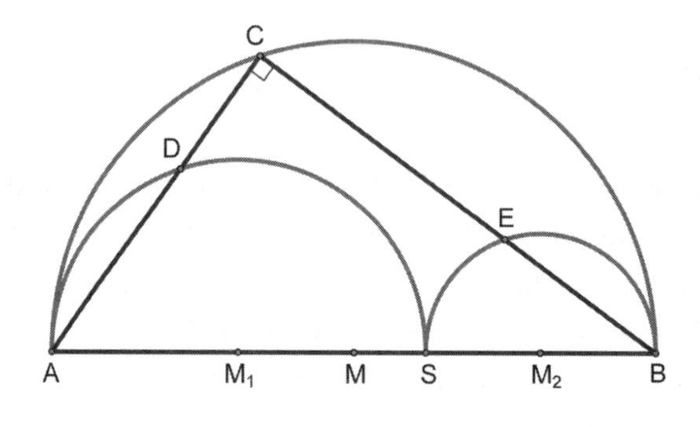


Figure 5-21

Symbolically, for figure 5-21, where we began with $\frac{AB}{AS} = \frac{AS}{BS} = \phi$, we then also have $\frac{AC}{AD} = \frac{AD}{CD} = \phi$ and $\frac{BC}{CE} = \frac{CE}{BE} = \phi$. Thus, we have created a tool that determines the golden ratio for a given line segment: Here in figure 5-21, the two line segments so partitioned are $AC$ and $BC$.

We also find that the semicircular arc lengths $b$, $b_1$, $b_2$ (fig. 5-20) are in the ratio of $\phi : 1 : \frac{1}{\phi}$, and the ratio of the respective areas of the semicircles is $\phi^2 : 1 : \frac{1}{\phi^2}$. This is not intuitively obvious, despite the special type of arbelos having the golden section.

It is also interesting to inspect the relationship of the perimeter and area of the arbelos with respect to the diameter ($AB$). Let us now embark on justifying these characteristics of this special arbelos.

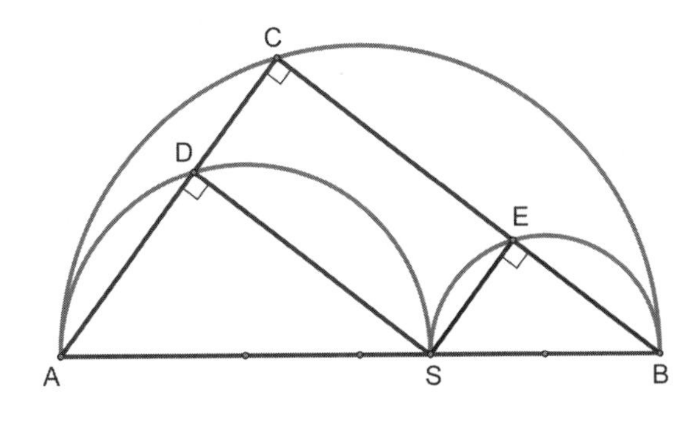


Figure 5-22

We can easily justify that point $D$ partitions the segment $AC$ into the same ratio as $S$ partitions $AB$, namely the golden ratio. Similarly, the point $E$ partitions chord $BC$ into the same ratio. We notice the right angles in figure 5-22, since these angles are inscribed in a semicircle. And we also have similar triangles: $\Delta ABC$, $\Delta ASD$, and $\Delta SBE$. Furthermore, since $CDES$ is a rectangle, we also have $CD = ES$ and $CE = DS$. Therefore, $\frac{AS}{SB} = \frac{AD}{SE} = \frac{AD}{DC} = \phi$. As we hinted at earlier, this argument can be repeated analogously for chord $BC$.

We can also appreciate the converse of this beautiful relationship. Suppose $AC$ is just any chord in the semicircle with diameter $AB$. Suppose further that $D$ determines the golden section of $AC$. Then the locus of all such points $D$ is a semicircle with diameter $AS$, where the point $S$ will determine the golden section of $AB$.

As an extra attraction, let us consider the arc lengths of the three semicircles: $\widehat{AB} = b = \pi r$, $\widehat{AS} = b_1 = \pi r_1$, and $\widehat{BS} = b_2 = \pi r_2$.

Then:

$$\frac{b}{b_1} = \frac{\pi r}{\pi r_1} = \frac{r}{r_1} = \frac{2r}{2r_1} = \frac{AB}{AS} = \phi, \text{ and } \frac{b_1}{b_2} = \frac{\pi r_1}{\pi r_2} = \frac{r_1}{r_2} = \frac{2r_1}{2r_2} = \frac{AS}{BS} = \phi.$$

We shall now consider the areas of the three semicircles:

$$AB\text{: Area}_b = \frac{\pi r^2}{2}, AS\text{: Area}_{b_1} = \frac{\pi r_1^2}{2}, \text{ and } BS\text{: Area}_{b_2} = \frac{\pi r_2^2}{2}.$$

Then, when we take the ratio of the areas of these semicircles, we once again find the golden ratio emerging.

$$\frac{\text{Area}_b}{\text{Area}_{b_1}} = \frac{\frac{\pi r^2}{2}}{\frac{\pi r_1^2}{2}} = \left(\frac{r}{r_1}\right)^2 = \left(\frac{2r}{2r_1}\right)^2 = \left(\frac{AB}{AS}\right)^2 = \phi^2$$

and

$$\frac{\text{Area}_{b_1}}{\text{Area}_{b_2}} = \frac{\frac{\pi r_1^2}{2}}{\frac{\pi r_2^2}{2}} = \left(\frac{r_1}{r_2}\right)^2 = \left(\frac{2r_1}{2r_2}\right)^2 = \left(\frac{AS}{BS}\right)^2 = \phi^2.$$

As for the last property we mentioned about the arbelos, we have for the perimeter:

$$\text{P}_{\text{Arbelos}} = b + b_1 + b_2 = \pi r + \pi r_1 + \pi r_2 = \pi(r + r_1 + r_2) = \pi(r + r) = 2\pi r = \pi AB.$$

The area of the arbelos can be obtained as follows:

$$\text{Area}_{\text{Arbelos}} = A - (A_1 + A_2) = \frac{\pi r^2}{2} - \frac{\pi r_1^2}{2} - \frac{\pi r_2^2}{2} = \frac{\pi(r_1 + r_2)^2}{2} - \frac{\pi r_1^2}{2} - \frac{\pi r_2^2}{2}$$

$$= \frac{2\pi r_1 r_2}{2} = \pi r_1 r_2 = \pi \cdot \frac{2r_1}{2} \cdot \frac{2r_2}{2} = \frac{\pi}{4} \cdot AS \cdot BS$$

$$= \frac{\pi}{4} \cdot \frac{1}{\phi} AB \cdot \frac{1}{\phi} AS = \frac{\pi}{4} \cdot \frac{1}{\phi^2} \cdot AB \cdot AS$$

$$= \frac{\pi}{4} \cdot \frac{1}{\phi^2} \cdot AB \cdot \frac{1}{\phi} \cdot AB = \frac{\pi}{4} \cdot \frac{1}{\phi^3} \cdot AB^2.$$

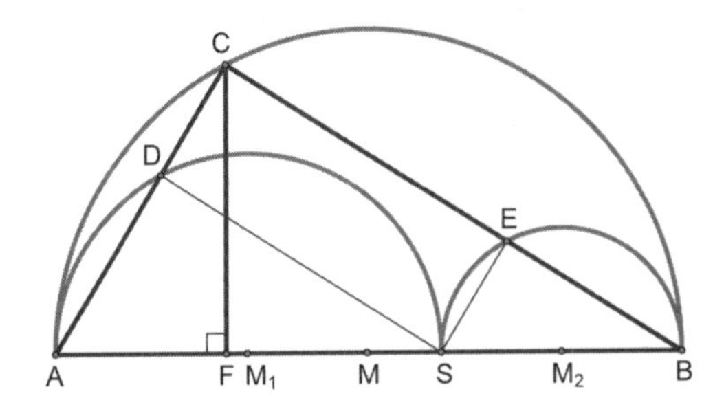


Figure 5-23

As a bonus to this curiosity, we can also say that if the legs of a right triangle (here, in fig. 5-23, we refer to right triangle *ABC*) are in the golden ratio, then the altitude, *CF*, to the hypotenuse, *AB*, determines the golden ratio as $\frac{CF}{AF}=\phi$ and $\frac{BF}{CF}=\phi$.

## CURIOSITY 11

The Chinese symbol yin and yang (fig. 5-24) depicts opposing forces and intends to show that they are interconnected in nature. This concept is key to much of Chinese philosophy and science. As you might expect by now, once again the golden section is embedded in this symbol.

Figure 5-24

The yin and yang design is comprised of two congruent semicircles on opposite sides of a diameter of a circle, whose diameter is twice that of the two smaller semicircles. (See fig. 5-25.)

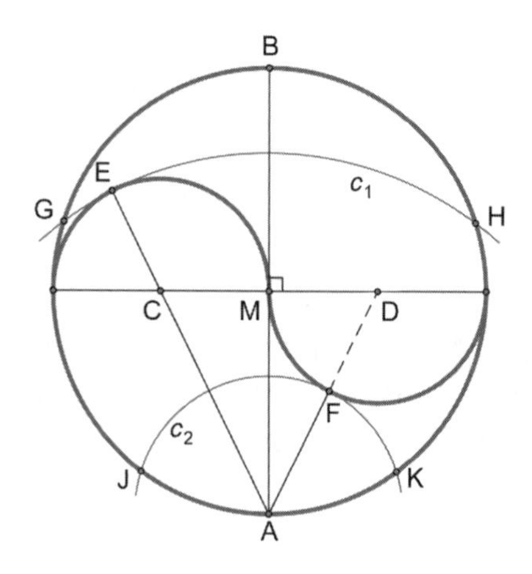


Figure 5-25

If we let the length of the radius of the larger circle (fig. 5-25) equal 1, then $AB=2$ and $CM=DM=\frac{1}{2}$. To this figure, we now add circles $c_1$ and $c_2$, each with point $A$ as center so that they are tangent to the congruent semicircles at points $E$ and $F$, respectively.

We then have $r_1=AE$ and $r_2=AF$. Since $CE$ and $AE$ are both perpendicular to the tangent to both circles (were it to be drawn) at $E$, they must coincide. This enables us to apply the Pythagorean theorem to $\Delta AMC$, which gives us

$$AC=\sqrt{AM^2+CM^2}=\sqrt{1^2+\left(\frac{1}{2}\right)^2}=\frac{\sqrt{5}}{2}.$$

Therefore, $r_1=AG=AE=AC+CE=\dfrac{\sqrt{5}}{2}+\dfrac{1}{2}=\phi$, and

$$r_2=AJ=AF=AD-DF=AC-DF=\frac{\sqrt{5}}{2}-\frac{1}{2}=\frac{1}{\phi}.$$

There, once again, appears the golden section.

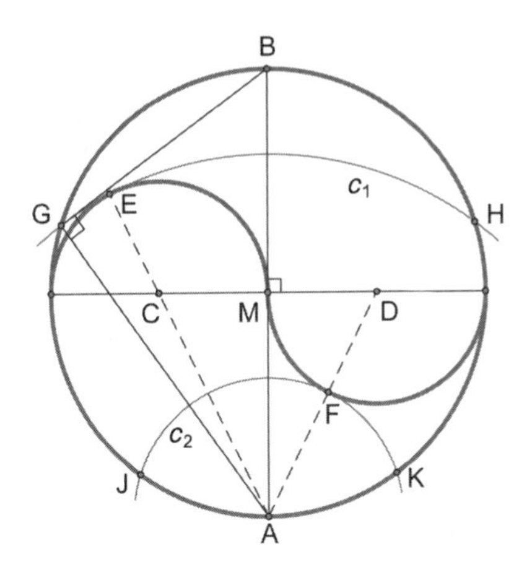


Figure 5-26

In figure 5-26, $\angle AGB$ is inscribed in a semicircle, and therefore it is a right angle. Thus, we can apply the Pythagorean theorem to right triangle $ABG$, to get

$$BG=\sqrt{AB^2-AG^2}=\sqrt{2^2-\phi^2}=\sqrt{4-\phi^2}=\frac{\sqrt{10-2\sqrt{5}}}{2}=\sqrt{\frac{5-\sqrt{5}}{2}}=\frac{\sqrt{\phi^2+1}}{\phi}.$$

In our investigation of the golden pentagon, we found that for a circumscribed circle with radius length 1, the side of the inscribed pentagon is

$$BG=r\cdot\frac{\sqrt{\phi^2+1}}{\phi}=1\cdot\frac{\sqrt{\phi^2+1}}{\phi}=\frac{\sqrt{\phi^2+1}}{\phi}.$$

Hence we have from within the yin and yang not only the golden section but also the golden pentagon, as shown in figure 5-27.

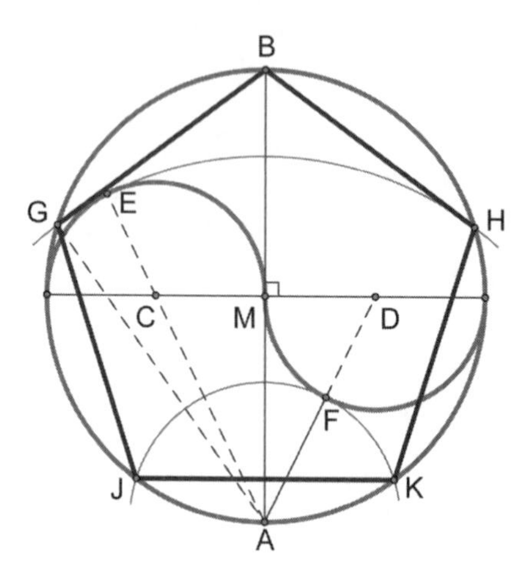


Figure 5-27

## CURIOSITY 12

This curiosity builds on the investigations we made in chapter 4 about the regular pentagon. Yet it requires one of the most powerful theorems—that is, sadly, too often neglected—from plane geometry. This relationship, first discovered by the Greek mathematician Claudius Ptolemy (ca. 90–ca. 168), provides us with a valuable relationship between the sides and diagonals of an inscribed quadrilateral. The theorem states that, for a quadrilateral inscribed in a circle, the sum of the products of the opposite sides equals the product of the diagonals.[4] In figure 5-28, we have inscribed quadrilateral *ABCD* and

$$AC \cdot BD = AB \cdot CD + BC \cdot AD, \text{ or}$$

$$e \cdot f = a \cdot c + b \cdot d.$$

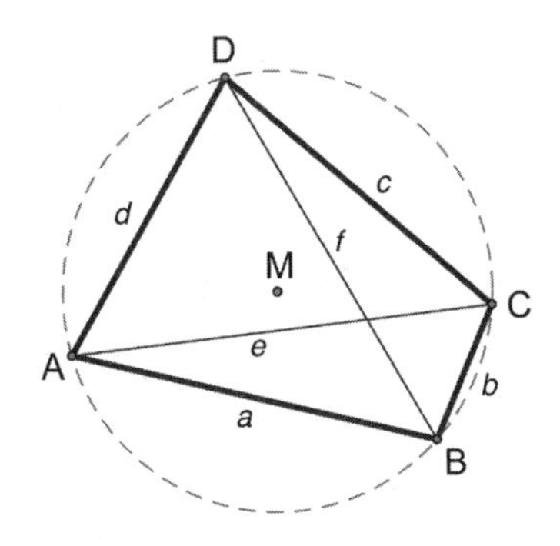


Figure 5-28

Now that we have mastered this lovely relationship, let's consider the regular pentagon *ABCDE* in figure 5-29 and focus on the isosceles trapezoid *ACDE*, with sides length *a*, bases length *a* and *d*, and diagonal length *d*. This trapezoid happens to be an inscribed quadrilateral, and therefore, we can apply Ptolemy's theorem (fig. 5-29):

$$AD \cdot CE = AC \cdot DE + AE \cdot CD$$

$$d \cdot d = d \cdot a + a \cdot a, \text{ or}$$

$$d^2 = da + a^2 = a(d + a).$$

We can rewrite this equation as a proportion, which then delivers for us the now-familiar result $\frac{d+a}{d} = \frac{d}{a} = \phi$.

Remember that in chapter 4 we already established that $\frac{AD}{AQ} = \frac{AQ}{DQ} = \frac{DQ}{QR} = \phi$.

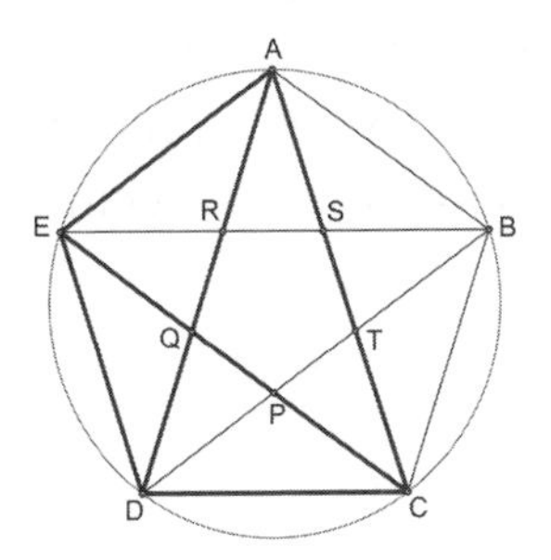

Figure 5-29

Figure 5-30

Suppose we now randomly select a point *X* on the pentagon's circumscribed circle (fig. 5-30), but not at one of the vertices of the pentagon. Applying Ptolemy's theorem to quadrilateral *ACDX* gives us

$$AD \cdot CX = d \cdot CX = AC \cdot DX + AX \cdot CD = d \cdot DX + AX \cdot a.$$

Therefore,

$$d \cdot CX = d \cdot DX + AX \cdot a,$$

which leads to

$$d \cdot CX - d \cdot DX = AX \cdot a,$$

or put another way:

$$\frac{d}{a} = \frac{AX}{CX - DX},$$

$$\text{which is } \frac{AX}{CX - DX} = \phi.$$

We leave other rich relationships to be found in this configuration to the reader.

## CURIOSITY 13

Most of the beauties involving this golden section are drawn on paper, yet one of them can be achieved by simply folding a knot with a strip of paper.[5] Just take a strip of paper, say, about one inch wide, and make a knot. Then very carefully flatten the knot as shown in figure 5-31. Notice the resulting shape appears to be a regular pentagon.

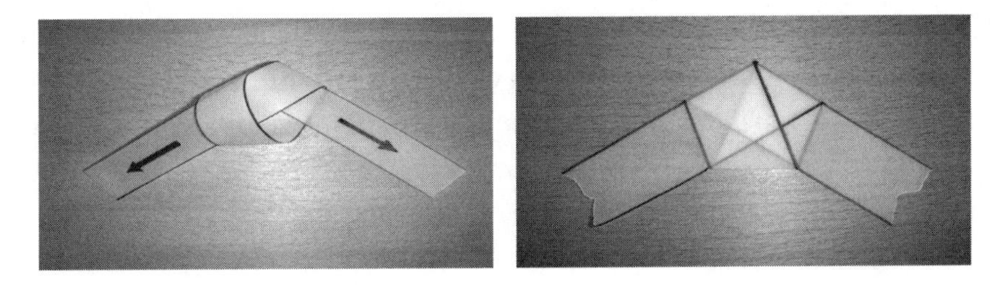

Figure 5-31. Photographs by I. Lehmann.

Figure 5-32 shows this in more detail.

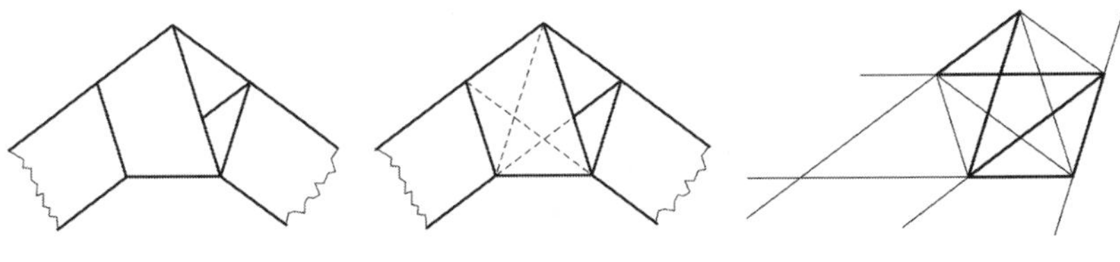

Figure 5-32

Through this paper-folding exercise, we can see the pentagon and the pentagram (with one part missing). This configuration allows us to visualize each of the diagonals of the pentagon parallel to one of its sides—since the sides of the paper strip are parallel. If you use relatively thin translucent paper and hold it up to a light, you ought to be able to see a pentagon with its diagonals. These diagonals intersect each other in the golden section.

If you now unfold this paper strip, you will have a parallelogram with four congruent isosceles trapezoids, each with three sides equal to the length of the sides of the pentagon, and the fourth side the length of the diagonals of the pentagon (fig. 5-33).

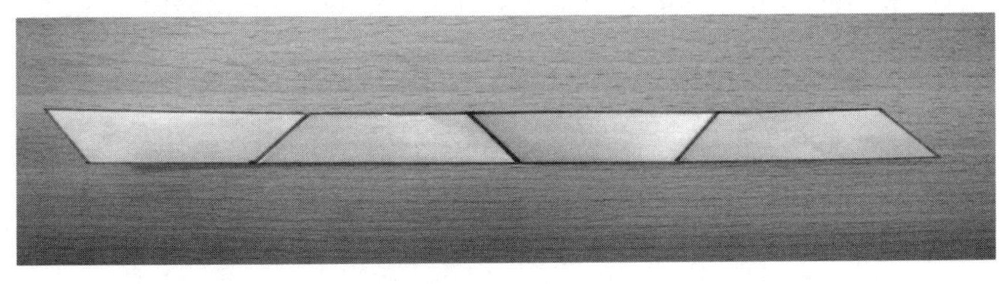

Figure 5-33. Photograph by I. Lehmann.

Elaborating on this a bit, in figure 5-34, we find that the shorter side of the parallelogram has the length of the side ($a$) of the pentagon, and the long side is the sum of twice a side ($a$) and twice a diagonal ($d = \phi a$) of the pentagon, that is, $2a + 2d = 2\phi^2 a$. The height ($h$) of the parallelogram is equal to the width of the paper strip.

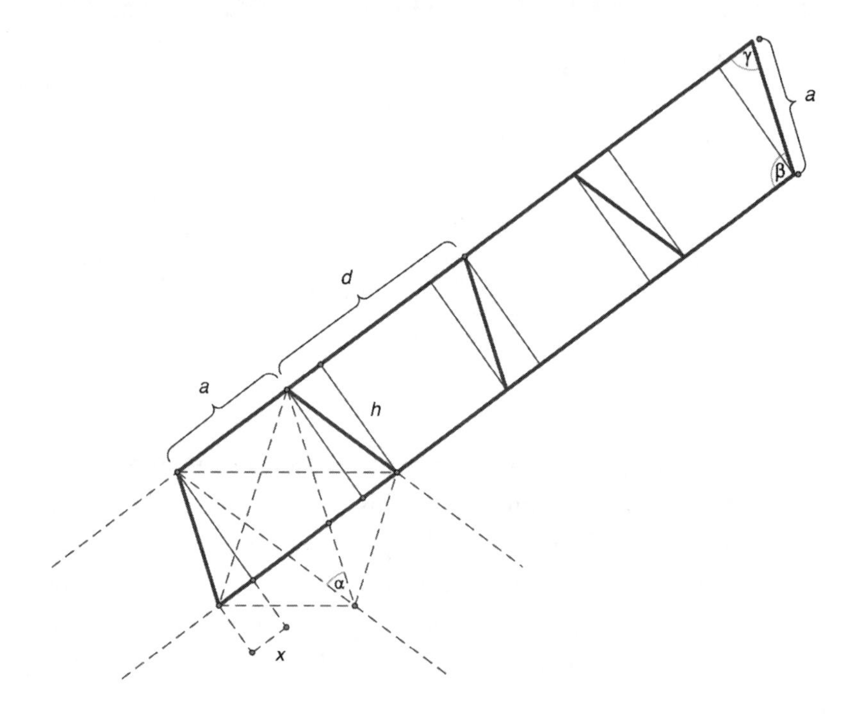


Figure 5-34

The angles are $\alpha = 36°$, $\beta = 108°$, and $\gamma = 72°$. (See chap. 4.) From the figure's symmetry, we get $a + 2x = d$. Therefore, $x = \frac{d-a}{2} = \frac{a}{2\phi}$.

We represent the width of the strip as $h$, the side length as $a$, and the pentagon's diagonal as $d$.

We then have $\sin\gamma = \sin 72° = \frac{h}{a}$ and

$$\sin 72° = \frac{\sqrt{\sqrt{5}\cdot\phi}}{2} = \frac{1}{2}\cdot\sqrt{\frac{5+\sqrt{5}}{2}},$$

from which we get

$$a=\frac{h}{\sin 72^\circ}=\frac{2\sqrt[4]{125}}{5\sqrt{\phi}}h=\frac{\sqrt{10}}{5}\sqrt{5-\sqrt{5}}\cdot h,$$

and

$$d=\frac{2\sqrt[4]{125}\cdot\sqrt{\phi}}{5}h=\frac{\sqrt{10}}{5}\sqrt{5+\sqrt{5}}\cdot h.$$

## CURIOSITY 14

Now we will examine a particularly curious problem. It was made popular by the English mathematician Charles Lutwidge Dodgson (1832–1898), who, under the pen name of Lewis Carroll, wrote *The Adventures of Alice in Wonderland*.[6] He posed the following problem: The square on the left in figure 5-35 has an area of 64 square units and is partitioned into quadrilaterals and triangles, and then these parts are rearranged and reassembled to form the rectangle on the right in figure 5-35.

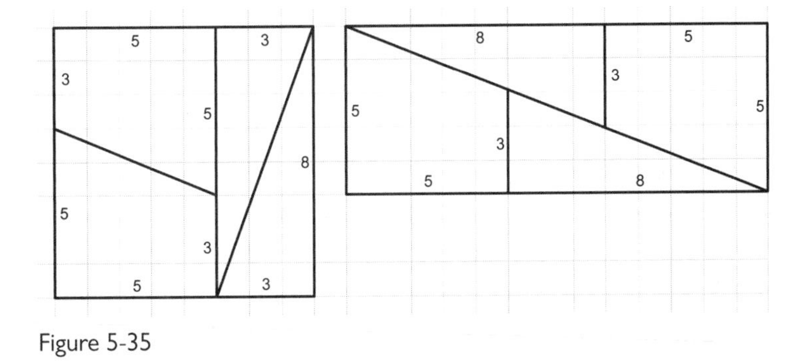

Figure 5-35

This rectangle has an area of $13\cdot 5=65$ square units. Where did this additional square unit come from? Think about it before reading further.

All right, we'll relieve you of the suspense. The "error" lies in the assumption that the rearranged triangles and quadrilaterals, when placed as in the right side of figure 5-35, will all line up along the drawn diagonal. This turns out not to be so. In fact, when put together properly, a "narrow" parallelogram is embedded here, and it has an area of one square unit (see fig. 5-36).

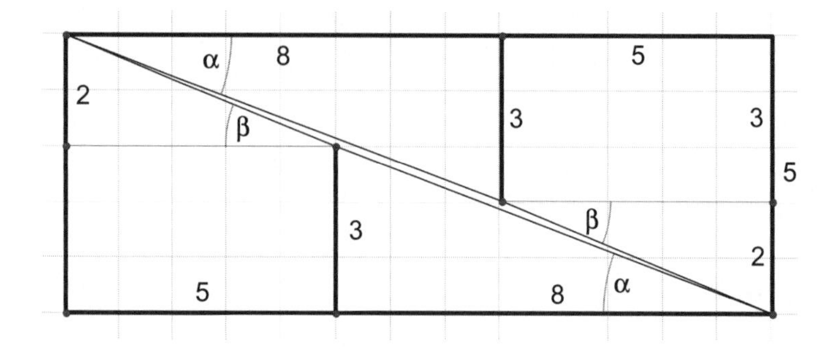


Figure 5-36

We can discover where the error lies by taking the tangent function of the angles marked $\alpha$ and $\beta$ so that we can discover their measures. Remember, they ought to be equal if they lie on the diagonal.[7]

$$\tan \alpha = \tfrac{3}{8}, \text{ then } \alpha \approx 20.6°;$$

$$\tan \beta = \tfrac{2}{5}, \text{ then } \beta \approx 21.8°.$$

The difference, $\beta - \alpha$, is merely 1.2°, yet it is enough to show that they are not on the diagonal.

You will notice that the segments above were 2, 3, 5, 8, and 13—all Fibonacci numbers. Moreover, you can prove[8] that $F_{n-1} F_{n+1} = F_n^2 + (-1)^n$, where $n \geq 1$. The rectangle has dimensions 5 and 13, and the square has a side length 8. These are the fifth, sixth, and seventh Fibonacci numbers: $F_5, F_6, F_7$.

This relationship tells us that

$$F_5 F_7 = F_6^{\,2} + (-1)^6$$

$$5 \cdot 3 = 8^2 + 1$$

$$65 = 64 + 1$$

This puzzle can then be done with any three consecutive Fibonacci numbers as long as the middle number is an even-numbered member of the Fibonacci sequence (i.e., in an even position). If we use larger

Fibonacci numbers, the parallelogram will be even less noticeable. Whereas if we use smaller Fibonacci numbers, then our eye cannot be deceived, as in figure 5-37.

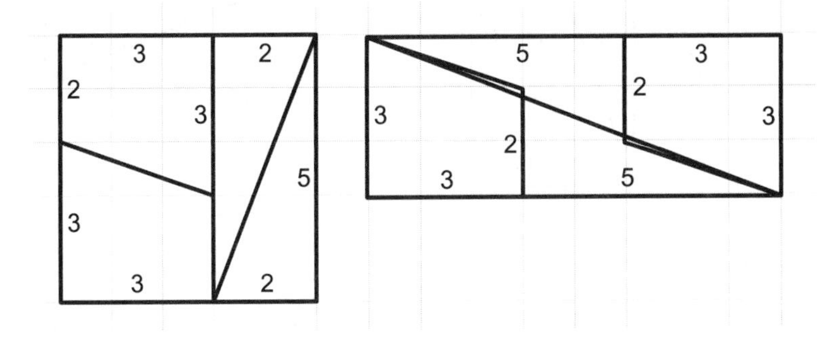

Figure 5-37

Here is the general form of the rectangle (fig. 5-38).

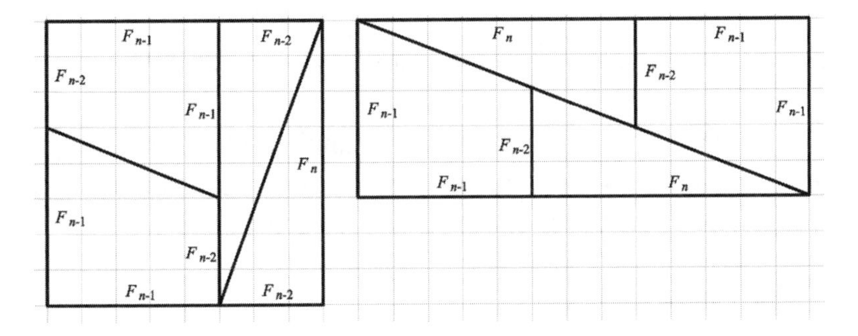


Figure 5-38

To do this properly without having the "missing area," the only such partitioning—amazingly enough—is with the golden ratio, $\phi$, as seen in figure 5-39.

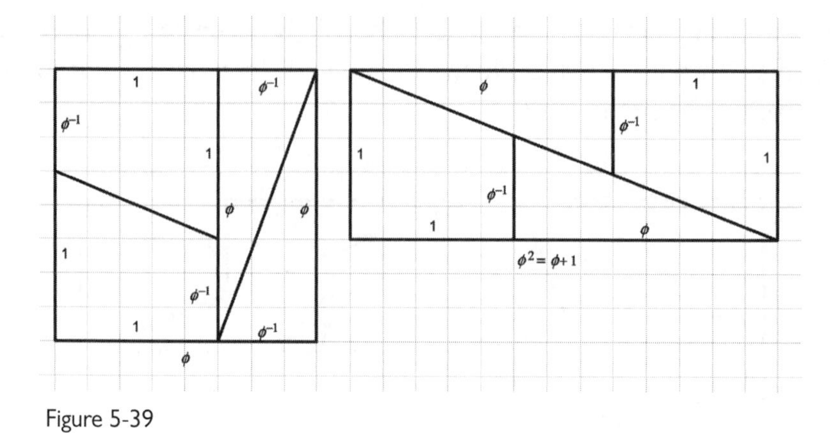


Figure 5-39

The areas of the rectangle and the square are equal here (fig. 5-39) as we shall show now:

The area of the square $= \phi \cdot \phi = \phi^2 = \phi + 1 = \dfrac{\sqrt{5}+3}{2} = 2.61803\ldots$, and

the area of the rectangle $= (\phi + 1) \cdot 1 = \phi + 1 = \dfrac{\sqrt{5}+3}{2} = 2.61803\ldots.$

Thus the areas of the square and the rectangle under this partition are equal. Once again, we find that the power of the golden ratio manifests itself in giving proper meaning to this dilemma.

## CURIOSITY 15

Although it may seem a bit contrived, this appearance of the golden section is quite surprising and requires us to make a "cross" composed of five congruent squares of side length 1 and cover it with a square that has side length $a$ and an area equal to that of the cross. The square should be placed in such a fashion that four small squares are formed in the corners, as shown in figure 5-40. These four small squares will have sides $b = \frac{1}{\phi}$, and the sides of the large square will have length $a = \sqrt{5}$. Furthermore, the golden section also appears in the sum of the areas of the portions of the cross that are *not* covered by the large square. This turns out to be $\frac{4}{\phi^2}$.

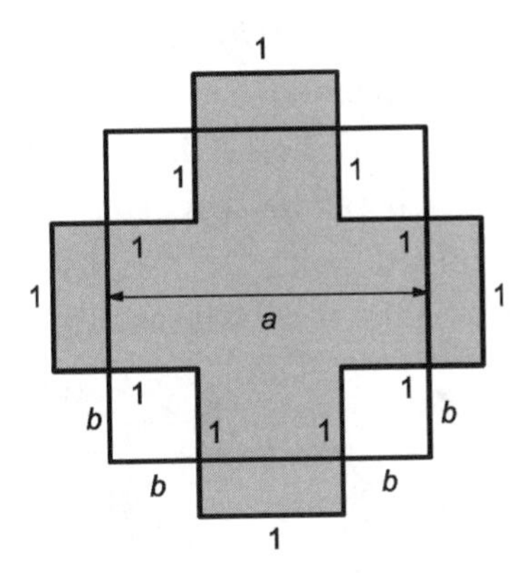


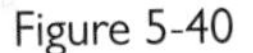
Figure 5-40

By inspecting figure 5-40, it is clear that if the area of the square is equal to the area of the cross, namely 5 (since it is comprised of five unit squares), then the side of the square is $a=\sqrt{5}$.

We can see in figure 5-40 that the side of the large square is $a=2b+1=\sqrt{5}$, which then leads to

$$b=\frac{\sqrt{5}-1}{2}=\frac{1}{\phi}.$$

Finally, to get the sum of the areas of the parts of the cross that are not covered by the square [i.e., four rectangles, each with dimensions $1\times(1-b)$], we do the following:

$$\text{Area}=4\cdot 1\cdot(1-b)=4\cdot\left(1-\frac{1}{\phi}\right)=4\cdot\left(1-\frac{\sqrt{5}-1}{2}\right)=2\cdot\left(3-\sqrt{5}\right)=\frac{4}{\phi^2}.$$

As we predicted, we arrive at a value involving the golden ratio—again at a time when you might have least expected it.

## CURIOSITY 16

There are times when the golden section just happens to be in a commonly seen design. Take the *Cross of Lorraine*, which was suggested by

General Charles de Gaulle (1890–1970) for the French flag (fig. 5-41) to represent the Free French as resistance to the Nazi occupation; it recalled Joan of Arc (ca. 1412–1431), who bore this symbol on her flag in battle against the English. Today, it can be seen as a 140-foot monument in de Gaulle's hometown of Colombey-les-Deux-Églises (fig. 5-42). It can also be seen (in partial form) on the coat of arms of Hungary.

Figure 5-41. Flag of Free France 1940–1944.

Figure 5-42. Monument Colombey, for Charles de Gaulle in Colombey-les-Deux-Églises.

This cross is constructed by thirteen congruent squares of side length 1, as shown in figure 5-43. If we now construct a line segment that divides the total cross area into two equal parts, then a most unexpected result appears. This area-dividing line segment $QPS$ partitions each of the small-square sides at its endpoints, $Q$ and $S$, into the golden ratio!

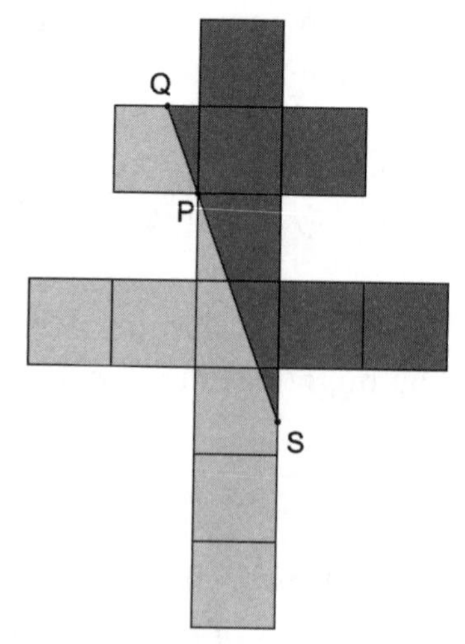


Figure 5-43

We will now set out to justify this remarkable occurrence—this strange and unexpected emergence of the golden section. Since the cross is comprised of thirteen unit squares, its total area is 13. The area of half the cross is comprised of $\Delta CQS$ plus four unit squares (see fig. 5-44).

We will seek the lengths of $BQ = x$ and $FS = y$, since that will help us determine if $Q$ and $S$ divide the sides of the unit squares in the golden ratio.

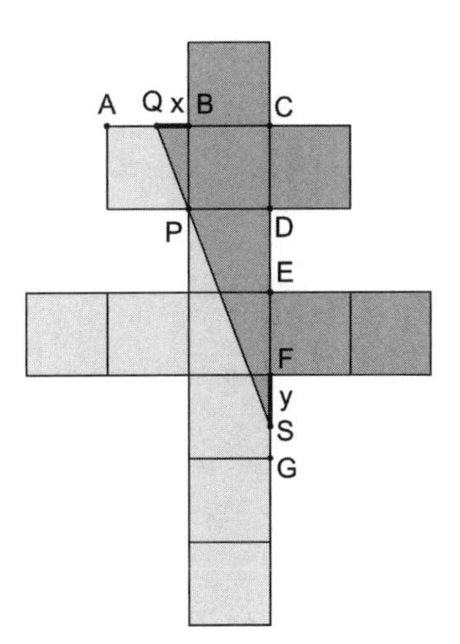


Figure 5-44

For the triangle $\Delta CQS$ we find $\text{Area}_{\Delta CQS} = \frac{13}{2} - 4 = \frac{5}{2}$. This area can also be represented by

$$\text{Area}_{\Delta CQS} = \frac{CQ \cdot CS}{2} = \frac{(BC + BQ)(CF + FS)}{2} = \frac{(1+x)(3+y)}{2}.$$

Equating these two expressions for the same area gives us the equation $\frac{(1+x)(3+y)}{2} = \frac{5}{2}$. This then leads us to the following algebraic steps:

$$(1+x) \cdot (3+y) = 5$$

$$1 \cdot (3+y) + x \cdot (3+y) = 5$$

$$x \cdot (3+y) = 5 - 3 - y = 2 - y\text{; therefore, } x = \frac{2-y}{3+y}.$$

Because $\Delta BQP$ and $\Delta DPS$ are similar, we get $\frac{BQ}{DP}=\frac{DP}{DS}=\frac{DP}{DF+FS}$, which, when expressed in terms of $x$ and $y$, is $\frac{x}{1}=\frac{1}{2+y}$, and so $x=\frac{1}{2+y}$.

We then equate the two values we found for $x$ and get $\frac{2-y}{3+y}=\frac{1}{2+y}$.

This, amazingly, leads us directly to the equation that yields the golden ratio: $y^2+y-1=0$, whose roots are: $y=-\phi$ (which we cannot use since it is negative), and

$$y=\frac{\sqrt{5}-1}{2}=\frac{1}{\phi}.$$

Then

$$x=\frac{1}{2+y}=\frac{1}{2+\frac{\sqrt{5}-1}{2}}=\frac{3-\sqrt{5}}{2}=\frac{1}{\phi^2}.$$

Note: The algebra that gave us this result is

$$\frac{3-\sqrt{5}}{2}=\frac{3-\sqrt{5}}{2}\cdot\frac{\sqrt{5}+3}{\sqrt{5}+3}=\frac{2}{\sqrt{5}+3}=\frac{1}{\frac{\sqrt{5}+3}{2}}=\frac{1}{\phi+1}=\frac{1}{\phi^2}.$$

Now that we have the values for $x$ and $y$, we can show that the points $Q$ and $S$ partition the segments $AB$ and $FG$, respectively, in the golden ratio. For segment $AB$:

$$\frac{AQ}{BQ}=\frac{1-x}{x}=\frac{1-\frac{3-\sqrt{5}}{2}}{\frac{3-\sqrt{5}}{2}}=\frac{\sqrt{5}+1}{2}=\phi;$$

and for segment $FG$:

$$\frac{FS}{GS}=\frac{FS}{FG-FS}=\frac{y}{1-y}=\frac{\frac{\sqrt{5}-1}{2}}{1-\frac{\sqrt{5}-1}{2}}=\frac{\sqrt{5}-1}{3-\sqrt{5}}=\frac{\sqrt{5}+1}{2}=\phi.$$

Therefore, we have shown that when the segment $PQS$ divides the cross into two equal areas, the points $Q$ and $S$ must partition the segments $AB$ and $FG$, respectively, in the golden ratio.

## CURIOSITY 17

To experience our next curiosity, we begin with a square and construct a line segment from one vertex to the midpoint of one side. We then come to the critical part of the construction, that is, to construct a circle tangent to the other two sides (as shown in fig. 5-45, the sides are $BC$ and $CD$) and to the line segment, $DE$, which we just drew inside the square. The golden ratio will now appear in several places. First, the line, $DK$, from a vertex of the square and through the center of the circle[9] intersects the side $BC$ at point $K$, and partitions it into the golden ratio: $\frac{CK}{BK} = \phi$; for convenience we use the side of the square as length 1.

Furthermore, the radius of the circle, $r = \frac{1}{\phi^2}$. As if this were not astounding enough, the circle's tangency points with the sides of the square provide the golden section of the sides, and the circle's third tangency point partitions the interior line segment in the the ratio $2 : \phi$.

Now let's inspect this in a bit more detail.

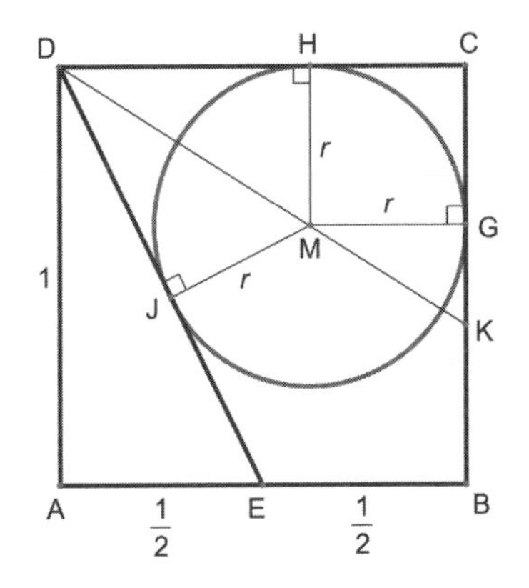


Figure 5-45

In figure 5-46 we have square $ABCD$, with $AE = BE = \frac{1}{2}$. Point $F$ is the intersection of the lines $CB$ and $DE$. Since $AB \parallel CD$, it follows that $\frac{BF}{CF} = \frac{BE}{CD} = \frac{1}{2}$, which gives us $\frac{BF}{BF+BC} = \frac{x}{x+1} = \frac{1}{2}$, and then we get $x = 1$.

In right triangle $CDF$ the Pythagorean theorem gives us $DF = \sqrt{5}$. Furthermore, because $\Delta AED \cong \Delta BEF$,

$$DE = EF = \frac{\sqrt{5}}{2}.$$

The radius, $r$, of the inscribed circle of $\Delta CDF$ can be found from the formula[10] $r = \frac{a+b-c}{2}$, where $a = CD$, $b = CF$, and $c = DF$ or in the following way:

Since $\Delta FBE \sim \Delta FCD$, we have $x = BF = 1$, and we have $CF = BC + BF = 1+1=2$.

Therefore, $FJ = DF - DJ = DF - DH = \sqrt{5} - (1-r) = \sqrt{5} - 1 + r$, and also $FG = CF - CG = 2 - r$.

Since $FJ = FG$, $\sqrt{5} - 1 + r = 2 - r$, whereupon it follows that

$$r = \frac{3-\sqrt{5}}{2} = \frac{1}{\phi^2}.$$

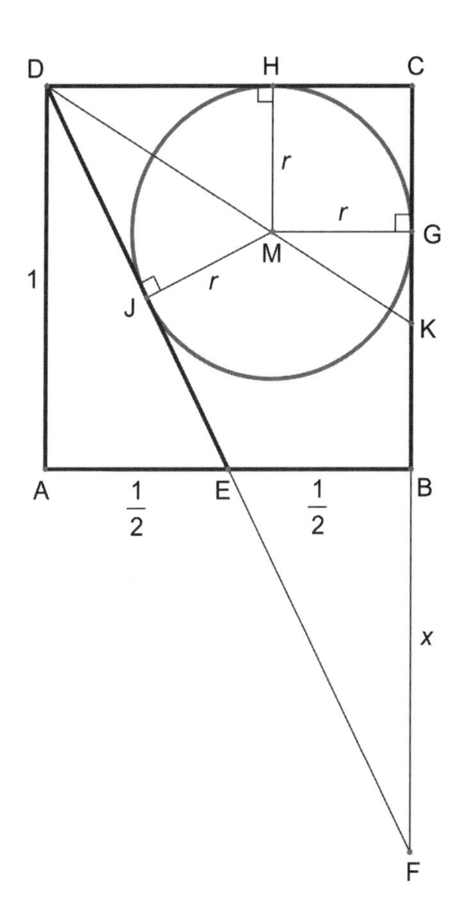


Figure 5-46

Recall that the center of the inscribed circle is the point of intersection of the angle bisectors of the triangle. When the Pythagorean theorem is applied to $\Delta DHM$ (fig. 5-46), we have

$$DM = \sqrt{DH^2 + HM^2} = \sqrt{(1-r)^2 + r^2} = \sqrt{5 - 2\sqrt{5}}.$$

The next few steps follow easily:

$$BG = DH = DJ = 1 - r = 1 - \frac{3-\sqrt{5}}{2} = \frac{\sqrt{5}-1}{2} = \frac{1}{\phi}$$

$$EJ = DE - DJ = \frac{\sqrt{5}}{2} - \frac{\sqrt{5}-1}{2} = \frac{1}{2}$$

$$\frac{BG}{CG} = \frac{1-r}{r} = \frac{1-\frac{1}{\phi^2}}{\frac{1}{\phi^2}} = \frac{\phi^2-1}{1} = \phi^2 - 1 = (\phi+1) - 1 = \phi$$

$$\frac{BC}{BG} = \frac{1}{1-r} = \frac{1}{1-\frac{1}{\phi^2}} = \frac{\phi^2}{\phi^2-1} = \frac{\phi^2}{\phi} = \phi,$$

which shows the partition into the golden ratio as we anticipated above.

Analogously, $\frac{DH}{CH} = \frac{1-r}{r} = \phi$, and $\frac{CD}{DH} = \frac{1}{1-r} = \phi$. Further to the list of $\phi$ appearances that we mentioned earlier, we can admire the following:

$$\frac{DJ}{EJ} = \frac{1-r}{\frac{1}{2}} = 2 \cdot (1-r) = 2 \cdot \left(1 - \frac{1}{\phi^2}\right) = 2 - 3 + \sqrt{5} = \sqrt{5} - 1 = \frac{2}{\phi}.$$

We now need to inspect the partitioning that point $K$ does to $BC$, where $K$ is the point at which the bisector of $\angle CDE$ meets $CF$. Point $K$ divides the side $CF$ of triangle $CDF$ proportional with the other two sides $CD$ and $DF$.

$$\frac{CK}{FK} = \frac{CD}{DF} = \frac{1}{\sqrt{5}}.$$

With $FK = CF - CK = 2 - CK$ we get

$$\frac{CK}{FK} = \frac{CK}{2 - CK} = \frac{1}{\sqrt{5}}.$$

Then $2 - CK = CK \cdot \sqrt{5}$, or $CK = \frac{1}{\phi}$. It follows that $BK = BC - CK = 1 - \frac{1}{\phi} = \frac{1}{\phi^2}$.

We are finally ready to set up the ratio we originally sought:

$$\frac{CK}{BK} = \frac{\frac{1}{\phi}}{\frac{1}{\phi^2}} = \phi,$$

which tells us that the angle bisector partitions the opposite side of the square into the golden ratio. When we look back at the simplicity of the original proposition—setting aside the computation that brought us here—again we have a truly wonderful appearance of the golden ratio.

## CURIOSITY 18

Here we have a little treat! Where you might least expect it, once again, the golden ratio pops up.

In figure 5-47, we have a square partitioned into four congruent trapezoids and a smaller square; all five parts have the same area. If the sides of the smaller (inner) square have length 1, then the sides ($a$) of the larger square will have length $\sqrt{5}$, and the height of each trapezoid will have length $x = \frac{1}{\phi}$.

In figure 5-48, we have a square partitioned into four congruent rectangles and a smaller square; again, all five parts have the same area. If the sides of the smaller (inner) square have length 1, then the the sides of the larger square ($a$) will have length $\sqrt{5}$. The widths (shorter sides) of each of the rectangles then have length $x = \frac{1}{\phi}$, and the longer sides have length $a - x = \phi$. These are rather unanticipated appearances of the golden section! Let's see why this is true.

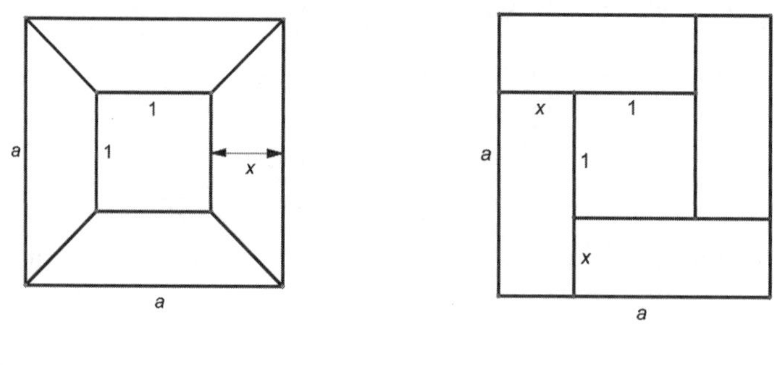


Figure 5-47 Figure 5-48

In both figures 5-47 and 5-48, $a=2x+1$, or $x=\frac{a-1}{2}$. Let's find the area of one of the trapezoids in figure 5-47. Remember, the area of a trapezoid is equal to one-half the product of the height and the sum of the bases.

$$\text{Area}_{\text{Trapezoid}}=x\left(\frac{a+1}{2}\right)=\frac{a-1}{2}\cdot\frac{a+1}{2}=\frac{a^2-1}{4}.$$

Since the area of one of the trapezoids is the same as the smaller square, we get: $\frac{a^2-1}{4}=1$, which is $a^2=5$, or $a=\sqrt{5}$. Then to find $x$, we use the earlier equation:

$$x=\frac{a-1}{2}=\frac{\sqrt{5}-1}{2}=\frac{1}{\phi}.$$

In figure 5-48, the area of the square is to be five times that of the inner square, or 5 times 1. Therefore, $a=\sqrt{5}$. Using our previously obtained value for $a$, we get $\sqrt{5}=a=2x+1$, whereupon it follows that

$$x=\frac{\sqrt{5}-1}{2}=\frac{1}{\phi}.$$

Then

$$a-x=\sqrt{5}-\frac{\sqrt{5}-1}{2}=\frac{\sqrt{5}+1}{2}=\phi.$$

Once again, the appearance of $\phi$ is clearly justified.

## CURIOSITY 19

Just as we were surprised with the unexpected appearance of $\phi$ in the previous curiosity, so, too, it is astonishing that the golden section will once again appear where we have no reason to expect it, as in the following configuration. Consider the rectangle *ABCD*, which in figure 5-49 has points *P* and *Q* on lines *AB* and *BC*, respectively, so that it is partitioned into four triangles: $\Delta DPQ$, $\Delta ADP$, $\Delta PBQ$, and $\Delta CDQ$, where the three shaded triangles have the same area. This produces an astonishing appearance of $\phi$, namely that points *P* and *Q* partition the sides *AB* and *BC* into the golden section.[11] Imagine, the location of points *P* and *Q* is determined to create equal areas and results in partitioning the line on which they lie in the golden section.

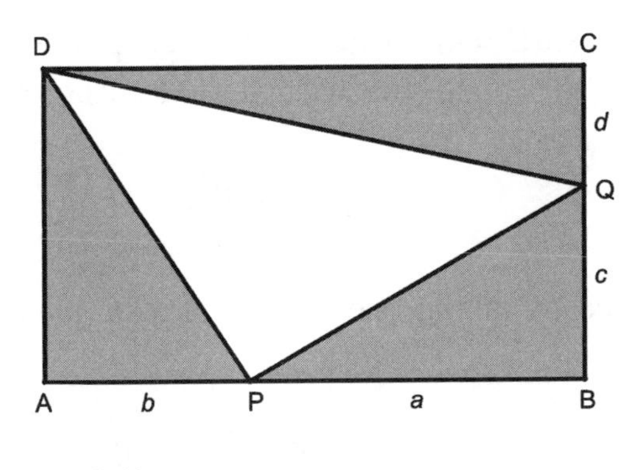

Figure 5-49

(Although the justification of this is rather simple, we shall provide it in the appendix so as not to disturb the flow of curiosities here.)

## CURIOSITY 20

This time we shall begin with lengths related to $\phi$, by constructing a triangle, $\Delta ABC$, with sides of lengths $a=\phi$, $b=\phi\sqrt{\phi}$, and $c=\phi+1$, and show that it will produce a right triangle. (See fig. 5-50.)

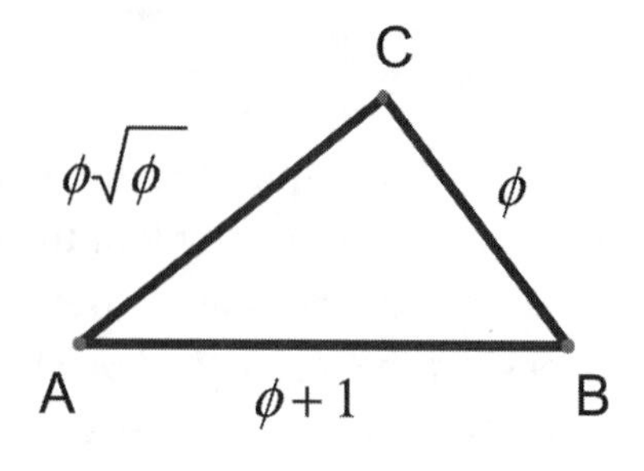

Figure 5-50

Using the converse of the Pythagorean theorem, we will show that this triangle is, in fact, a right triangle, since[12]

$$(\phi\sqrt{\phi})^2+\phi^2=\phi^3+\phi^2=\phi^2(\phi+1)=(\phi+1)(\phi+1)=(\phi+1)^2,$$

and therefore, $AC^2+BC^2=AB^2$, which tells us that $\angle ACB$ is a right angle.

Furthermore, the area of this triangle (in terms of $\phi$) is

$$\text{Area}_{\Delta ABC}=\frac{\sqrt{\phi^5}}{2},$$

which is obtained by taking one-half the product of the legs of this right triangle:

$$\frac{1}{2}(\phi)\left(\phi\sqrt{\phi}\right)=\frac{1}{2}\phi^2\sqrt{\phi}=\frac{1}{2}\sqrt{\phi^4}\sqrt{\phi}=\frac{\sqrt{\phi^5}}{2}.$$

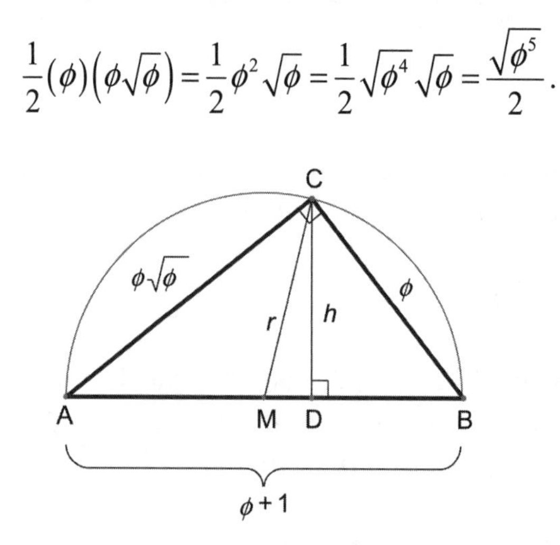

Figure 5-51

If, as in figure 5-51,[13] we have $BD=1$ and we construct the median $(CM)$ and the altitude $(CD)$ to the hypotenuse $AB$, we find the distance between them is

$$DM=BM-BD=\frac{(\phi+1)}{2}-1=\frac{\phi-1}{2}=\frac{\phi(\phi-1)}{2\phi}$$
$$=\frac{\phi^2-\phi}{2\phi}=\frac{(\phi+1)-\phi}{2\phi}=\frac{1}{2\phi}=\frac{\sqrt{5}-1}{4}.$$

There are some more little features worth citing, namely that the areas of the two triangles determined by the altitude $CD$ also can be expressed in terms of the golden ratio:

$$\text{Area}_{\Delta ACD}=\frac{\sqrt{\phi^3}}{2}, \text{ and Area}_{\Delta BCD}=\frac{\sqrt{\phi}}{2}.$$[14]

This shows that $\text{Area}_{\Delta ACD}=\phi\cdot\text{Area}_{\Delta BCD}$.

And, last, but not least, the ratio of the side lengths is

$$AB:AC:BC=(\phi+1):\phi\sqrt{\phi}:\phi=\phi^{\frac{1}{2}}:1:\phi^{-\frac{1}{2}}.$$

## CURIOSITY 21

Our next curiosity will build on the previous one. We shall begin by taking the configuration that we produced in Curiosity 20 and represent the lengths in a somewhat different way, yet keeping the values the same. Thus, we have the lengths as shown in figure 5-52:

$$a=BP=\phi,$$
$$b=AP=\phi\sqrt{\phi}=\phi^{\frac{3}{2}},$$
$$c=AB=\phi+1=\phi^2.$$

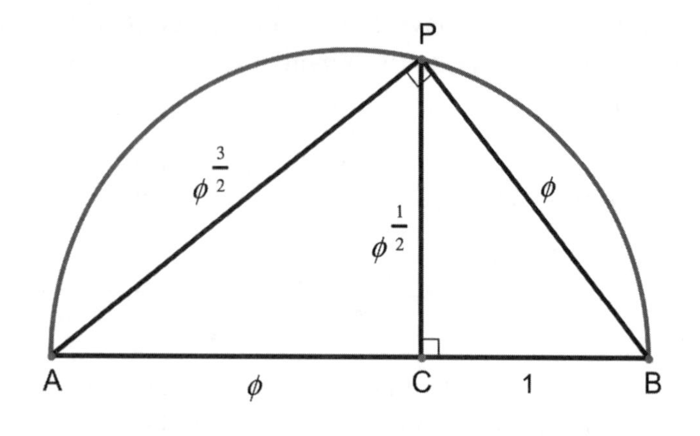


Figure 5-52

We will now draw a series of perpendiculars and parallels in succession. The perpendicular line to $AB$ at $B$ ($=B_1$) intersects the line through $A$ and $P$ ($=P_1$) at $P_2$, the parallel line to $B_1P_1$ through $P_2$ intersects $AB$ at $B_2$. (See fig. 5-53.)

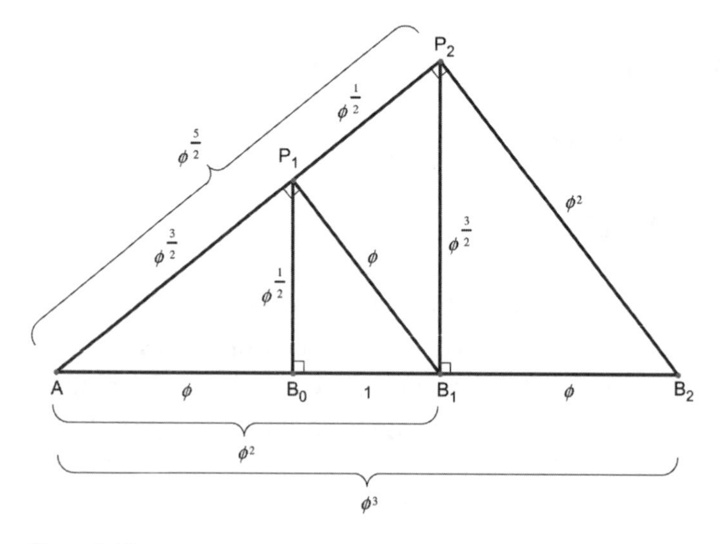


Figure 5-53

Here, $B_0B_1 = \phi^0 = 1$

$$P_1P_2 = B_0P_1 = \phi^{\frac{1}{2}}$$

$$AB_0 = B_1P_1 = B_1B_2 = \phi^1 = \phi$$

$$AP_1 = B_1P_2 = \phi^{\frac{3}{2}}, \text{ and } AB_1 = B_2P_2 = \phi^2$$

$$AP_2 = \phi^{\frac{5}{2}}, \text{ and } AB_2 = \phi^3.$$

This process can then continue as shown in figure 5-54.

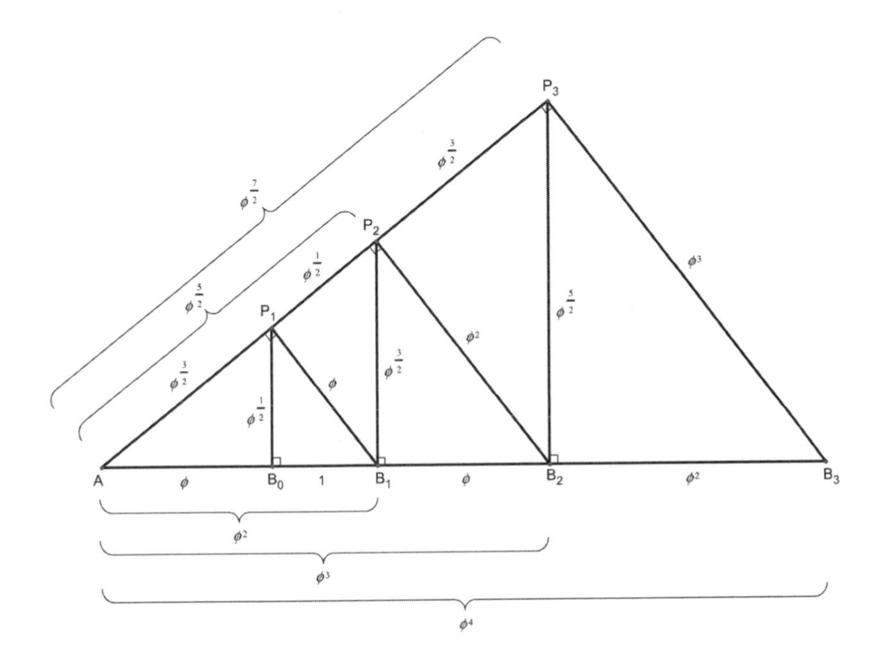

Figure 5-54

As one might have expected by now, we have a geometric representation of the powers of the golden ratio and at the same time the Fibonacci numbers (see also chap. 4, pp. 92, 127):

$$AB_1 = AB_0 + B_0B_1 \quad \Rightarrow \quad \phi^2 = 1\phi + 1$$

$$AB_2 = AB_1 + B_1B_2 \quad \Rightarrow \quad \phi^3 = 2\phi + 1$$

$$AB_3 = AB_2 + B_2B_3 \quad \Rightarrow \quad \phi^4 = 3\phi + 2$$

$$AB_4 = AB_3 + B_3B_4 \quad \Rightarrow \quad \phi^5 = 5\phi + 3$$

$$AB_5 = AB_4 + B_4B_5 \qquad \Rightarrow \qquad \phi^6 = 8\phi + 5$$

$\ldots$

$$AB_n = AB_{n-1} + B_{n-1}B_n \qquad \Rightarrow \qquad \phi^n = F_n\phi + F_{n-1} \quad (\text{for } n > 0).$$

Once again the golden ratio provides us with yet another example of the close connection between algebra and geometry. Were we to do this in reverse, we would have the segments as noted in figure 5-55.

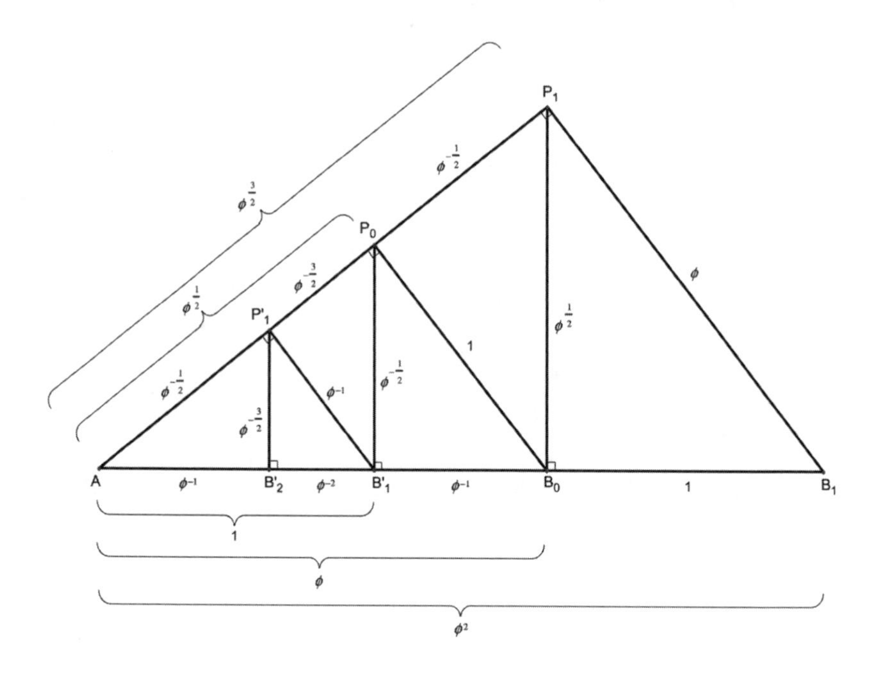

Figure 5-55

This would give us the sequence 1, $\phi^{-1}$, $\phi^{-2}$, $\phi^{-3}$, $\phi^{-4}$, $\phi^{-5}$, $\phi^{-6}$, …, for which the analogous construction of the triangles produces $\phi = \phi^{-1} + \phi^{-2} + \phi^{-3} + \phi^{-4} + \phi^{-5} + \phi^{-6} + \ldots$.

We can also construct a rectangular "spiral," where the $n^{\text{th}}$ side lengths are these powers of $\phi$ (fig. 5-56). The spiral approaches the limit point determined by the intersection $B_0B_2'$ and $B_1'B_3'$.

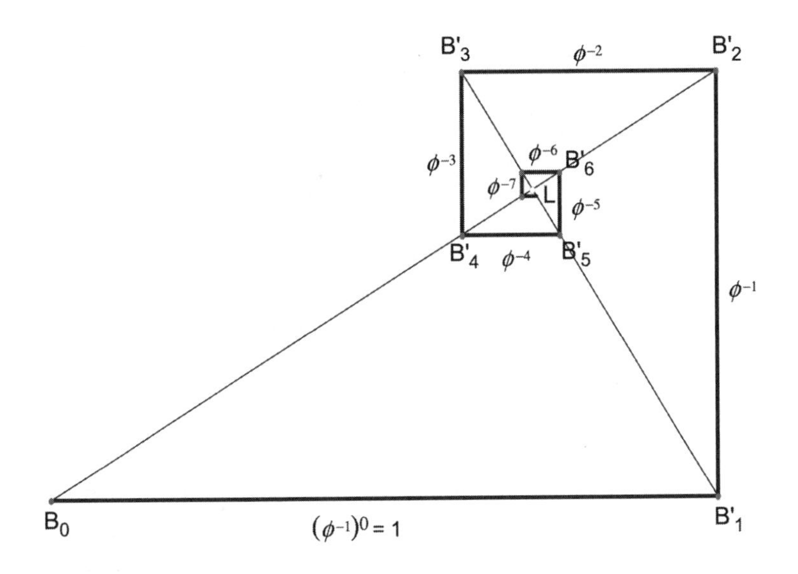


Figure 5-56

The length of this rectangular "spiral" is finite, namely $1+\phi^{-1}+\phi^{-2}+\phi^{-3}+\phi^{-4}+\ldots=1+\phi=\phi^2$.

## CURIOSITY 22

Yet again we have a simple situation in which the golden ratio appears quite unexpectedly! We begin with a parallelogram *ABCD* with an acute angle of measure 60°, and then two isosceles triangles are formed as shown in figure 5-57. The two parallelograms, *ABCD* and *DEBF*, are similar if their corresponding sides are in the same ratio, $(a+x):a=a:x$. This should remind us of the golden ratio, where the ratio is then $\phi:1$.

Then the ratio of the areas is the square of this ratio of similitude, namely $\text{Area}_{ABCD}:\text{Area}_{EBFD}=\phi^2:1$. This is easily justified.

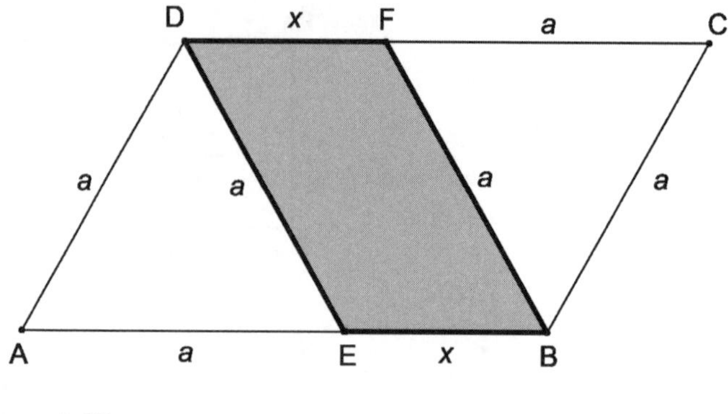


Figure 5-57

From the similarity of the parallelograms, we have $\frac{AB}{BC} = \frac{DE}{BE}$, which is also  $\frac{a+x}{a} = \frac{a}{x}$. This will lead us to a now-familiar quadratic equation: $x^2 + ax - a^2 = 0$, where the positive root is $x = \frac{1}{\phi}a$. Therefore, the ratio of the corresponding sides of the similar parallelograms is $a : x = \phi : 1$, and the ratio of the areas of the two parallelograms is $\text{Area}_{ABCD} : \text{Area}_{EBFD} = \phi^2 : 1$.

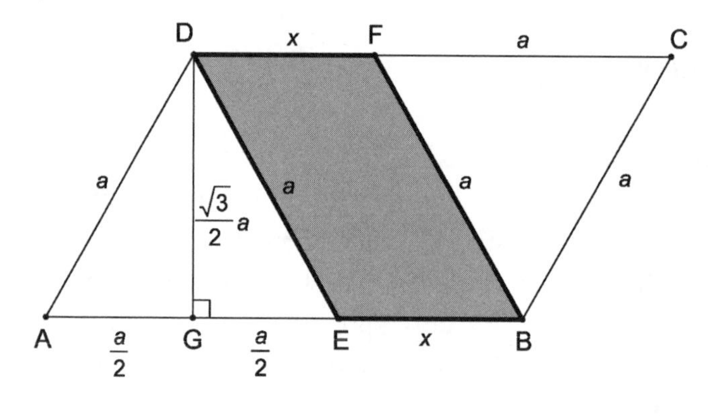


Figure 5-58

Alternatively, we can also show the ratio of the areas of the two parallelograms independently. The height, $DG$, of $\Delta ADE$ is

$$h = \frac{\sqrt{3}}{2}a,$$

which also serves as the height of each of the two parallelograms. Therefore, we can calculate the area of each of the parallelograms using the formula: the product of the base and the height.

$$\text{Area}_{ABCD} = AB \cdot DG = (a + x) \cdot h$$

$$= \left(a + \frac{1}{\phi}a\right) \cdot \frac{\sqrt{3}}{2}a = \phi \cdot \frac{\sqrt{3}}{2} \cdot a^2 = \frac{\sqrt{3} \cdot \left(\sqrt{5}+1\right)}{4} \cdot a^2.$$

$$\text{Area}_{EBFD} = BE \cdot DG = x \cdot h = \frac{1}{\phi}a \cdot \frac{\sqrt{3}}{2}a = \frac{\sqrt{3} \cdot \left(\sqrt{5}-1\right)}{4} \cdot a^2.$$

Then the ratio of the areas of the two parallelograms is

$$\text{Area}_{ABCD} : \text{Area}_{EBFD} = \left(\phi \cdot \frac{\sqrt{3}}{2} \cdot a^2\right) : \left(\frac{1}{\phi} \cdot \frac{\sqrt{3}}{2} \cdot a^2\right) = \frac{\phi}{\frac{1}{\phi}} = \phi^2 \cdot$$

This can also be written as $\text{Area}_{ABCD} : \text{Area}_{EBFD} = \phi^2 : 1 = (\phi + 1) : 1.$

## CURIOSITY 23

The trapezoid presents us with a curious occurrence of the golden section. In figures 5-59 and 5-60, we have trapezoids *ABCD*, one isosceles (fig. 5-60) and one not isosceles (fig. 5-59). The line segment *FE*, which joins points *E* and *F* on the sides of the trapezoids, is parallel to the bases. The bases have lengths 3*b* and *b*, as shown. The length of

$$FE = c = \sqrt{\frac{a^2 + b^2}{2}}.$$

We call *FE* the *root mean square*,[15] and it has the property that divides the original trapezoid into two trapezoids of equal area. Of particular interest to us here is that this segment, *FE*, partitions the two sides of the trapezoid in the golden section.

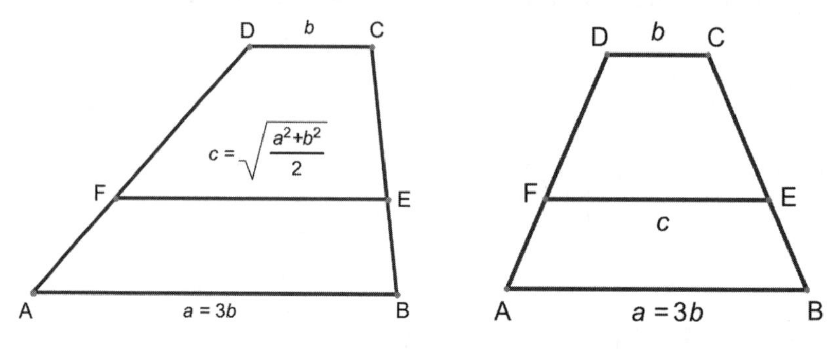


Figure 5-59 Figure 5-60

So far, the following is given: The parallel bases are $AB=a$, $CD=b$, and $a=3b$. Also

$$EF=c=\sqrt{\frac{a^2+b^2}{2}}.$$

Since $b<a$, we then have $b<c<a$, as

$$c=\sqrt{\frac{9b^2+b^2}{2}}=\sqrt{5}b<3b.$$

This justifies that the line segment *EF* actually does exist, since its length lies between the lengths of the bases. We shall now refer to figures 5-61 and 5-62, where $DG \parallel BC$ and $BG=CD=b$. Then because of the similarity of $\Delta ADG$ and $\Delta DFH$, we get

$$\frac{AD}{DF}=\frac{AG}{FH}=\frac{2b}{FE-EH}=\frac{2b}{c-b}=\frac{2b}{b\sqrt{5}-b}=\frac{2}{\sqrt{5}-1}=\frac{\sqrt{5}+1}{2}=\phi.$$

This allows us to conclude that $\frac{DF}{AF}=\phi$, and once again we arrived at our golden ratio!

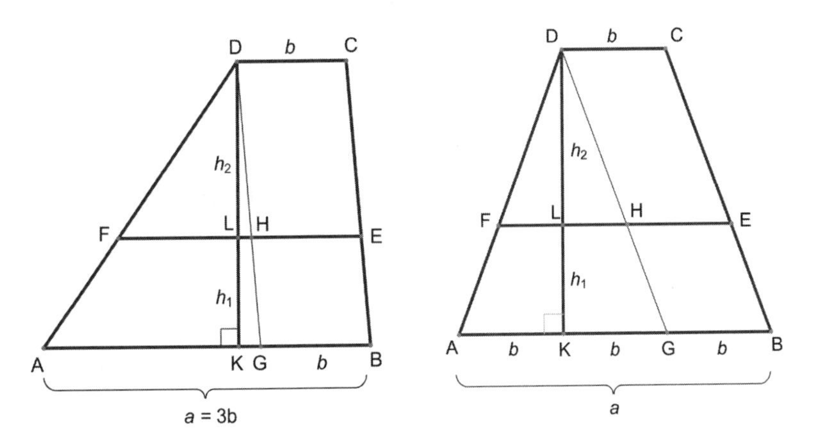


Figure 5-61

Figure 5-62

The altitudes of the trapezoid are also partitioned into the golden section. In figures 5-61 and 5-62, we note that the altitudes $DK = KL + DL = h_1 + h_2$ and $DL = h_2$ give us the golden ratio as $h_2 : h_1 = \phi : 1$. We leave to the reader to show that the area of the original trapezoid is, in fact, divided in half by $FE$.

For the ambitious reader, we offer a concise procedure to construct this configuration in the appendix.

## CURIOSITY 24

Here we will consider an isosceles trapezoid that can have an inscribed circle (one tangent to each of its four sides), and where the trapezoid's circumscribed circle has its larger base as the circumcircle's diameter. This special trapezoid has the golden ratio embedded in it.

We will use the isosceles trapezoid $ABCD$ (shown in fig. 5-63) with sides $AB = a$, $BC = b$, $CD = c$, and $AD = b$. It has an inscribed and a circumscribed circle, where $AB \parallel CD$ and $AB = a$, which is the diameter of the circumscribed circle $c_0$. Then the radius of the circumscribed circle is $r_0 = \frac{a}{2}$, and $\varepsilon = \angle BM_0C$. We then have the following unexpected properties: $b = \frac{a}{\phi}$ and $c = a(\frac{\phi - 1}{\phi^2})$, and the radius of the inscribed circle is

$$r_i = \frac{a\sqrt{\phi}}{2\phi^2}.$$

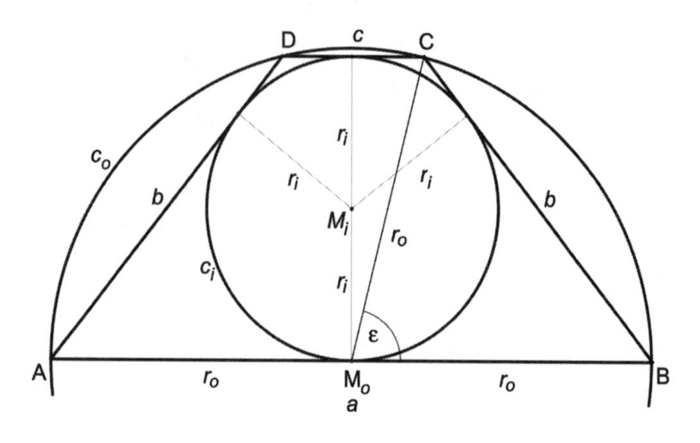


Figure 5-63

As a bonus, we also can show that the golden ratio also appears in relation to an angle, namely when the sine function is applied to $\angle BM_0C = \varepsilon$, we find that $\sin\frac{\varepsilon}{2} = \frac{1}{\phi}$.

(The justification for these appearances of the golden ratio can be found in the appendix.)

## CURIOSITY 25

Sometimes a sighting of the golden section is not only unexpected but also *not* intuitively obvious. Here we begin with a right pyramid[16] with a rectangular base. All lateral sides are, therefore, isosceles triangles as shown in figure 5-64. A plane containing one side ($BC$) of the base and intersecting the opposite lateral face in line segment $EF$ divides the volume of the pyramid in half. The fascinating thing here is that the points $E$ and $F$ partition the lateral edges, $AS$ and $DS$, respectively, in the golden section, that is, $\frac{AS}{ES} = \frac{\phi}{1}$ and $\frac{ES}{AE} = \frac{\phi}{1}$. The same holds true for point $F$ with regard to $DS$, namely $\frac{DS}{FS} = \frac{\phi}{1}$ and $\frac{FS}{DF} = \frac{\phi}{1}$.

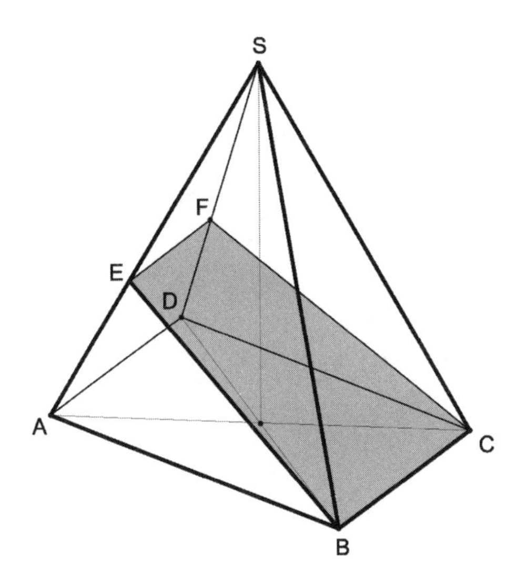


Figure 5-64

## CURIOSITY 26

This curiosity may appear to be a bit contrived, yet it shows the strange ways that the golden section appears where it may be least expected. We begin with a triangle $\Delta ABC$, where $AB=2$, $BC=1$, and $AB \perp BC$, as shown in figure 5-65. We will call the point $C=C_0$, for convenience, and it will allow us to make a generalization at the end of this curiosity. We now bisect $\angle ACB$ with a line segment that intersects $AB$ at point $C_{-1}$. At the point $C$ (or $C_0$), we will construct a perpendicular to $C_{-1}C_0$, which will intersect the line $AB$ at point $C_1$. Then at point $C_1$, we will construct a perpendicular to $C_0C_1$ to meet $BC$ at $C_2$. Repeating the process, we have at $C_2$ the perpendicular to $C_1C_2$ intersecting $AB$ at $C_3$. We then get points $C_4, C_5, \ldots, C_n$ (where $n \geq 0$) in the same fashion.[17] The result is that

$$\frac{BC_{-1}}{BC_0}=\frac{BC_0}{BC_1}=\frac{BC_1}{BC_2}=\frac{BC_2}{BC_3}=\frac{BC_3}{BC_4}=\frac{BC_4}{BC_5}= \ldots =\frac{BC_{n-1}}{BC_n}=\frac{1}{\phi}$$

and

$$BC_n = \phi \cdot BC_{n-1} = \phi^n \text{ (for } n=-1, 0, 1, 2, 3, \ldots).$$

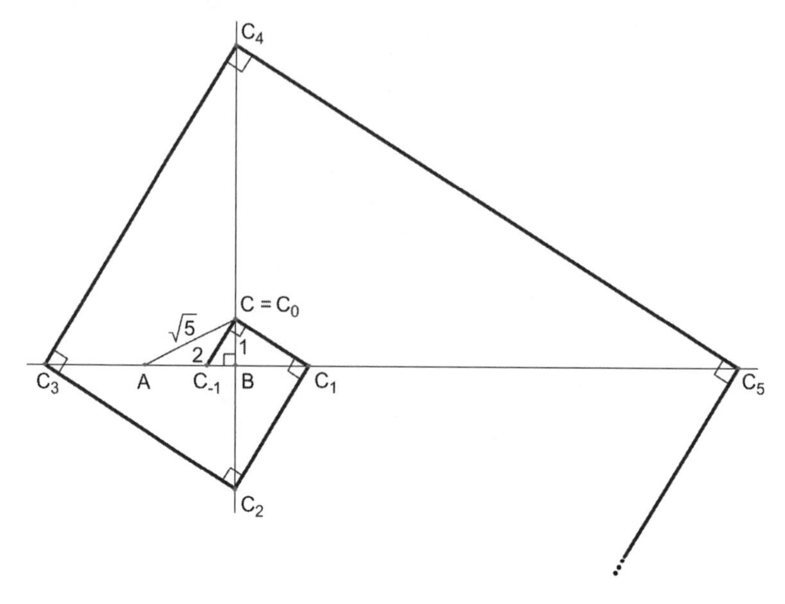


Figure 5-65

To see why this holds true, we shall go back to the original triangle $\Delta ABC$, where $AB=2$, $BC=BC_0=1$, and $AC=\sqrt{5}$. If we let $BC_{-1}=x$, then $AC_{-1}=2-x$. Since $CC_{-1}$ is an angle bisector,

$$\frac{AC}{AC_{-1}}=\frac{BC}{BC_{-1}}, \text{ or } \frac{\sqrt{5}}{2-x}=\frac{1}{x},$$

which then gives us $x=\frac{1}{\phi}$. We then have $BC_{-1}=\phi^{-1}$, $BC_0=\phi^0(=1)$ and $BC_1=\phi^1$.

In general, the right triangles of the form $\Delta BC_nC_{n+1}$ are all similar to each other. Therefore,

$$\frac{BC_0}{BC_{-1}}=\frac{BC_1}{BC_0}=\frac{BC_2}{BC_1}=\frac{BC_3}{BC_2}=\frac{BC_4}{BC_3}=\frac{BC_5}{BC_4}=\ldots=\frac{BC_n}{BC_{n-1}}=\frac{\phi}{1},$$

for $n=0, 1, 2, 3, \ldots$.

With

$$\frac{BC_{-1}}{BC_0}=\frac{1}{\phi}=\phi^{-1},$$

we get $BC_0 = \phi BC_{-1}$. In general, for $n =$ 0, 1, 2, 3, …
$BC_n = \phi \cdot BC_{n-1} = \phi^n$.

Without the distraction of the auxilary line in figure 5-65, this sort of "spiral" can be a bit more attractive—as shown in figure 5-66.

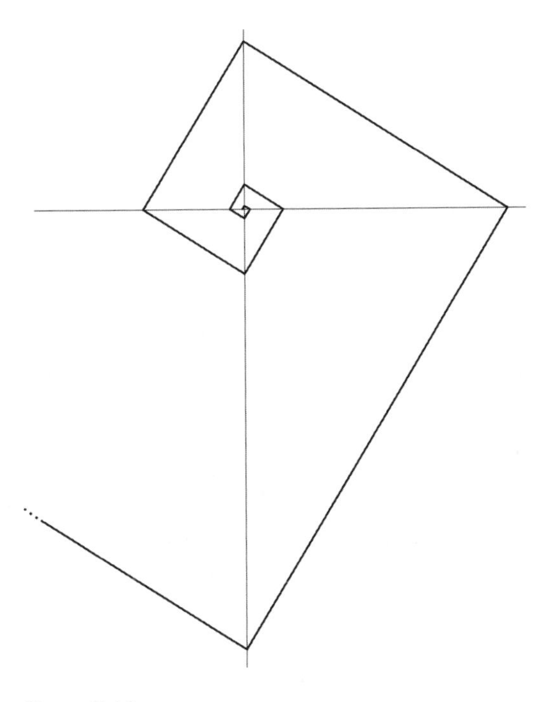

Figure 5-66

## CURIOSITY 27

Our next curiosity will not be geometric but rather use a pattern that should not prove to be too strange, although it may look so at the start. We will consider a sequence of 1s and 0s as shown here:

1 0 1 1 0 1 0 1 1 0 1 1 0 1 0 1 1 0 1 0 1 1 0 1 1 0 1 0 1 1 0 1 ….

The construction of this sequence is quite simple. We begin with a 1. Then the next step is to replace the 1 with 10, as you can see in the listing below. Then, in each succeeding step, we replace the 1s with 10 and the 0s with 1.

## GOLDEN STRINGS

1
10
101
10110
10110101
1011010110110
101101011011010110101
1011010110110101101011011010110110
. . . .

Another way of looking at this sequence development is to take the third generation (101) and add to it the previous generation (10), to get 101 10. To get the next generation, we take the previous one, 101 10, and add to it its predecessor, to get 10110 101. In general terms, to get the $n$th generation, we take the $(n-1)$ generation and add to it the $(n-2)$ generation. This sequence is often called the *golden string.*

By now you may wonder what this has to do with the golden ratio. Consider the function $y=f(x)=\phi x$.

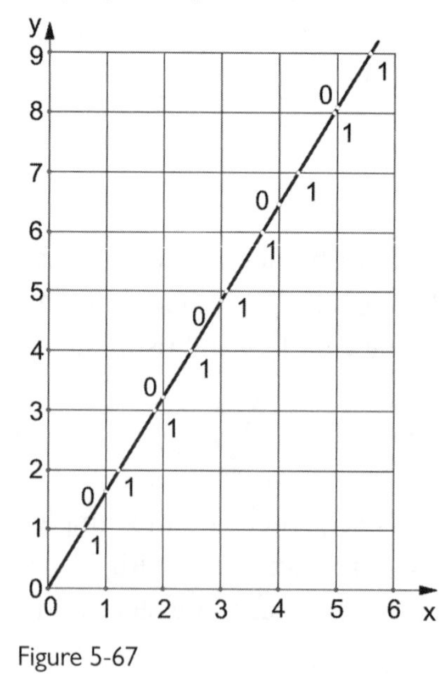


Figure 5-67

Let us graph this equation ($y=\phi x$), as shown in figure 5-67. The graph is a straight line containing no lattice point. Let us indicate each point that this line crosses a horizontal line with a 1, and each time it crosses a vertical line with a 0. Beginning after the origin, we will list the numbers along the line, as shown in figure 5-67. We then get the following: 10110101101101, which is the golden sequence. The graph of the golden ratio was able to generate this sequence.

The procedure for creating the golden string may remind you of the Fibonacci numbers. Consider the table in figure 5-68.

| Golden Strings | Number of 0s and 1s |
|---|---|
| 1 | 1 |
| 10 | 1+1 |
| 101 | 1+2 |
| 10110 | 2+3 |
| 10110101 | 3+5 |
| 1011010110110 | 5+8 |
| 101101011011010110101 | 8+13 |
| 1011010110110101101011011010110110 | 13+21 |
| ... | ... |

Figure 5-68

Counting the number of 0s and 1s reminds us of the Fibonacci numbers. Furthermore, if we take the ratio of the numbers of 0s to the number of 1s, we get the golden ratio—again!

Another strange aspect of the golden string can be seen with the following instructions. One might say that the sequence is self-reproducing. To demonstrate this, we will begin with the golden string:

1011010110110101101011011010110110 ....

We shall focus on all the 10s in the string as underlined below:

$$\underline{10}1\underline{10} \mid \underline{10}1\underline{10}1\underline{10} \mid \underline{10}1\underline{10} \mid \underline{10}1\underline{10}1\underline{10} \mid \underline{10}1\underline{10}1\underline{10} \ldots.$$

We next replace each of these 10s with a 2, as shown here:

$$\underline{2}1\underline{2} \mid \underline{2}1\underline{2}1\underline{2} \mid \underline{2}1\underline{2} \mid \underline{2}1\underline{2}1\underline{2} \mid \underline{2}1\underline{2}1\underline{2} \ldots.$$

Next, we will replace all the 2s with a 1 and all the 1s with a 0, which gives us

101 | 10101 | 101 | 10101 | 010101…,
or when written together: 1011010110110101101010….

Yes, this is our original string self-generated as we claimed at the outset.

## CURIOSITY 28

A nice recreational activity in arithmetic is to see if one can represent the natural numbers using only four 4s. This sort of exercise is shown below for the first twenty natural numbers and zero.

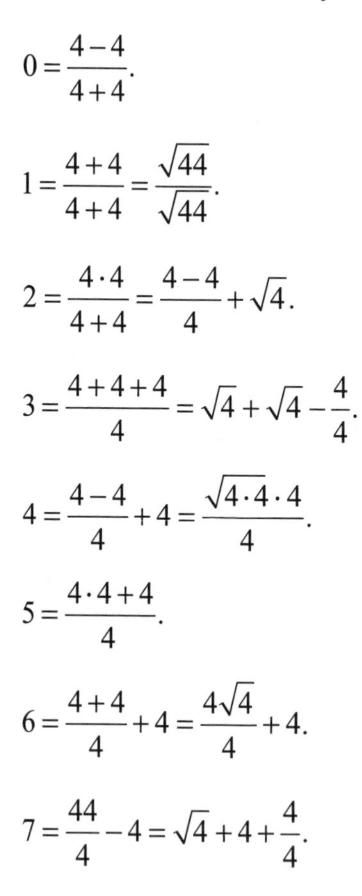

$$0=\frac{4-4}{4+4}.$$

$$1=\frac{4+4}{4+4}=\frac{\sqrt{44}}{\sqrt{44}}.$$

$$2=\frac{4\cdot 4}{4+4}=\frac{4-4}{4}+\sqrt{4}.$$

$$3=\frac{4+4+4}{4}=\sqrt{4}+\sqrt{4}-\frac{4}{4}.$$

$$4=\frac{4-4}{4}+4=\frac{\sqrt{4\cdot 4}\cdot 4}{4}.$$

$$5=\frac{4\cdot 4+4}{4}.$$

$$6=\frac{4+4}{4}+4=\frac{4\sqrt{4}}{4}+4.$$

$$7=\frac{44}{4}-4=\sqrt{4}+4+\frac{4}{4}.$$

$$8 = 4 \cdot 4 - 4 - 4 = \frac{4(4+4)}{4}.$$

$$9 = \frac{44}{4} - \sqrt{4} = 4\sqrt{4} + \frac{4}{4}.$$

$$10 = 4 + 4 + 4 - \sqrt{4}.$$

$$11 = \frac{4}{4} + \frac{4}{.4}.$$

$$12 = \frac{4 \cdot 4}{\sqrt{4}} + 4 = 4 \cdot 4 - \sqrt{4} - \sqrt{4}.$$

$$13 = \frac{44}{4} + \sqrt{4}.$$

$$14 = 4 \cdot 4 - 4 + \sqrt{4} = 4 + 4 + 4 + \sqrt{4}.$$

$$15 = \frac{44}{4} + 4 = \frac{\sqrt{4} + \sqrt{4} + \sqrt{4}}{.4}.$$

$$16 = 4 \cdot 4 - 4 + 4 = \frac{4 \cdot 4 \cdot 4}{4}.$$

$$17 = 4 \cdot 4 + \frac{4}{4}.$$

$$18 = \frac{44}{\sqrt{4}} - 4 = 4 \cdot 4 + 4 - \sqrt{4}.$$

$$19 = \frac{4 + \sqrt{4}}{.4} + 4.$$

$$20 = 4 \cdot 4 + \sqrt{4} + \sqrt{4}.$$

By now, one would anticipate that this must also be possible for many other numbers. You might want to continue this list. However, when we come to our main subject here, the golden ratio, we would not expect to be able to represent this number, since it is not a natural number. Well, again the golden ratio surprises us with its ubiquity. Here is the golden ratio expressed using four 4s (4!, or 4 factorial, is defined as $1 \cdot 2 \cdot 3 \cdot 4$):

$$\frac{\sqrt{4}+\sqrt{4!-4}}{4}=\phi.$$

Yes, this is precise!

You can show that this is equivalent to

$$\frac{1+\sqrt{5}}{2}=\phi$$

as follows:

$$\frac{\sqrt{4}+\sqrt{4!-4}}{4}=\frac{\sqrt{4}+\sqrt{24-4}}{4}=\frac{\sqrt{4}+\sqrt{4}\sqrt{5}}{4}=\frac{2+2\sqrt{5}}{4}=\frac{1+\sqrt{5}}{2}.$$

In this chapter, we have tried to demonstrate how the appearances of the golden ratio can seem to be practically limitless. Often rather unrelated situations have the golden ratio embedded. We hope that the reader will be motivated to search for other hidden golden ratio occurrences.

## LAST, BUT NOT LEAST . . .

In the mathematical community, $\pi$ lovers celebrate March 14 as $\pi$-day, since its short form is 3-14. And at 1:59, they will be jubilant! (Can you guess why?). Fittingly, we should now celebrate $\phi$ on January 6, whereupon $\phi$ enthusiasts will be particularly jubilant at 18:03 o'clock!

## Chapter 6

# The Golden Ratio in the Plant Kingdom[1]

The spiral patterns of the sunflower, fir cone, and the pineapple have fascinated plant biologists for hundreds of years, and the attempt to account for their appearance is still an exciting field of research today (called *phyllotaxis*). These considerations provide excellent examples of how simple mathematical description and modeling can contribute to our understanding of complex plant growth processes. One of the more interesting examples is the informative application of Fibonacci numbers to analyze certain repetitive or regular patterns in nature, especially in the plant kingdom.

When one considers the enormous variety of growth forms in the plant kingdom, it seems even more astounding that the Fibonacci numbers are found so abundantly. For example, if we count the number of clockwise or counterclockwise spirals in a sunflower or in a pineapple, then usually we will find two successive Fibonacci numbers, $F_n$. This is true even in instances where we would hardly expect them, such as in the flower heads (*capitulum*) of dandelions. After all the sailing dandelion seeds have been dispersed, a Fibonacci spiral pattern (fig. 6-1d and 6-2a) with 34 clockwise and 55 counterclockwise spirals can be seen. In the case of the crassulacean succulent (*Aeonium tabuliforme*) (fig. 6-3), 5 clockwise and 8 counterclockwise spirals can be clearly observed.

Figure 6-1a Figure 6-1b Figure 6-1c Figure 6-1d

Growth phases of the dandelion flower head (Photos: Mascolus).

Figure 6-2a Figure 6-2b Figure 6-2c

Fibonacci spiral patterns:

(a) Dandelion capitulum (b) Marguerite (daisy) flower head (c) Pineapple scale pattern

In the following discussion, we will attempt to establish a universal law to explain the frequent occurrences of the Fibonacci numbers in the plant kingdom.

Figure 6-3. Fibonacci spiral pattern with five clockwise and eight counter-clockwise spirals of the crassulacean succulent (*Aeonium tabuliforme*).

## FIBONACCI NUMBERS AND THE GOLDEN ANGLE

The Fibonacci numbers have a close connection to the *golden angle*, which is defined as follows (see p. 136): The golden angle is attained by dividing the circumference of a circle in the golden ratio. In this way, two angles are created that are here defined, respectively, as the large and small golden angles as follows: $\frac{360°}{\phi} = 222.4\ldots°$ and $360° - \frac{360°}{\phi} = 360°(2-\phi) = 137.5\ldots°$ (fig. 6-4).

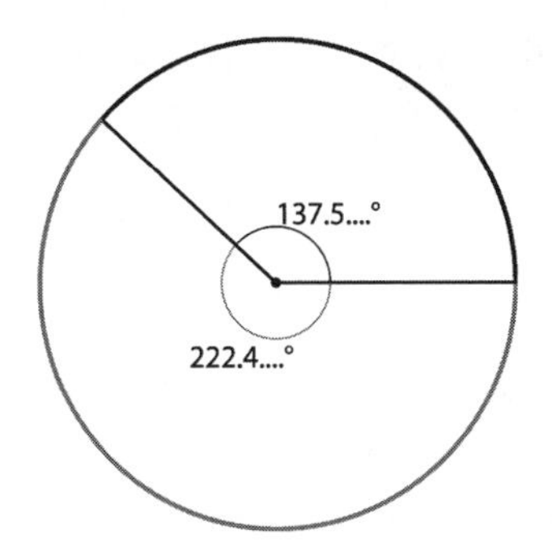


Figure 6-4. The golden angle.

An approximation of the golden angle is often observed in nature as the angle between successive leaves or the *divergence angle* (fig. 6-5). The golden angle is already apparent early in the history of plant development.

Figure 6-5. The golden angle (137.5...°) as approximation of the divergence angle.

The association between the golden angle and the Fibonacci numbers was empirically proven for the first time in 1830 by the German geologist, botanist, and poet Karl Friedrich Schimper (1803–1867)[2] (see fig. 6-6).

| Divergence (°) | Plant |
|---|---|
| $\frac{1}{2}\cdot 360 = 180$ | lime |
| $\frac{1}{3}\cdot 360 = 120$ | beech |
| $\frac{2}{5}\cdot 360 = 144$ | oak |
| $\frac{3}{8}\cdot 360 = 135$ | pear |
| $\frac{5}{13}\cdot 360 = 138.4\ldots$ | almond |

Figure 6-6. The divergence angle of selected plants.[3]

We established earlier (chap. 3) that

$$\frac{F_n}{F_{n+1}} \to \frac{1}{\phi}.$$

Therefore, the sequence of fractions of the divergence angle,

$$\frac{F_n}{F_{n+2}},$$

has the limiting value

$$\frac{F_n}{F_{n+2}} = \frac{F_n}{F_n + F_{n+1}} = \frac{1}{1+\frac{F_{n+1}}{F_n}} \to \frac{1}{1+\phi} = \frac{1}{\phi^2} = 2 - \phi.$$

This equates exactly to the golden angle:

$$(2 - \phi) \cdot 360° = \left(2 - \frac{\sqrt{5}+1}{2}\right) \cdot 360° = 137.507\ldots°.$$

In 1979, the central role that the golden angle played in phyllotaxis was impressively illustrated through computer simulations by H. Vogel in his paper "A Better Way to Construct the Sunflower Head."[4] Vogel made two model assumptions about the distribution of the florets in the sunflower head (i.e., in the *capitulum*):

1. The divergence angle is constant.
2. The packing is compact.

The constancy of the divergence angle means that successive establishments are developed with the constant angle $\alpha$; also the compact packing requires that the increase in the area of the capitulum is the same as the area of the newly established growth.[5] With the computer model, the influence of the divergence angle $\alpha = \lambda \cdot 360°$ can be assessed for various lambda ($\lambda$) parameters (fig. 6-7).

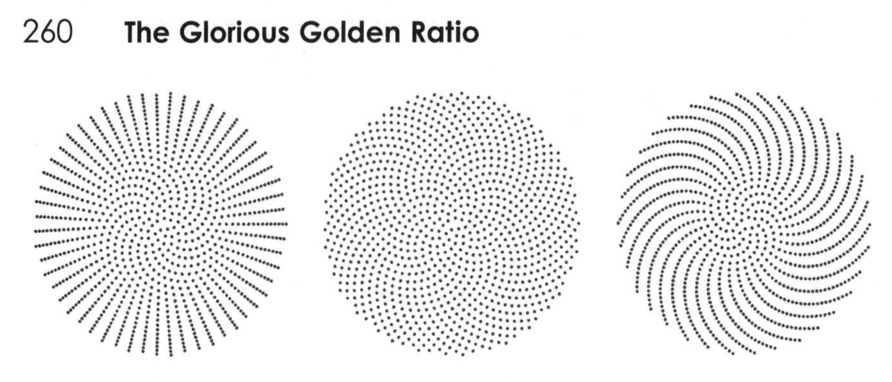

Figure 6-7. The three spiral patterns were generated with the Vogel model for various $\lambda$ values:

Left $\lambda = \frac{21}{55} = \mathbf{0.3\overline{18}}$, Middle $\lambda = 2 - \phi \approx 0.381966$, and Right $\lambda = 0.3825$.

The connection between the divergence angle of the real number $\lambda$ and the number of visible spirals or *contact parastichies* is determined by the development of the continued fraction (see the appendix).

The convergent denominators of the golden angle are precisely the Fibonacci numbers, which explains the above occurrences and coincides with the number of spirals in the same rotational direction.

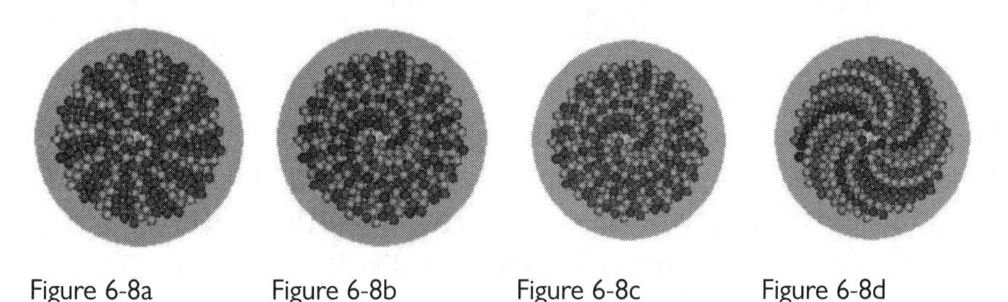

Figure 6-8a Figure 6-8b Figure 6-8c Figure 6-8d

Spiral patterns generated with the Vogel model for the golden angle. The Fibonacci spiral pattern has been made visible through coloring every fifth, eighth, thirteenth, and twenty-first spiral.

## PARASTICHY NUMBERS, DIVERGENCE ANGLE, AND GROWTH

It is an empirical fact that during plant development, the growth $h$ (the vertical interval between leaf nodes) lessens. This can be verified by taking a walk through the garden and inspecting the plants (fig. 6-9).

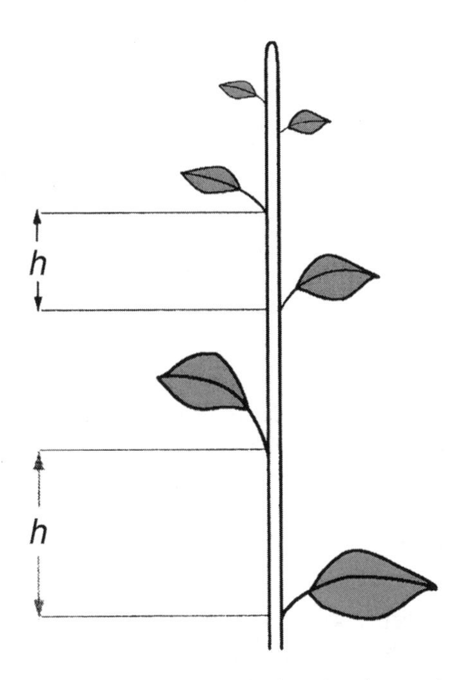


Figure 6-9. Reduction in growth between leaf nodes during plant development.

This is the source of the alternation in the spiral pattern and can be explained more easily as a cylindrical lattice. For this purpose, we observe the points of a helix on a cylinder surface with cylinder radius $R$, growth $h$, and divergence angle $\alpha = 360° \cdot \lambda$ ($\lambda$ is the determinant divergence for this angle). If a cylinder lattice is normalized to $C = 2\pi \cdot R = 1$ and rolled out into a plane, we obtain a plane point lattice, which is explicitly characterized by $(h, \lambda)$ (fig. 6-10).[6] The biological sequence of new growth adheres to the natural numbers, whereby the youngest is depicted as number 1 and the second youngest with the number 2, and so on. From a geometric viewpoint, it is not the age of the growth that is important but rather its relation to neighboring growth.

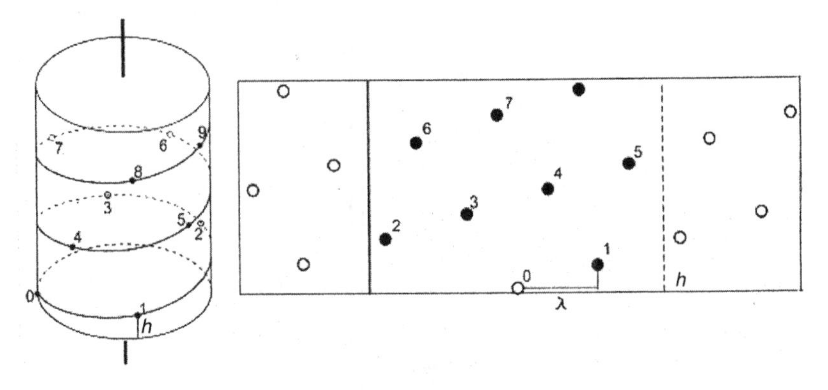


Figure 6-10. If we imagine the cylinder as a roller and roll it out in the plane, we create a point lattice, which is defined by assigning the growth rate $h$ (the vertical interval between points) and the divergent angle $\alpha = 360° \bullet \lambda$. It is sufficient to observe the foundational strip $T$ between $-0.5$ and $0.5$, because the whole lattice is created through displacement of $T$.

For the lattice in figure 6-11, the points 2 and 3 are the immediate neighbors of the origin. This creates 2 counterclockwise-rotating contact parastichies on the cylinder and 3 clockwise-rotating contact parastichies, respectively. It is said that the parastichies pair (2, 3) belongs to the lattice. Through a reduction in growth of the interval, point 5 replaces point 2 as the second neighbor of the origin. That is to say, the parastichies pair changes the lattice from (2, 3) to (5, 3).

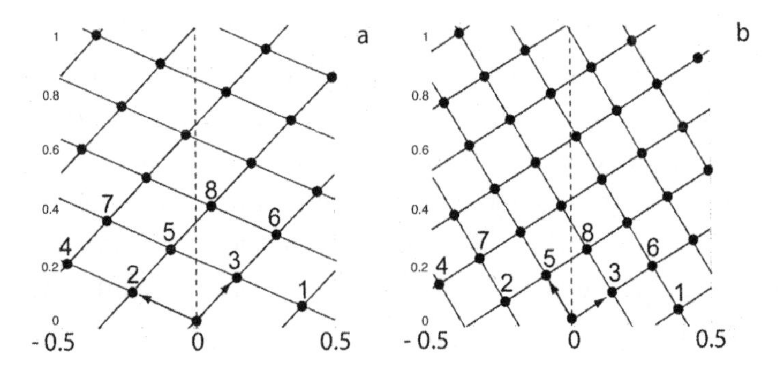


Figure 6-11. The cylinder lattices above were generated with a divergence $\lambda = 2 - \phi$. For the left lattice (a) a growth of $h = 0.05$ and for the right lattice a growth of $h = 0.03$ was chosen.

Although a point lattice can be constructed for every parameter pair ($h$, $\lambda$), the lattices that have their origin in biological growth processes are often subject to severe constraints. Because the phyllotactic lattice is often idealized as a tangential circle, these lattices are rhombic (fig. 6-10b), which means that both generating vectors are of the same length. These considerations are fundamental for the so-called sphere packing model that the Dutch botanist G. van Iterson (1878–1972) introduced in his doctoral thesis in 1907.[7] This allows us to explain the interrelationship between the golden angle and the Fibonacci numbers in the case of an ideal (i.e., generated with fixed divergence angle) and of consistently changing divergence angles, according to the Van Iterson diagram (see fig. 6-14).

## CAUSAL MODEL OF PHYLLOTAXIS

One explanation that has been proposed for the spiral pattern, as a functional principle or *morphological adaptation*, is that it allows for optimal light exploitation, which enables maximal photosynthetic activity.[8] However, this can be disregarded because, on the one hand, these patterns are also found in saltwater algae, which are kept in motion by the continuous water currents and therefore have no specific advantage in being arranged according to the golden angle; and on the other hand, these patterns are also found in scale insects (Placentalia) and seeds, which obviously have nothing to do with light exploitation.[9] Furthermore, it is difficult to separate cause from effect.

The idea of a black box in which over millions of years evolution has written the code for the angles of the leaf primordia is also not very helpful in explaining the functional value of this universal phenomenon.

The following model, on the other hand, ensures that the generation of the Fibonacci spiral pattern is solely a result of biologically plausible principles.

The mathematical modeling of the spiral phyllotaxis must at least replicate the following two biological processes: the processes in the growing tip that lead to the generation of the primordia in specific locations and the physical interactions of the primordia during their alignment on the hypothecium.

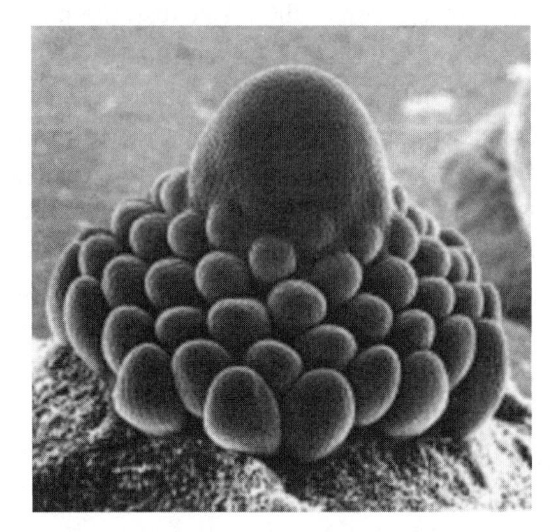

Figure 6-12. Spiral inception of numerous individual stamen primordia on a conical floral apex. (REM-Foto from Erbar & Leins. Reprinted with permission from Professor Erbar.)

The *shoot apical meristem* is characterized by a union of cells with a high cell density and cell division rate. The apical ring is situated at the base of the apical meristem, which is where new biological primordia for leaves or flowers are initiated. These can be seen as small balls in figure 6-12. The location of the primordia initiation is decisively regulated by the plant hormone auxin (Greek *auxein* = growth, enrichment).

E. J. H. Corner stated, "The spiral pattern of the apical meristem . . . is one of the biggest wonders of the botanic."[10] A simple causal model for this wonder, which was proposed by J. N. Ridley[11] after preliminary work from I. Adler,[12] is based on the contact-pressure hypothesis of the Swiss botanist Simon Schwendener (1829–1919).[13] Mechanical forces were first recognized by the German mathematician Johannes Kepler (1571–1630)[14] as the main factor leading to specific organic forms and patterns. In this way, he explained that the rhombic form of pomegranate seeds is due to the pressure contact on the seeds during growth. As a result of these pressure forces, tightly packed rhombic seed structures are generated. Hubert Airy (1838–1903) conjectured in 1873 that in an embryonic state, the plant has a large advantage from the compact packing condition: "In the bud we see at once, what must be the use of leaf-order. It is the economy of space, whereby the bud is entire to itself and presents the least surface to outward danger and vicissitudes of temperature."[15]

The compact packing of the leaf primordia makes the hypothesis of pressure-force related forms in the early development stages plausible. Ridley's simulation of the contact force model contains the following steps:

**Ridley Algorithm**

1. **Generation** of a new primordium
2. **Interaction** of the primordia
3. **Expansion** of the primordia

The understanding of the position regulation of the primordia is of particular significance. It is generally accepted[16] today that historically, after 1868, the position of the primordia initiation can be empirically explained using the hypothesis presented by the German botanist Wilhelm Hofmeister (1824–1877): *A new primordium is initiated in the position of the apex ring, which has the largest interval of all already-existing primordia.*[17]

Simulations of an improved Ridley model have shown that for large parameter areas, the spiral pattern is generated exactly as the one most often seen in nature.[18] For the sunflower, this is with a frequency of 82 percent Fibonacci spirals and 14 percent Lucas spirals.[19]

Similar results were obtained by the French physicists Stéphane Douady and Yves Couder in their famous experiment from 1992.[20] Small ferromagnetic balls were dripped into an oil pool while continuously lowering the frequency of additions, and they were then slowly drawn to the outside by an external magnet field. A regular Fibonacci spiral pattern was generated. Since then, this pattern building has also been observed in many other nonbiological systems.

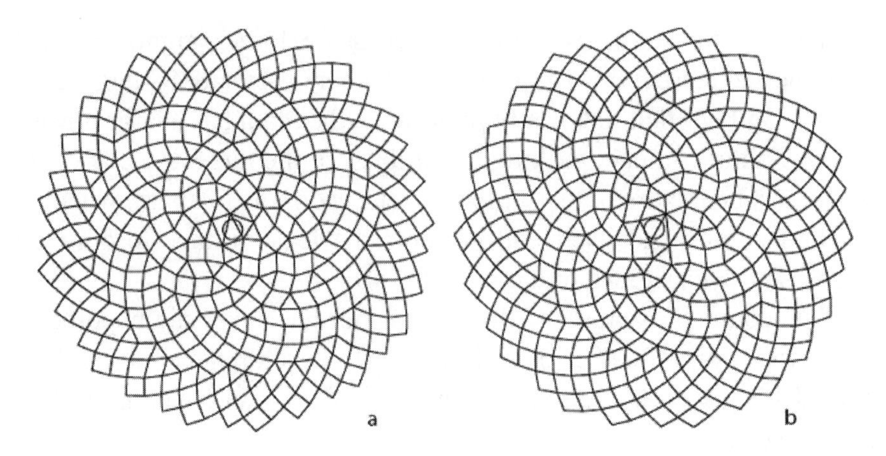


Figure 6-13. Spiral patterns generated with the Ridley algorithm (a) Fibonacci spirals and (b) Lucas spirals.

In 2002, Pau Atela and his colleagues, Christophe Golé and Scott Hotton,constructed a dynamic system that proved that the fix points of this system make up exactly the stabile lattice, which in $(d, h)$-parameter space are given by a truncated Van Iterson diagram (fig. 6-13).[21] Through the influence of the contact pressure, the growth of a phyllotactic pattern during decreasing elongation can be described by a downward zigzag path. The path that starts earliest, and has the largest sphere of influence, is the Fibonacci path $(1, 1)\rightarrow(1, 2)\rightarrow(2, 3)\rightarrow\ldots\rightarrow(m, n)\rightarrow(n, m+n)\rightarrow\ldots$, which becomes closer and closer to the straight line $x=2-\phi$. Therefore, it is no wonder that the Fibonacci numbers are so often found in the plant kingdom. The golden ratio, on the other hand, is a mathematical construction that works in nature as is impressively shown by the Van Iterson diagram (fig. 6-14).

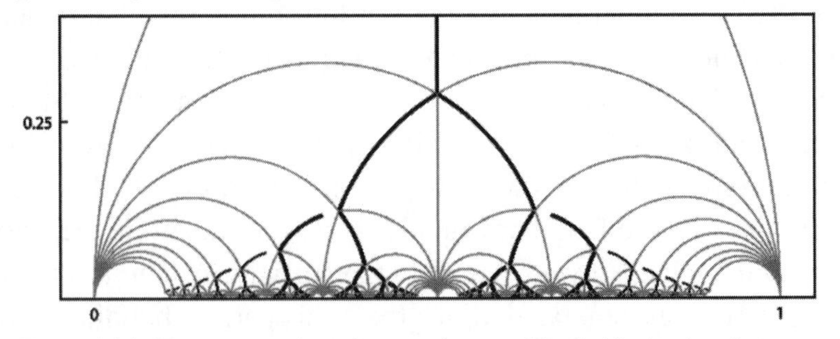


Figure 6-14. The truncated van Iterson diagram (Grafic Neukirchner).

As we have seen, the golden section with its partner the Fibonacci numbers is embedded in nature. You might want to search for the many other manifestations of these mathematical aspects in nature.

# Chapter 7

# The Golden Ratio and Fractals[1]

When mentioning the golden ratio, perhaps the geometrical figures that most promptly come to mind are regular pentagons, because of the relationship between their sides and diagonals. Or maybe even the famous golden rectangle. But another realm in which the golden ratio plays an important role is in the construction of some fractals.

One of the easiest ways to understand the nature of fractals is to observe trees. The way in which each branch of a tree bifurcates into smaller branches in order to create a fork constitutes the basic idea we try to replicate when creating fractals: that is, repeatedly adding to a geometric figure reduced copies of itself, or in some cases, replacing parts of the figure by those reduced copies, according to a determined rule.

We can imitate a tree, or create a fractal tree, according to a very simple geometric rule: We start with a trunk (a segment), and at one of its endpoints we create a bifurcation by placing two reduced copies of the trunk. At the other endpoint of the two new stems, we will repeat the rule and create other bifurcations, as shown on figure 7-1.

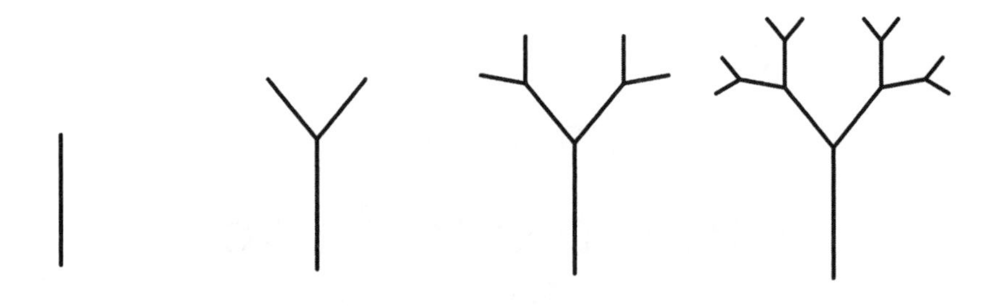

Figure 7-1

The factor of reduction and the angle at which the branches will be placed are a matter of choice.

Trees clearly show the idea of self-similarity, one of the most remarkable characteristics of fractals: An object is made up of several smaller, perhaps overlapping, copies of itself.

As we repeat that geometric rule more and more times, and the number of copies in the figure increases, making it more "crowded," it is often the case that some of those parts will overlap. For example, compare the trees in figures 7-2 and 7-3. Both were obtained by fourteen repetitions of the bifurcation procedure described above. For the one in figure 7-2 we used a reduction factor of $\frac{4}{7}$, while for figure 7-3 we used a reduction factor of $\frac{5}{7}$. We can see that with the reduction of the branches in figure 7-2, the resulting tree has plenty of room for more branches without having them overlap. This is not true of the tree in figure 7-3. The factor $\frac{5}{7}$ does not reduce the branches enough to provide room for growth without overlap.

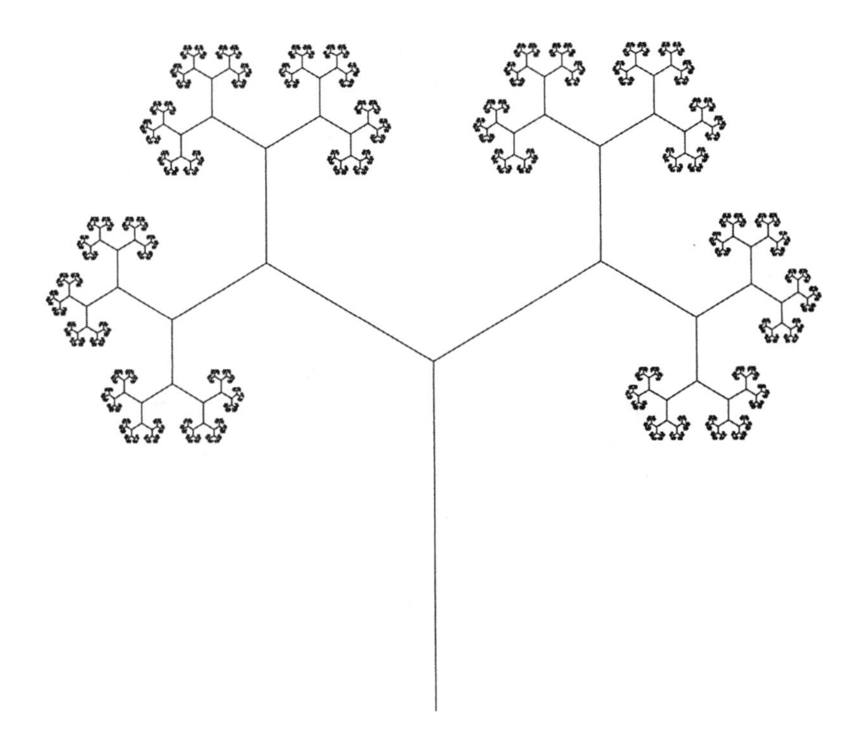

Figure 7-2. Tree obtained by fourteen iterations of a 120°-bifurcation with reduction factor $\frac{4}{7}$.

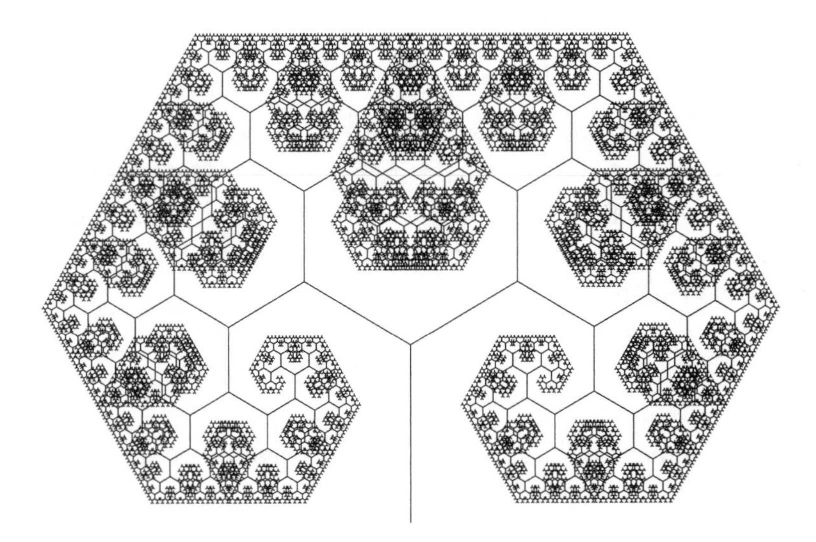

Figure 7-3. Tree obtained by fourteen iterations of a 120°-bifurcation with reduction factor $\frac{5}{7}$.

A natural question that one may then ask when constructing such fractals is, What choices of angles or reduction factors will yield overlapping figures, and which ones will not?

Amazingly, the pursuit of an answer to that question will lead us to the now-familiar golden ratio.

Let us look more closely to the bifurcation procedure illustrated by figure 7-1. We start out with a segment *k*, whose length we will stipulate to be one unit and will label $l_0$. We then branch that segment into two other segments, which are reduced copies of *k*. The factor of reduction is our choice, and we will call it *f*. In order to have branches spread evenly around the bifurcation point, we will use angles measuring 120° between the branches (fig. 7-4).

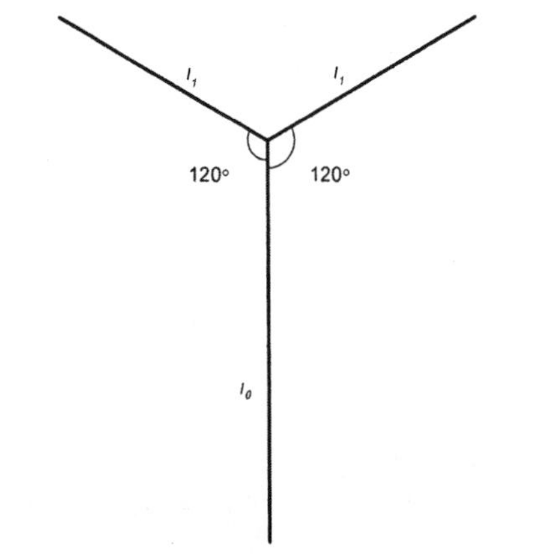


Figure 7-4. The first stage of the tree fractal.

We will label $l_1$ the length of the new segments, since these were obtained by one iteration of the copying procedure. The two new segments have length:

$$l_1 = l_0 \cdot f = 1 \cdot f = f.$$

When we iterate the bifurcation procedure again, we will have four new segments, of length

$$l_2 = l_1 \cdot f = f \cdot f = f^2.$$

A third iteration will produce segments of length $l_3 = f^3$, and so forth. In general, the segment generated by the $n$th iteration will have length $f^n$ (see fig. 7-5).

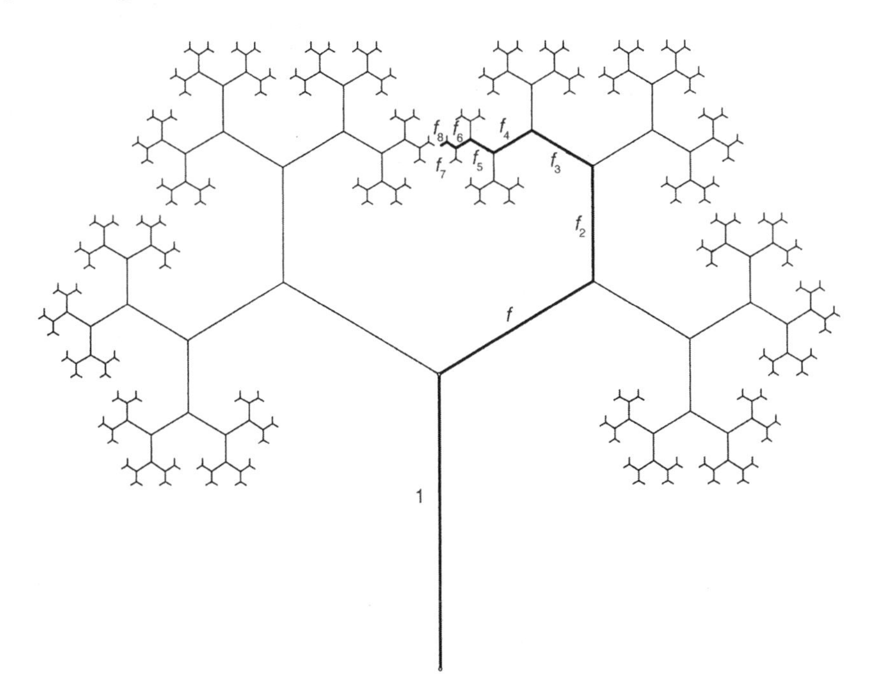


Figure 7-5. If we start the fractal tree with a segment of length $l_0 = 1$ and use a reduction factor $f$, the segment generated by the $n$th iteration will have length $f^n$.

As we noted earlier, whether or not the branches overlap will depend on our choice of $f$. We have seen from figures 7-2 and 7-3 that when $f$ equals $\frac{4}{7}$, or approximately 0.57, no overlapping occurs, but if we choose $f$ to be $\frac{5}{7}$, or approximately 0.71, the branches of the tree will overlap. This leads us to conjecture that there should be a real number between $\frac{4}{7}$ and $\frac{5}{7}$ that will make the branches of the tree lightly touch, with no overlaps, when used as a reduction factor. Let us try to find such a number.

We want the zigzag made by the segments of lengths $f^3, f^4, f^5, f^6, \ldots$ to fit exactly between the two parallel axes $o$ and $p$ shown as dashed lines in figure 7-6.

What is the distance between those two parallel lines?

We can see from figure 7-7 that it will be the projection of the segment of length $f$ onto line $r$, that is, $f \sin 60°$.

If we also "flatten" the zigzag horizontally, we get the equation we want to be true in order to have branches touching but not overlapping:

$$f^3 \sin 60° + f^4 \sin 60° + f^5 \sin 60° + f^6 \sin 60° + \ \ldots = f \sin 60°.$$

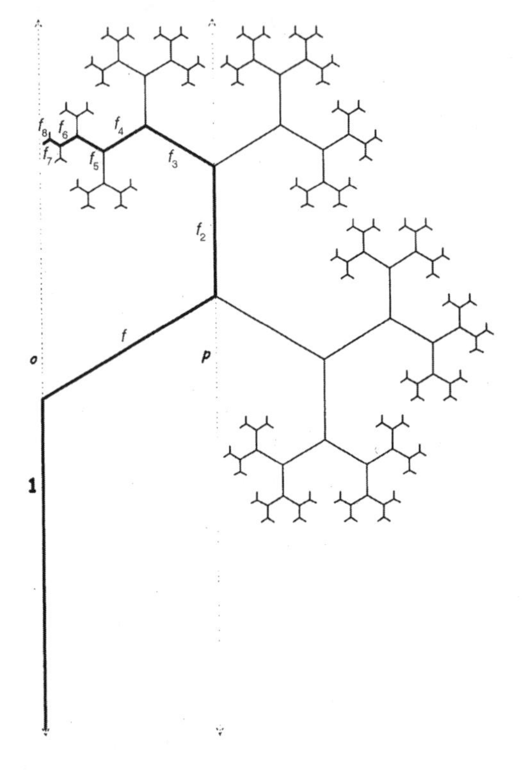

Figure 7-6

We can simplify this equation by dividing both its sides by $\sin 60°$:

$$f^3 + f^4 + f^5 + f^6 + \ \ldots = f.$$

The left side of this equation can be rewritten as: $f^3(1 + f + f^2 + f^3 + \ \ldots)$.

The summation in parentheses is the geometric series, which converges to $\frac{1}{1-f}$ for values of $f$ between 0 and 1, which is the case of our reduction factor, since by figures 7-2 and 7-3 we know that it must be between the values $\frac{4}{7}$ and $\frac{5}{7}$.

We arrive then at the equation $f^3 \frac{1}{1-f} = f$.

Can we find values of $f$ that satisfy this equation? So far the equation does not look very familiar, but we can simplify it further. If we divide both sides by $f$, we get

$$f^2 \frac{1}{1-f} = 1.$$

Finally, let's multiply both sides by $1-f$:

$$f^2 = 1-f.$$

This equation looks more familiar. It is the quadratic equation $f^2+f-1=0$, whose roots are

$$f_{1,2} = -\frac{1}{2} \pm \frac{\sqrt{5}}{2}; \quad f_1 = \frac{\sqrt{5}}{2} - \frac{1}{2} = \frac{\sqrt{5}-1}{2} = \frac{1}{\phi}; \text{ and}$$

$$f_2 = -\frac{\sqrt{5}}{2} - \frac{1}{2} = -\frac{\sqrt{5}+1}{2} = -\phi\ .$$

$$\frac{1}{\phi} = \frac{\sqrt{5}-1}{2} \approx 0.61803 \text{ and } -\phi = -\frac{\sqrt{5}+1}{2} = \frac{-1-\sqrt{5}}{2} \approx -1.61803,$$

that is, the reciprocal and the opposite of the golden ratio.

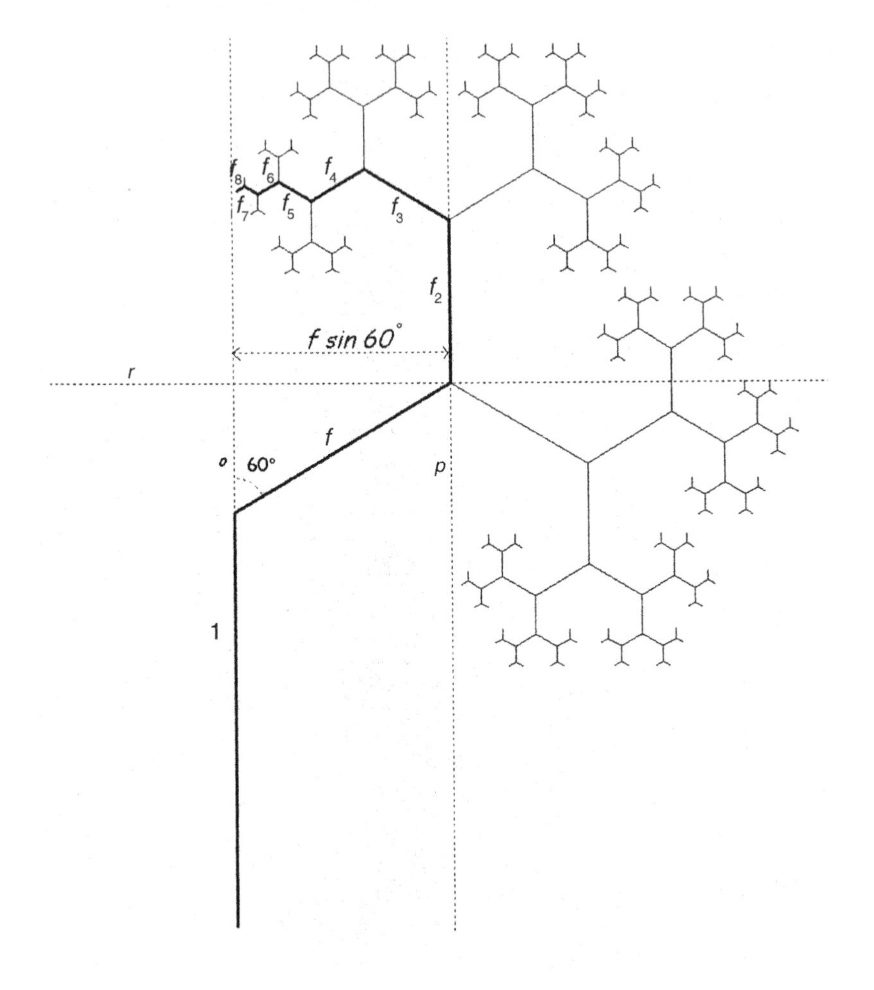

Figure 7-7

And there we find our amazing golden number again as the optimal solution for our aesthetic demand. A fractal tree constructed with the reciprocal of the golden ratio as the reduction factor will have branches covering up as much space as they can, and branches getting as close to each other as they can, until they touch, though not covering other branches.

Other beautiful fractals can be obtained by using the golden ratio in one way or another. The square fractal, for example, is constructed by starting with a square and adding reduced copies of it at each corner.[2] At each subsequent step, reduced copies are added to each one of the three free corners of the new squares. Figure 7-8 shows such a fractal in the ninth stage of its construction. The reduction factor used is a $\frac{4}{9}$ linear reduction. That is, the side of the squares created at a particular stage measure $\frac{4}{9}$ of the length of the side of squares at the preceding stage.

Figure 7-8. Square fractal, nine stages, reduction factor of $\frac{4}{9}$.

If we use the golden ratio as the ratio between the sides of squares in the square fractal, the resulting picture is a perfectly crafted tapestry. The squares snuggle perfectly, and as the iterations progress we can see many golden rectangles being delineated in the resulting picture (fig. 7-9). As in the case with the tree fractal, the golden ratio is the ratio we find if we want optimal fit in the square fractal.

Figure 7-9. Golden square fractal, nine stages, reduction factor is $\frac{1}{\phi} \approx 0.61803$.

Another way in which we can combine the golden ratio and fractals is by deliberately using in our constructions geometric figures that we know entail the golden ratio in their measurements. Three such figures are the regular pentagon, the isosceles triangle with base angles measuring 36°, and the isosceles triangle with base angles measuring 72°—which we call, respectively, the obtuse and the acute golden triangles (see chap. 4). We can also use the fact that we can dissect each of these figures into a combination of other regular pentagons and golden triangles, as shown in figures 7-10 through 7-13.

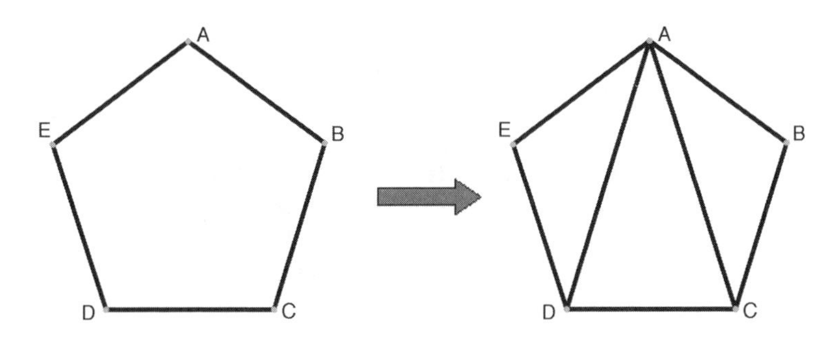


Figure 7-10

A regular pentagon can be dissected into two obtuse golden triangles and one acute golden triangle as shown in figure 7-10.

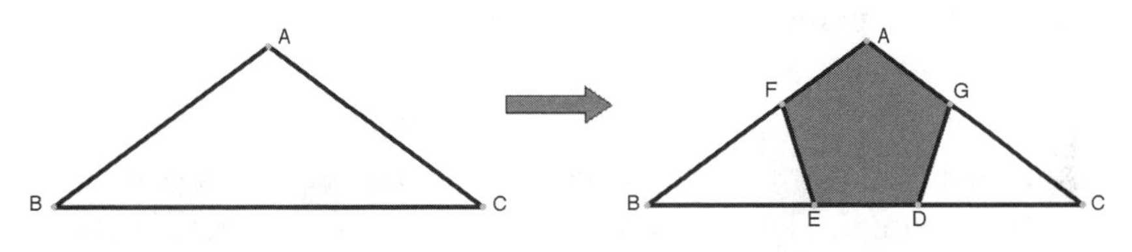


Figure 7-11

An obtuse golden triangle (fig. 7-11) can be dissected into one regular pentagon and two acute golden triangles. Notice that the points *D* and *E* are found by marking off a length equal to that of *AB* on segment *BC* from points *B* and *C*, respectively.

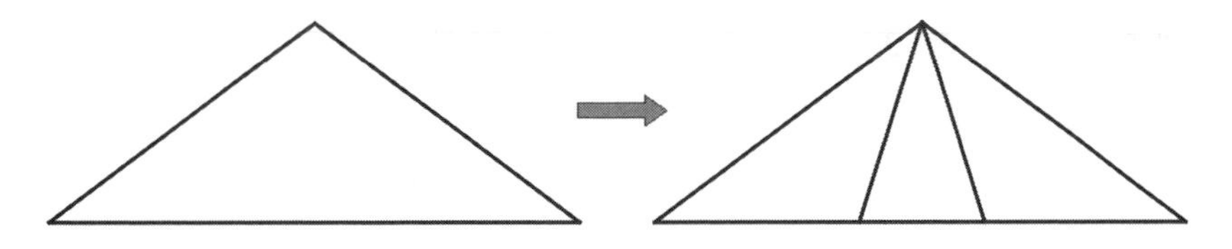

Figure 7-12

Obtuse golden triangles can also be dissected into three other golden triangles, two obtuse and one acute, as shown in figure 7-12.

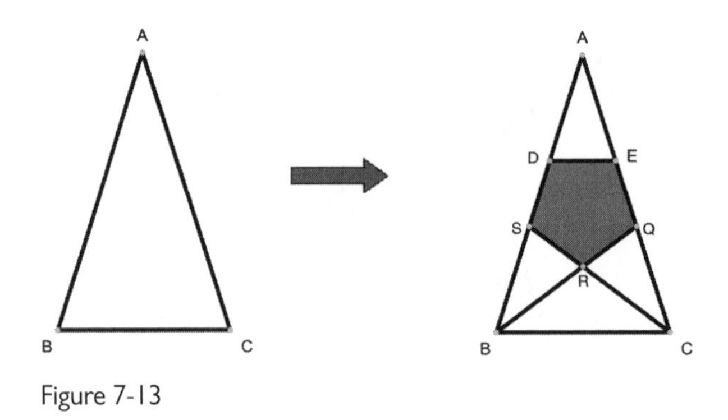


Figure 7-13

An acute golden triangle (fig. 7-13) can be dissected into one regular pentagon, three acute golden triangles, and one obtuse golden triangle. Notice that the points $Q$ and $S$ are found by bisecting the 72° angles $ABC$ and $BCA$, respectively.

We can choose one of these figures to start our construction, then decide on a way to partition it, one that can be iterated over and over. As a first example, let's start with an obtuse golden triangle and the dissection shown in figure 7-12.

We can stipulate that our rule will be to divide the obtuse golden triangle in that manner and subsequently to remove the acute golden triangle in the middle. Each iteration will consist of applying the same rule to every obtuse golden triangle at any stage of the construction. Figure 7-14 shows the result of five iterations of this rule.

Figure 7-14

This construction fits nicely into a pentagonal shape if we add to it rotated copies of itself, such as in figure 7-15.

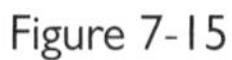
Figure 7-15

In figure 7-16, we generate a fractal from an acute golden triangle. Careful inspection will show the connection between the buildup from figure 7-16 to figure 7-17.

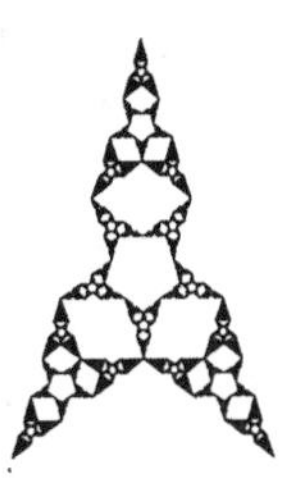

Figure 7-16

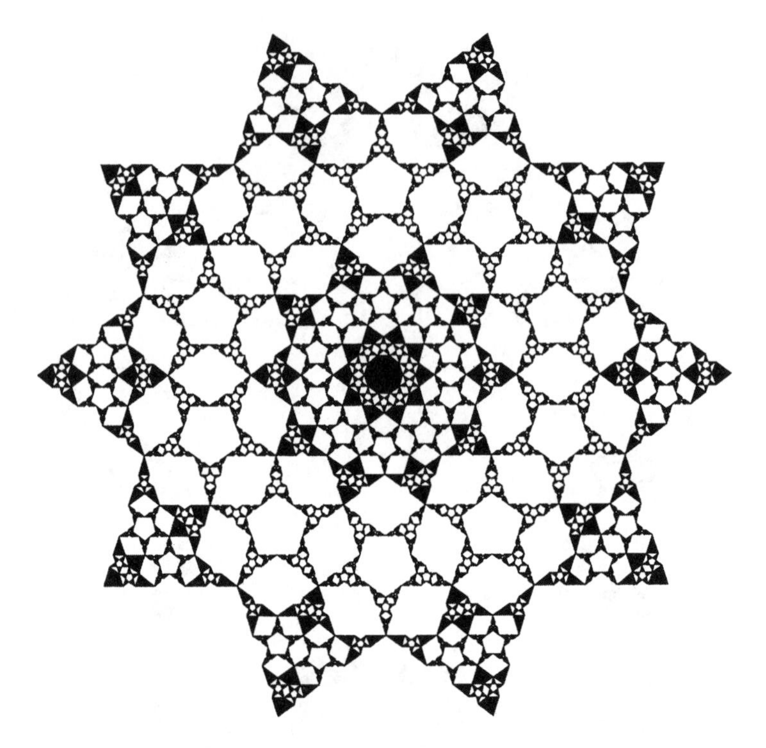

Figure 7-17

The next fractal we will consider is built around a pentagon. The process of construction is detailed in figure 7-18.

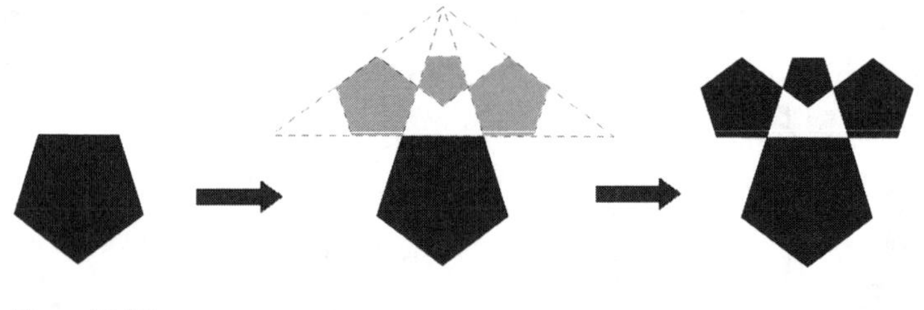

Figure 7-18

In figure 7-19, we have this process iterated three times. Notice that in the fourth stage of the construction (third iteration) some pentagons start overlapping others.

In figure 7-20 we see the results of the first five iterations. Once again we rotated the figure around a point to create symmetry.

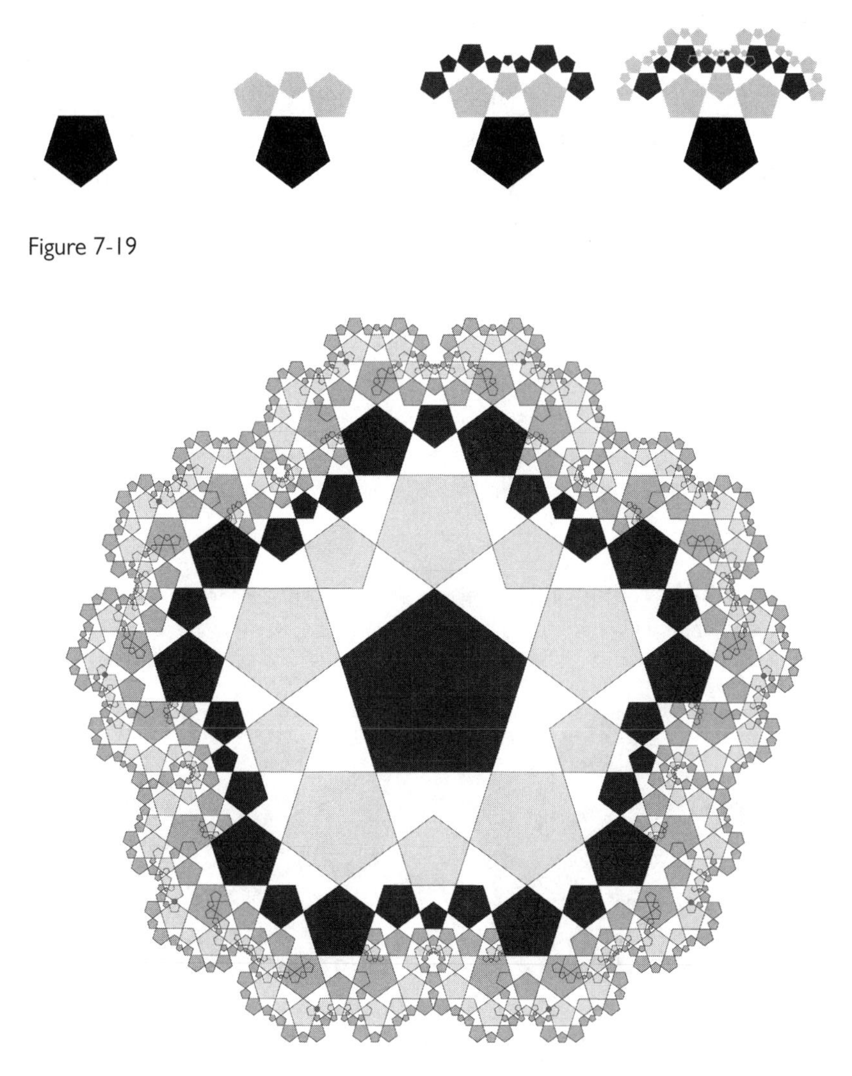

Figure 7-19

Figure 7-20

Another fractal involving pentagons—and consequently the golden ratio—is the pentaflake. This construction is said to have been first thought of by German artist Albrecht Dürer (1471–1528). We start with a regular pentagon. We construct its diagonals and find their points of intersection. Those points will be the vertices of a new regular pentagon (fig. 7-21). To construct the pentaflake, we use auxiliary circles to mark off two points on each side of our original pentagon. Figure 7-21 shows this procedure for one of the sides. Figure 7-22 shows the complete construction. This construction will be used as the generator of the fractal. At each stage of the fractal's construction, we will apply this rule to every pentagon at that stage. Figure 7-23 shows the first three iterations in the construction of the pentaflake.

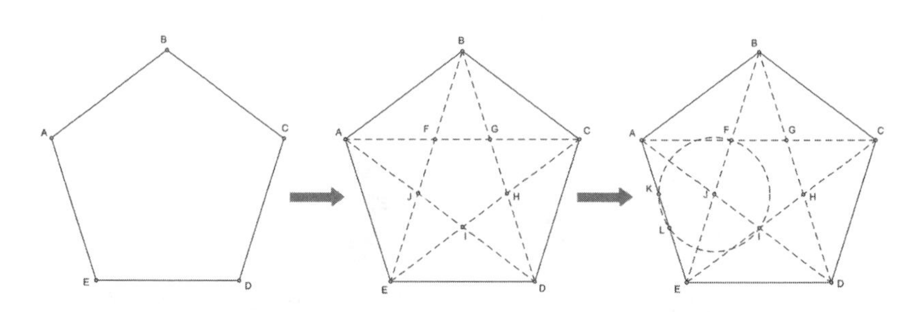


Figure 7-21. The diagonals of regular pentagon *ABCDE* form a new regular pentagon, *FGHIJ*.

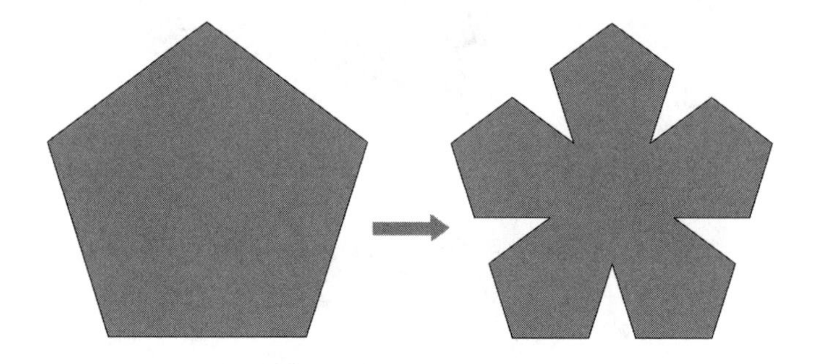

Figure 7-22. The generative procedure for the construction of the pentaflake. At each stage this procedure will be applied to every pentagon at that stage.

Figure 7-23. The results of the first three iterations in the construction of the pentaflake.

How can we be sure that these constructions are fractals? Besides the presence of self-similarity, one thing that characterizes fractals is the fact that their dimension can be an irrational number.

This last sentence may not make sense unless we revise our concept of *dimension*. There are many ways in which dimension of a geometric object can be defined. What most people have heard about dimension is that a point has dimension zero, a line has dimension one, plane figures have dimension two, and solids are three-dimensional. With only that in mind, it may be impossible to conceive of an object having dimension that is an irrational number. Therefore, to understand this affirmation, we will briefly extend our concept of dimension.

The concept of dimension we will use is also called similarity dimension. It is calculated by observing what happens to a figure once we dilate it by a linear factor $f$. We will try to understand the idea by examining objects whose dimensions we already know: a segment, a two-dimensional figure, and a three-dimensional figure. Then, once we figure out the process that originates those numbers, we will use it to calculate dimensions of fractal objects.

Let's start with a line segment of length $l$. We will then make a dilated copy of it. We can choose the dilation factor $f$ to be any number, for example, 2. In this case, the copy will have length $2l$ (fig. 7-24).

The key thing now to calculate the dimension is to determine how many self-similar copies of the original figure can be found in the dilated figure. Obviously, in this case, we have two copies of the original segment in the dilated segment.

Figure 7-24. Doubling the length of a segment yields two copies of the original segment.

We will now see what happens if we dilate a square by the same factor *f*. One important thing to keep in mind is that *f* is a factor of *linear* dilation. That is, if we choose it to be 2, we are going to double the *lengths* for the new figure, not the areas. In this case, that means we will double the length of the side of the square (fig. 7-25).

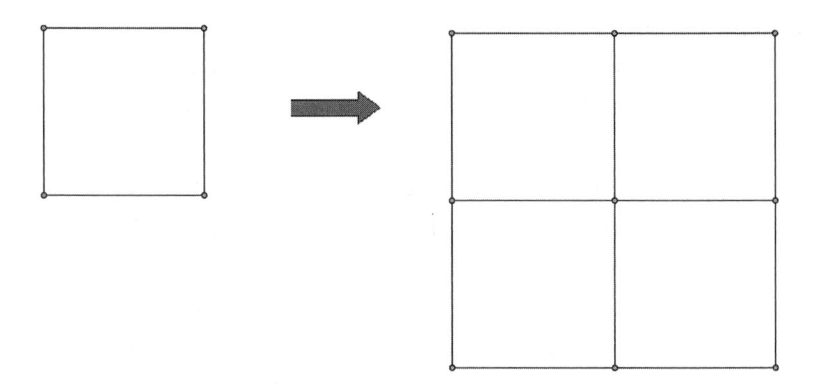

Figure 7-25. Dilating a square by a linear factor of 2 gives us four copies of the original square.

We can see that a linear dilation of 2 will yield four copies of the original figure, in this case.

Finally, let's examine what happens once we dilate a cube, which we assume to have three dimensions, by a linear factor *f*.

Figure 7-26 shows that when we double the side of a cube, the new cube has eight copies of the original cube in itself.

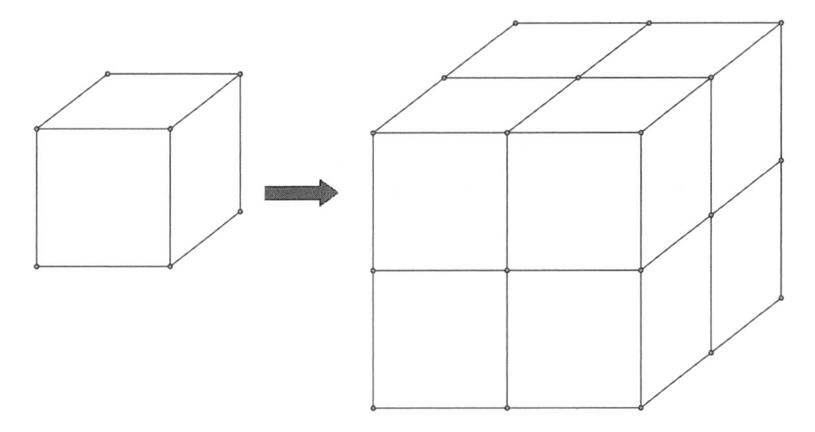

Figure 7-26. A cube with the side double the length of that of the side of the original cube will entail eight copies of the original in itself.

The results we obtained are summarized in a table (fig. 7-27).

| Figure | Known dimension | Linear factor used for dilation ($f$) | Number of copies obtained |
|---|---|---|---|
| segment | 1 | 2 | 2 |
| square | 2 | 2 | 4 |
| cube | 3 | 2 | 8 |

Figure 7-27

From this table you can notice the number of copies obtained can be rewritten as a power of $f$, and the exponents are precisely the dimension we commonly have heard of: 1 in the case of the segment, 2 for the square, and 3 for the cube (fig. 7-28).

| Figure | Linear factor used for dilation ($f$) | Number of copies obtained |
|---|---|---|
| segment | 2 | $2^1$ |
| square | 2 | $2^2$ |
| cube | 2 | $2^3$ |

Figure 7-28

So if we define *dimension* to be the exponent obtained when we write the number of self-similar copies as a power of the linear dilation factor, we get results that match our previous, informal, notion of dimension. This is just a way of simplifying a more formal definition of dimension, known as box-counting dimension, which we will not cover here. But notice that this simplification can only be used in self-similar figures.

Let us write these observations algebraically. We will use the following notation:

Dimension $= d$

Number of self-similar copies $= N$

Linear dilation factor $= f$

Using the variables above, we can write the following equation: $N = f^d$.

Let us use this definition of dimension to calculate the dimension of the fractal in figure 7-14.

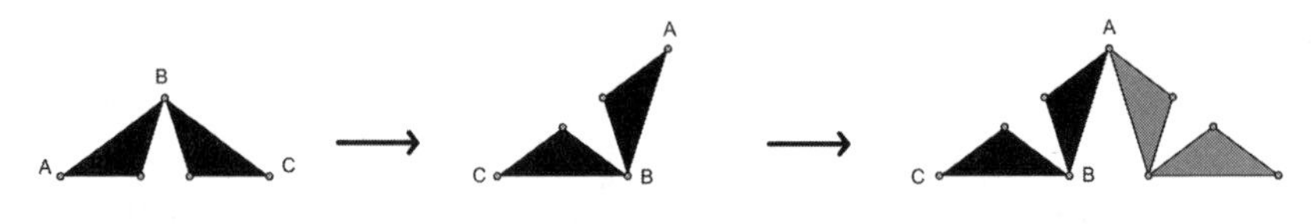


Figure 7-29

Figure 7-29 shows that a linear dilation of $\phi$ gives us two copies of the original figure. This can be seen if we notice that triangle $ABC$ in the figure is a golden triangle, so its sides $AB$ and $AC$ are to each other in the golden ratio.

Using our formula for dimension we have that this fractal has dimension:

$$N = f^d$$

$$2 = \phi^d.$$

Since our unknown $d$ is the exponent, to find its value we need to apply the logarithmic function to both sides of the equation:

$$\log 2 = \log(\phi^d).$$

Using a known property of logarithms, the equation becomes

$$\log 2 = d \log \phi.$$

The value of $d$ is obtained if we divide both sides of the equation by $\log \phi$:

$$d = \frac{\log 2}{\log \phi} \approx \frac{0.3010}{0.2090} \approx 1.44.$$

So $d$ is irrational. Having an irrational dimension is a common trait among fractals.

The number also lies between 1 and 2 (it is approximately equal to 1.44). What does it mean to have dimension greater than 1 but less than 2? A dimension equal to 1 is characteristic of segments, objects that have only length. Two-dimensional objects, on the other hand, have an area. A dimension of 1.44 seems to suggest that our fractal has more than just length, but not quite an area.

One might at this point argue that the objects in figure 7-29 have an area. But we have to remember that those illustrations represent just initial stages in the construction of the fractal. The actual fractal is the set of points that would be obtained after an infinite number of iterations of the generative procedure.

But since we can calculate the area at a specific stage, let us calculate a few of those and infer what the area of the fractal would be by examining the change pattern we will find.

Let us also calculate the perimeter of the fractal, that is, the length of its boundary.

Figure 7-30 shows the calculations for the first ten stages of the fractal.

| Stage | Length of short side of triangle | Length of long side of triangle | Number of triangles | Perimeter of each triangle | Area of each triangle | Total perimeter | Total area |
|---|---|---|---|---|---|---|---|
| 0 | 1 | 1.6180 | 1 | 3.6180 | 0.4755 | 3.6180 | 0.4755 |
| 1 | 0.6180 | 1 | 2 | 2.2361 | 0.1816 | 4.4721 | 0.3633 |
| 2 | 0.3820 | 0.6180 | 4 | 1.3820 | 0.0694 | 5.5279 | 0.2775 |
| 3 | 0.2361 | 0.3820 | 8 | 0.8541 | 0.0265 | 6.8328 | 0.2120 |
| 4 | 0.1459 | 0.2361 | 16 | 0.5279 | 0.0101 | 8.4458 | 0.1620 |
| 5 | 0.0902 | 0.1459 | 32 | 0.3262 | 0.0039 | 10.4396 | 0.1237 |
| 6 | 0.0557 | 0.0902 | 64 | 0.2016 | 0.0015 | 12.9041 | 0.0945 |
| 7 | 0.0344 | 0.0557 | 128 | 0.1246 | 0.0006 | 15.9503 | 0.0722 |
| 8 | 0.0213 | 0.0344 | 256 | 0.0770 | 0.0002 | 19.7157 | 0.0552 |
| 9 | 0.0132 | 0.0213 | 512 | 0.0476 | 0.0001 | 24.3699 | 0.0421 |

Figure 7-30

The table was constructed cognizant of the fact that if we start with an obtuse golden triangle in which the shorter side measures one unit length, the length of the longer side will be equal to $\phi$. At each subsequent stage, the lengths of the sides will be reduced by a factor of $\frac{1}{\phi}$. The area of each triangle was calculated with the help of the Pythagorean theorem and the formula for area of a triangle. The total perimeter and total area at a particular stage are the sums of the perimeters and areas of all triangles at that stage, respectively.

The graphs in figures 7-31 and 7-32 help us see that while the perimeter of the fractal increases at each stage, the area decreases. In the long run, the fractal will have an infinite perimeter but an area equal to zero. No wonder it has dimension greater than 1, but less than 2. As we suspected, the dimension of 1.44 means that our fractal has more than just length (dimension 1), but not quite an area (dimension 2).

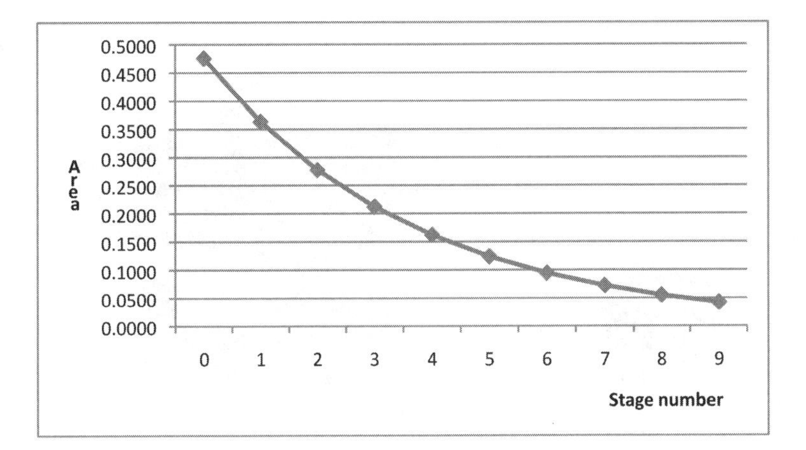


Figure 7-31

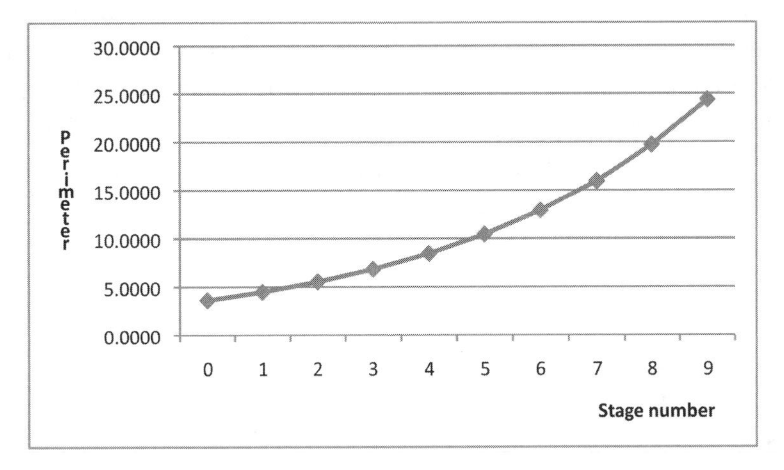


Figure 7-32

What would the dimension of the pentaflake be? We see that at each stage of its construction, the pentaflake gets more and more "holes." This suggests that its dimension is more than 1 and less than 2. Let us see if calculations confirm this conjecture.

Figure 7-33 shows that when we arrange six copies of a pentagon to form a pentaflakelike figure, the corresponding linear factor of dilation is $1+\phi$. Calculating the dimension the same way we did for the previous fractal, we will find that the dimension of the pentaflake is $\frac{\log 6}{\log(1+\phi)}$, which is an irrational number approximately equal to 1.86.

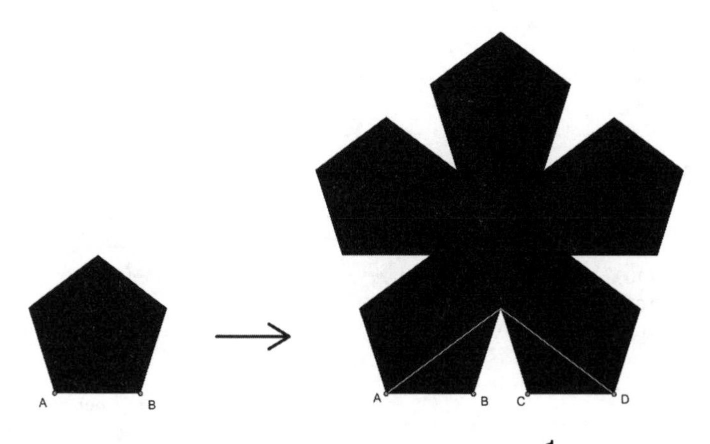


Figure 7-33. If $AB = 1$, $AD = 1 +$ $AB = 1$, $AD = 1 + \frac{1}{\Phi} + 1 = 1 + \phi$.

Does the fact that the golden ratio was employed in the construction of the fractals in this chapter make them more visually appealing than other fractals? That may not be the case. But the study of the mathematical relationships in these fractals and the fact that the golden ratio plays such a strong role in them can certainly generate awe.

# Concluding Thoughts

As we have reached the end of our journey, you must be thoroughly convinced that the golden section is a truly extraordinary phenomenon in mathematics. It appears both by design and by chance. Although we have to take the clues from history, and as best our modern minds can reconstruct the pieces, we can see that this relationship has permeated all aspects of society: structurally, aesthetically, biologically, and mathematically, which has given us an enormous range of areas to explore. The ratio's history is fascinating, and we traced it from ancient times to its more recent manifestations. Were the ancients aware of this ratio in all cases, or are we speculating that they did to some degree? Whatever the case may be, it is truly delightful to view its past permutations in our search for this ratio. You might well find other situations where this ratio emerges. The possibilities are practically boundless.

By now you know how to construct the ratio by partitioning a line segment, constructing a golden rectangle, a golden triangle, and a regular pentagon—all of which exhibit the golden section clearly. However, we have also examined other geometric configurations that in some fashion exhibited the golden section—many of which were quite unexpected appearances. Yet with each of these unexpected sightings of the golden section, there was usually an introduction to what for many were some new geometric relationships beyond the golden section. It is our hope that this type of exploration enriches one's encounter with geometry—something sorely lacking from traditional high school geometry.

The numerical value of the golden ratio is fascinating largely because of its ubiquity. Perhaps its most well-known connection to another structure in mathematics is its connection to the Fibonacci numbers—that is, the ratio of two consecutive Fibonacci numbers approaches either the golden ratio or its reciprocal, depending on the order of the ratio. This brings us to the most unusual relationship of the numerical value of the golden ratio, namely it is the only number that differs from its reciprocal by 1, that is, $\phi = \frac{1}{\phi} + 1$. This led us to a value of $\phi$ that is irrational and that in turn has opened up yet another area of further investigation.

Aside from its appearance in architecture and art, the golden section may be found throughout the plant kingdom. You will quite likely now be looking for other golden ratio specimens in the biological world. We just provided you an open door from which to peek at the garden of possibilities in this arena.

The closing chapter, showing the golden section in the field of fractals, can be seen as both mathematical and artistic. This rounds out our appreciation for this most famous ratio in mathematics. Since it relates in some form to almost everything in the field of mathematics, it is truly a ratio that has earned the title of *golden*. So now go for it, and expand on your introduction to the golden ratio!

Appendix

# Proofs and Justifications of Selected Relationships

## FOR CHAPTER 1:

### Derivation of the Quadratic Formula

The quadratic equation, $ax^2 + bx + c = 0$ (where $a > 0$), can be solved for $x$ in the following way:

$ax^2 + bx + c = 0.$

$x^2 + \frac{b}{a}x + \frac{c}{a} = 0.$

$x^2 + \frac{b}{a}x\ (+\frac{b^2}{4a^2} - \frac{b^2}{4a^2}) + \frac{c}{a} = 0.$

$(x + \frac{b}{2a})^2 - \frac{b^2}{4a^2} + \frac{c}{a} = 0.$ [Add $\frac{b^2}{4a^2} - \frac{c}{a}$ to both sides of the equation.]

$(x + \frac{b}{2a})^2 = \frac{b^2}{4a^2} - \frac{4ac}{4a^2} = \frac{b^2 - 4ac}{4a^2}.$ [Take the square root of both sides.]

$|x + \frac{b}{2a}| = \sqrt{\frac{b^2 - 4ac}{4a^2}} = \frac{\sqrt{b^2 - 4ac}}{2a}.$ [Note the absolute value.]

$x_{1,2}=-\frac{b}{2a}\pm\frac{\sqrt{b^2-4ac}}{2a}=\frac{-b\pm\sqrt{b^2-4ac}}{2a}$. Therefore,

$$x_{1,2}=\frac{-b\pm\sqrt{b^2-4ac}}{2a}.$$

## FOR CHAPTER 3:

### Proof of $\phi^n=F_n\phi+F_{n-1}$, with $n\geq 1$ and $F_0=0$.

We begin by showing that the statement to be proved by mathematical induction is true for $n=1$.

Yes, it holds true: $\phi^1=F_1\phi+F_0=1\cdot\phi+0=\phi$.

It is also true for the cases of $n=2, 3, 4, 5$, as shown below:

$$\phi^2=\phi+1=1\cdot\phi+1,$$
$$\phi^3=\phi\phi^2=\phi(\phi+1)=\phi^2+\phi=\phi+1+\phi=2\phi+1,$$
$$\phi^4=\phi\phi^3=2\phi^2+\phi=2\phi+2+\phi=3\phi+2,$$
$$\phi^5=\phi\phi^4=\phi(3\phi+2)=3\phi^2+2\phi=3\phi+3+2\phi=5\phi+3.$$

What now remains is that we accept its truth for $k$: $\phi^k=F_k\phi+F_{k-1}$, and must show it is then also true for $k+1$, namely $\phi^{k+1}=F_{k+1}\phi+F_k$.

By multiplying the first equation by $\phi$, we get: $\phi^{k+1}=F_k\phi^2+F_{k-1}\phi$

Since $\phi^2=\phi+1$, we have

$$\phi^{k+1}=F_k\phi^2+F_{k-1}\phi=F_k(\phi+1)+F_{k-1}\phi=(F_k+F_{k-1})\phi+F_k=F_{k+1}\phi+F_k,$$

which we were required to show.

### On Continued Fractions

A continued fraction is a fraction in which the denominator contains a mixed number (a whole number and a proper fraction). We can take an improper fraction such as $\frac{13}{7}$ and express it as a mixed number: $1\frac{6}{7}=1+\frac{6}{7}$. Without changing the value, we could then write this as

$$1+\frac{6}{7}=1+\frac{1}{\frac{7}{6}},$$

which in turn could be written (again without any value change) as

$$1+\cfrac{1}{1+\cfrac{1}{6}}.$$

This is a continued fraction. We could have continued this process, but when we reach a unit fraction (as in this case, the unit fraction is $\frac{1}{6}$), we are essentially finished.

So that you can get a better grasp of this technique, we will create another continued fraction. We will convert $\frac{12}{7}$ to a continued fraction form. Notice that at each stage, when a proper fraction is reached, take the reciprocal of the reciprocal (e.g., change

$$\frac{2}{5} \text{ to } \cfrac{1}{\cfrac{5}{2}},$$

as we will do in the example that follows), which does not change its value:

$$\frac{12}{7}=1+\frac{5}{7}=1+\cfrac{1}{\cfrac{7}{5}}=1+\cfrac{1}{1+\cfrac{2}{5}}=1+\cfrac{1}{1+\cfrac{1}{\cfrac{5}{2}}}=1+\cfrac{1}{1+\cfrac{1}{2+\cfrac{1}{2}}}.$$

If we break up a continued fraction into its component parts (called *convergents*),[1] we get closer and closer to the actual value of the original fraction.

First convergent of $\frac{12}{7}$: $1.$

Second convergent of $\frac{12}{7}$: $1+\frac{1}{1}=2.$

Third convergent of $\frac{12}{7}$: $1+\cfrac{1}{1+\cfrac{1}{2}}=1+\frac{2}{3}=1\frac{2}{3}=\frac{5}{3}.$

Fourth convergent of $\frac{12}{7}$: $1+\cfrac{1}{1+\cfrac{1}{2+\cfrac{1}{2}}}=\frac{12}{7}.$

The above examples are all *finite* continued fractions, which are equivalent to rational numbers (those that can be expressed as simple fractions). It would then follow that an irrational number would result in an *infinite* continued fraction. That is exactly the case. A simple example of an infinite continued fraction is that of $\sqrt{2}$. Although we show it here, we will actually generate it just a bit further on.

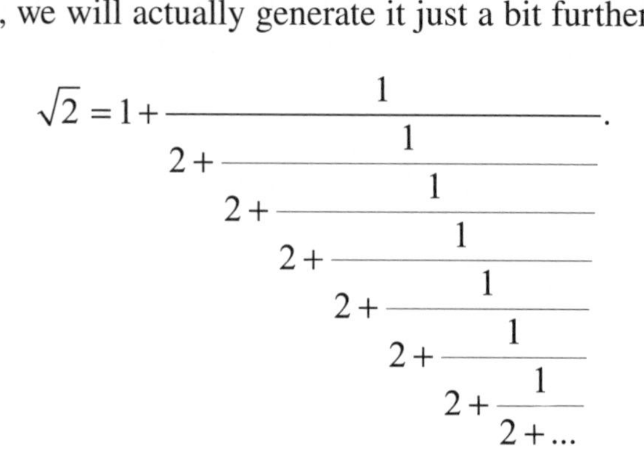

$$\sqrt{2}=1+\cfrac{1}{2+\cfrac{1}{2+\cfrac{1}{2+\cfrac{1}{2+\cfrac{1}{2+\cfrac{1}{2+\cfrac{1}{2+\dots}}}}}}}.$$

We have a short way to write a long (in this case infinitely long!) continued fraction: $[1; 2, 2, 2, 2, 2, 2, 2, \ldots]$, or when there are these endless repetitions, we can even write it in a shorter form as $[1; \overline{2}]$, where the bar over the 2 indicates that the 2 repeats endlessly.

In general, we can represent a finite continued fraction as

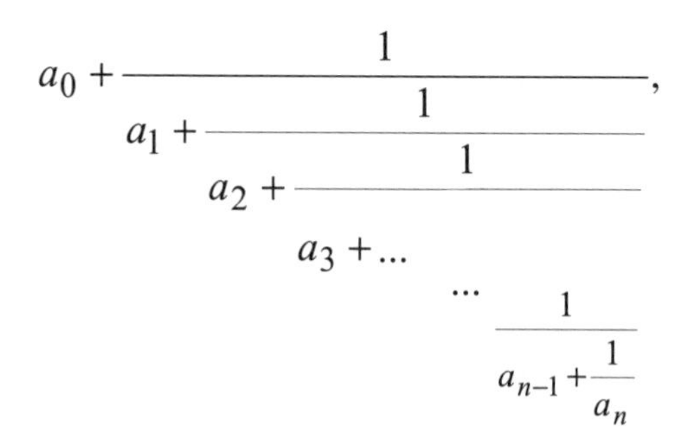

$$a_0+\cfrac{1}{a_1+\cfrac{1}{a_2+\cfrac{1}{a_3+\dots \cfrac{\ddots}{\cfrac{1}{a_{n-1}+\cfrac{1}{a_n}}}}}},$$

where $a_i$ are real numbers and $a_i \neq 0$ for $i>0$. We can write this in a shorter fashion as $[a_0; a_1, a_2, a_3, \ldots, a_{n-1}, a_n]$, but as an infinite continued fraction as $[a_0; a_1, a_2, a_3, \ldots, a_n, \ldots]$.

As we said before, we will generate a continued fraction equal to $\sqrt{2}$. Begin with the identity $\sqrt{2}+2=\sqrt{2}+2$.

Factor the left side and split the 2 on the right side: $\sqrt{2}(1+\sqrt{2})=1+\sqrt{2}+1$.

Divide both sides by $1+\sqrt{2}$ to get

$$\sqrt{2}=1+\frac{1}{1+\sqrt{2}}=[1;\,1,\sqrt{2}\,].$$

Replace $\sqrt{2}$ with $\sqrt{2}=1+\frac{1}{1+\sqrt{2}}$ and simplify the terms:

$$\sqrt{2}=1+\cfrac{1}{1+(1+\cfrac{1}{1+\sqrt{2}})}=1+\cfrac{1}{2+\cfrac{1}{1+\sqrt{2}}}=[1;\,2,\,1,\sqrt{2}\,].$$

Continue this process. The pattern now becomes clear.

$$\sqrt{2}=1+\cfrac{1}{2+\cfrac{1}{2+\cfrac{1}{1+\sqrt{2}}}}=[1;\,2,\,2,\,1,\sqrt{2}\,]\text{, and so on.}$$

Eventually we conclude with the following:

$$\sqrt{2}=1+\cfrac{1}{2+\cfrac{1}{2+\cfrac{1}{2+\dots}}}=[1;\,2,\,2,\,2,\,\dots\,]$$

Thus we have a periodic continued fraction for $\sqrt{2}$ (that is, $\sqrt{2}=[1;\,2,\,2,\,2,\dots\,]=[1;\,\overline{2}]$).

There are continued fractions equal to some famous numbers such as Euler's $e$ ($e=2.7182818284590452353\dots$)[2] and the famous $\pi$ ($\pi=3.14159265358979323 84\dots$):

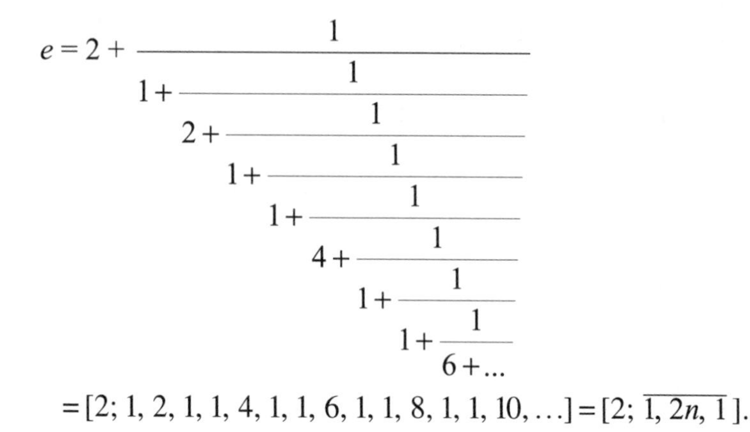

$$e = 2 + \cfrac{1}{1+\cfrac{1}{2+\cfrac{1}{1+\cfrac{1}{1+\cfrac{1}{4+\cfrac{1}{1+\cfrac{1}{1+\cfrac{1}{6+\ldots}}}}}}}}$$

$$= [2; 1, 2, 1, 1, 4, 1, 1, 6, 1, 1, 8, 1, 1, 10, \ldots] = [2; \overline{1, 2n, 1}].$$

Here are two ways that $\pi$ can be approximated as a continued fraction.[3]

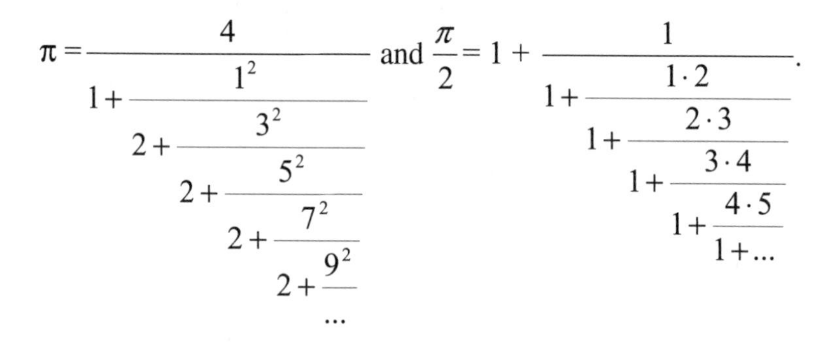

$$\pi = \cfrac{4}{1+\cfrac{1^2}{2+\cfrac{3^2}{2+\cfrac{5^2}{2+\cfrac{7^2}{2+\cfrac{9^2}{\ldots}}}}}} \quad \text{and} \quad \frac{\pi}{2} = 1 + \cfrac{1}{1+\cfrac{1\cdot 2}{1+\cfrac{2\cdot 3}{1+\cfrac{3\cdot 4}{1+\cfrac{4\cdot 5}{1+\ldots}}}}}.$$

Sometimes we have continued fractions representing these famous numbers that do not seem to have a distinctive pattern:

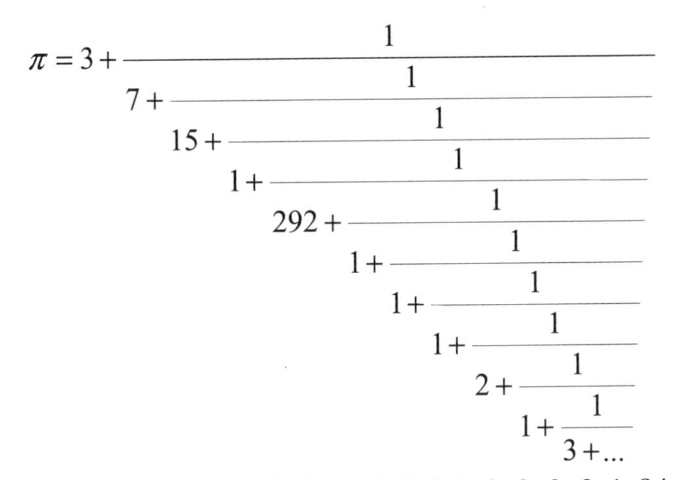

$\pi = [3; 7, 15, 1, 292, 1, 1, 1, 2, 1, 3, 1, 14, 2, 1, 1, 2, 2, 2, 2, 1, 84, 2, \ldots]$.

We have now set the stage for the golden ratio. Can we express this Fibonacci-related ratio ($\phi = 1.6180339887498948482\ldots$) as a continued fraction?

## Proof of the Binet Formula

Following you will find a simple way to express the Binet formula:

$$F_n = \frac{\phi^n - \psi^n}{\sqrt{5}},$$

where $\phi = \frac{1+\sqrt{5}}{2}$ and $\psi = \frac{1-\sqrt{5}}{2}$.

Recall the relationships that exist between $\phi$ and $\psi$ (since $\psi = -\frac{1}{\phi}$):

$$\phi + \psi = 1,$$
$$\phi - \psi = \sqrt{5}, \text{ and}$$
$$\phi\psi = -1.$$

The proof will be done using mathematical induction.

We begin by noting that

$$F_0 = \frac{1-1}{\sqrt{5}} = 0 \text{ and } F_1 = \frac{1+\sqrt{5}-(1-\sqrt{5})}{2\sqrt{5}} = 1,$$

that is, for $n=0$ and $n=1$, the Binet formula is correct.

Therefore, we assume that it is true for $n-2$ and $n-1$.

Because of the recursive formula, we have $F_n = F_{n-1} + F_{n-2}$, and we must therefore show that

$$\frac{\phi^{n-1}-\psi^{n-1}}{\sqrt{5}} + \frac{\phi^{n-2}-\psi^{n-2}}{\sqrt{5}} = \frac{\phi^{n}-\psi^{n}}{\sqrt{5}}.$$

Thus it suffices that $\phi^{n-1}+\phi^{n-2}=\phi^n$ and $\psi^{n-1}+\psi^{n-2}=\psi^n$:

$$\phi^{n-1}+\phi^{n-2} = \left(\frac{1+\sqrt{5}}{2}\right)^{n-1} + \left(\frac{1+\sqrt{5}}{2}\right)^{n-2}$$

$$= \left(\frac{1+\sqrt{5}}{2}\right)^{n-2}\left(\frac{1+\sqrt{5}}{2}+1\right) = \left(\frac{1+\sqrt{5}}{2}\right)^{n-2}\left(\frac{3+\sqrt{5}}{2}\right)$$

$$= \left(\frac{1+\sqrt{5}}{2}\right)^{n-2}\left(\frac{1+\sqrt{5}}{2}\right)^{2} = \left(\frac{1+\sqrt{5}}{2}\right)^{n} = \phi^n.$$

Thus, $\phi^{n-1}+\phi^{n-2}=\phi^n$, as required. The corresponding result for $\psi$ is proved in a similar way. The two together conclude the induction.

## FOR CHAPTER 4:

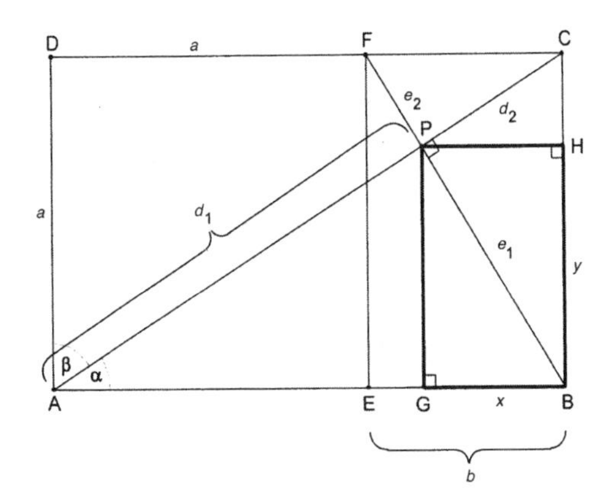


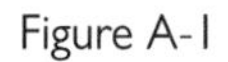
Figure A-1

*Development of the ratio:*

Applying the Pythagorean theorem to $\Delta ABC$:

$$d = d_1 + d_2 = AC$$

$$= \sqrt{AB^2 + BC^2} = \sqrt{(a+b)^2 + a^2} = a \cdot \sqrt{2 + \frac{2}{\phi} + \frac{1}{\phi^2}} = a \cdot \sqrt{\frac{5+\sqrt{5}}{2}}.$$

$\Delta ABP \sim \Delta ABC$; therefore, $\frac{AP}{AB} = \frac{AB}{AC}$, or $\frac{d_1}{a+b} = \frac{a+b}{d}$, therefore,

$$d_1 = \frac{(a+b)^2}{d} = \frac{a^2\left(1+\frac{1}{\phi}\right)^2}{a\sqrt{2+\frac{2}{\phi}+\frac{1}{\phi^2}}},$$

$$\text{or } d_1 = a \cdot \sqrt{\frac{5+2\sqrt{5}}{5}}.$$

$\Delta BCP \sim \Delta ABC$; therefore, $\frac{CP}{BC} = \frac{BC}{AC}$, or $\frac{d_2}{a} = \frac{a}{d}$.

We then get $d_2 = \dfrac{a^2}{d} = a \cdot \dfrac{1}{\sqrt{2 + \dfrac{2}{\phi} + \dfrac{1}{\phi^2}}}$,

or $d_2 = a \cdot \sqrt{\dfrac{5 - \sqrt{5}}{10}}$.

Analogously, we have segments $e_1$ and $e_2$ along the diagonal $BF$, which enables us to have $\Delta BCP \sim \Delta ABC$, with $\frac{BP}{BC} = \frac{AB}{AC}$, or $\frac{e_1}{a} = \frac{a+b}{d}$.

Therefore, $e_1 = \dfrac{a(a+b)}{d} = \dfrac{a^2\left(1 + \dfrac{1}{\phi}\right)}{a\sqrt{2 + \dfrac{2}{\phi} + \dfrac{1}{\phi^2}}}$.

Thus, $e_1 = a \cdot \sqrt{\dfrac{5 + \sqrt{5}}{10}}$.

Furthermore, $\Delta CFP \sim \Delta ABC$, with $\frac{FP}{CP} = \frac{BC}{AB}$, with $\frac{e_2}{d_2} = \frac{a}{a+b}$,

or $e_2 = \dfrac{ad_2}{a+b} = \dfrac{a^2 \dfrac{1}{\sqrt{2 + \dfrac{2}{\phi} + \dfrac{1}{\phi^2}}}}{a\left(1 + \dfrac{1}{\phi}\right)}$,

which, simplified, gives us $e_2 = a \cdot \sqrt{\dfrac{5 - 2\sqrt{5}}{5}}$.

Whereupon we can establish the lengths $x$ and $y$ on sides $AB$ and $BC$ as follows:

For $\Delta BGP \sim \Delta ABC$, with $\frac{BG}{BP} = \frac{BC}{AC}$, or $\frac{x}{e_1} = \frac{a}{d}$, we get

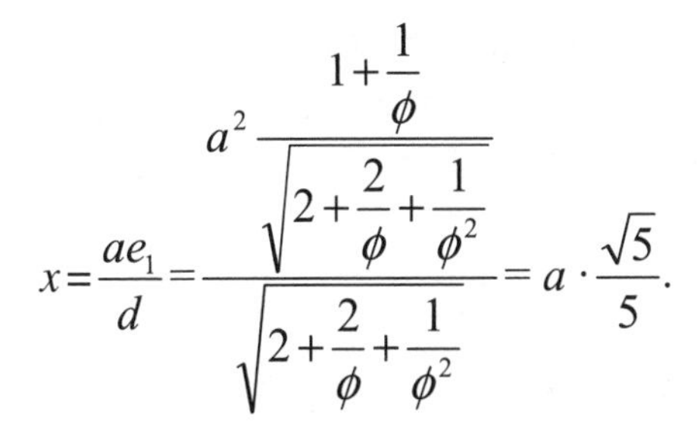

$$x = \frac{ae_1}{d} = \frac{a^2 \dfrac{1+\dfrac{1}{\phi}}{\sqrt{2+\dfrac{2}{\phi}+\dfrac{1}{\phi^2}}}}{\sqrt{2+\dfrac{2}{\phi}+\dfrac{1}{\phi^2}}} = a \cdot \frac{\sqrt{5}}{5}.$$

Then $x = a \cdot \frac{\sqrt{5}}{5}$.

Applying the Pythagorean theorem to $\Delta BHP$, we get

$$y = BH = \sqrt{BP^2 - HP^2} = \sqrt{e_1{}^2 - x^2} = a\sqrt{\frac{5+\sqrt{5}}{10} - \frac{5}{25}},$$

or $y = a \cdot \frac{5+\sqrt{5}}{10}$.

Thus we now have $AG = AB - BG = a + b - x =$

$$a + \frac{a}{\phi} - a \cdot \frac{\sqrt{5}}{5} = a \cdot \frac{5+3\sqrt{5}}{10}.$$

$$CH = BC - BH = a - y = a - a \cdot \frac{5+\sqrt{5}}{10} = a \cdot \frac{5-\sqrt{5}}{10}.$$

$$EG = BE - BG = b - x = \frac{a}{\phi} - a \cdot \frac{\sqrt{5}}{5} = a \cdot \left(\frac{\sqrt{5}-1}{2} - \frac{\sqrt{5}}{5}\right) = a \cdot \frac{3\sqrt{5}-5}{10}.$$

Thus the following segment lengths give us the golden ratio:

$$\frac{a+b}{a} = \frac{a}{b} = \frac{d_1}{e_1} = \frac{d_2}{e_2} = \frac{a+b-x}{y} = \frac{y}{x} = \frac{a-y}{b-x} = \phi = \frac{\sqrt{5}+1}{2}.$$

Now finally we have $\frac{d_1}{d_2} = \frac{e_1}{e_2} = \phi^2 = \phi + 1 = \frac{\sqrt{5}+3}{2}$.

## To Prove That the Maximum Area of the Shaded Region Formed by Two Congruent Perpendicular Rectangles Is Obtained When They Are Golden Rectangles[4]

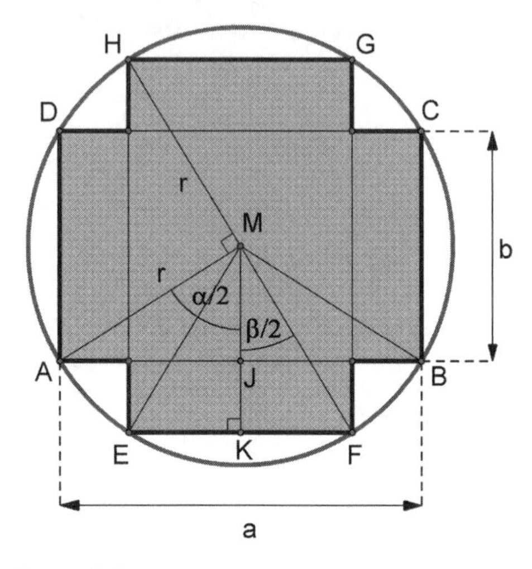


Figure A-2

We have rectangles that give us $AB=CD=FG=EH=a$, $AD=BC=EF=GH=b$, $AM=BM=EM=FM=r$, and as marked in figure A-2: $\alpha=\angle AMB$ and $\beta=\angle EMF$. By symmetry, $\beta=\angle EMF=\angle AMD$. Therefore, $\alpha+\beta=\angle AMB+\angle AMD=180°$, since $BD$ is the diagonal of the rectangle $ABCD$.

The shaded region in figure 4-13 is actually composed of the original rectangle $ABCD$ and two rectangles with side lengths $EF$ and $JK$.

The shaded region (in fig. 4-13) is actually composed $=ab+2\cdot\frac{a-b}{2}\cdot b=ab+(a-b)b=2ab-b^2$.

Applying the Pythagorean theorem to $\Delta AJM$, we get $AM^2=AJ^2+JM^2$, which then gives us

$$r^2=\frac{a^2}{4}+\frac{b^2}{4}, \text{ or } a=\sqrt{4r^2-b^2}.$$

$$\text{In } \Delta AJM\text{: } \sin\frac{\alpha}{2}=\frac{AJ}{AM}=\frac{\frac{a}{2}}{r}=\frac{a}{2r}\text{; therefore, } a=2r\sin\frac{\alpha}{2}.$$

In $\Delta EKM$: $\sin\frac{\beta}{2}=\frac{EK}{EM}=\frac{\frac{b}{2}}{r}=\frac{b}{2r}$ ; therefore, $b=2r\sin\frac{\beta}{2}$, and $\cos\frac{\beta}{2}=\frac{MK}{EM}=\frac{\frac{a}{2}}{r}=\frac{a}{2r}$.

With $\alpha=180°-\beta$, or in another form $\frac{\alpha}{2}=90°-\frac{\beta}{2}$, and $\sin(90°-\frac{\beta}{2})=\cos\frac{\beta}{2}$, we can get $\sin^2\frac{\beta}{2}=\frac{1-\cos\beta}{2}$ .

The area of the shaded region $=2ab-b^2=2b\sqrt{4r^2-b^2}-b^2$

$$=2\cdot 2r\sin\frac{\beta}{2}\sqrt{4r^2-\left(2r\sin\frac{\beta}{2}\right)^2}-\left(2r\sin\frac{\beta}{2}\right)^2=2\cdot 2r\sin\frac{\beta}{2}\cdot 2r\sqrt{1-\sin^2\frac{\beta}{2}}-4r^2\sin^2\frac{\beta}{2}$$

$$=4r^2\left(2\sin\frac{\beta}{2}\cdot\cos\frac{\beta}{2}-\sin^2\frac{\beta}{2}\right)=4r^2\left(\sin\beta+\frac{\cos\beta}{2}-\frac{1}{2}\right).$$

The factor $4r^2$ has no effect on the maximum area of the shaded region. Therefore, we shall focus our attention on the remaining factor: $f(\beta)=\sin\beta+\frac{\cos\beta}{2}-\frac{1}{2}$ [where the area of the shaded region $=4r^2\cdot f(\beta)$], and it is this we must maximize.

Differentiate $f$ and then set it equal to 0 to get

$$f'(\beta)=\frac{df}{d\beta}=\cos\beta-\frac{\sin\beta}{2}=0$$

$$\text{or } \cos\beta=\frac{\sin\beta}{2} \text{ or } \frac{\sin\beta}{\cos\beta}=\tan\beta=2 \text{ or } \beta=\arctan 2.$$

It is necessary for us to maximize, and for that we need to show for $0<\beta<180°$ the value $\beta=\arctan 2\approx 1.107$ (radians) $\approx 63.4°$.

We have

$$2=\tan\beta=\frac{\sin\beta}{\cos\beta}=\frac{2\sin\frac{\beta}{2}\cos\frac{\beta}{2}}{\cos^2\frac{\beta}{2}-\sin^2\frac{\beta}{2}}=\frac{2\cdot\frac{a}{2r}\cdot\frac{b}{2r}}{\left(\frac{a}{2r}\right)^2-\left(\frac{b}{2r}\right)^2}=\frac{2ab}{a^2-b^2}, \text{ also } 2(a^2-b^2)=2ab \text{ bzw. } a^2-ab-b^2=0\,.$$

That is,

$$\frac{a}{b}=\phi \text{ bzw. } \frac{b}{a}=\frac{2r\sin\frac{\beta}{2}}{2r\cos\frac{\beta}{2}}=\tan\frac{\beta}{2}=\frac{\sqrt{5}-1}{2}=\frac{1}{\phi}\text{ bzw. } \beta=2\arctan\frac{1}{\phi}.$$

The second derivative, $f''(\beta)=-\frac{\cos\beta}{2}-\sin\beta$, is at this point less than zero, $f''(\arctan 2)=f''(\arctan\frac{1}{3}+\frac{\pi}{4})=-\frac{\sqrt{5}}{2}<0$, so that the maximum is at $\beta=\arctan 2=2\arctan\frac{1}{\phi}\approx 63.4°$.

Because $\frac{a}{b}=\phi$, we have a golden rectangle.

The area of the shaded region $=2ab-b^2=(2\sqrt{5}-2)\cdot r^2\approx 2.472135954\cdot r^2$.

The shaded region covers an area of approximately 78.7 percent of the area of the circle.

The ratio of the areas is $\phi$ to $\frac{4}{\pi}$. That is,

$$\frac{Area_{\text{Circle}}}{Area_{\text{Shaded region}}}=\frac{\pi r^2}{\left(2\sqrt{5}-2\right)r^2}=\frac{\pi}{4}\cdot\frac{\sqrt{5}+1}{2}=\frac{\pi}{4}\cdot\phi=\frac{\phi}{\frac{4}{\pi}}.$$

## To Show That the Golden Ratio Is Present in Parts of the Pentagon and the Pentagram

We can use a number of approaches. Here we offer two such.

### Option 1:

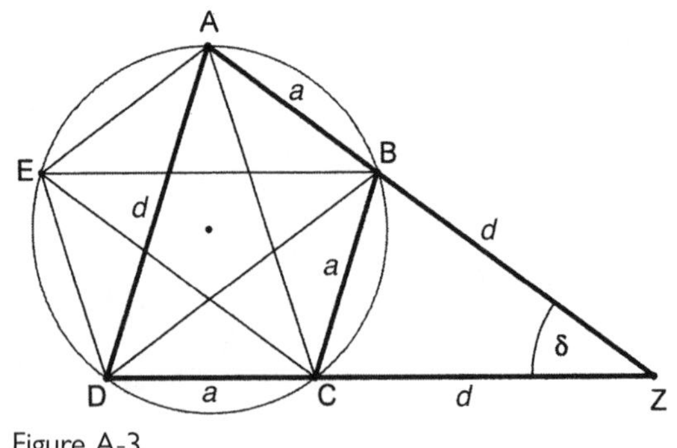


Figure A-3

For $\Delta ADZ \sim \Delta BCZ$, we get $\frac{AD}{BC}=\frac{DZ}{CZ}=\frac{CD+CZ}{CZ}=\frac{CD+AC}{AC}$, or, using the length designations from figure A-3, we can write this[5] as

$$\frac{d}{a}=\frac{a+d}{d},$$

which gives us $d^2-ad-a^2=0$, or

$$\left(\frac{d}{a}\right)^2-\frac{d}{a}-1=0.$$

If we replace $\frac{d}{a}$ by $x$, we arrive at the (by now) well-known golden ratio equation: $x^2-x-1=0$, where we know that $x=\phi=\frac{d}{a}$, or $d=a\phi$.

## Option 2:

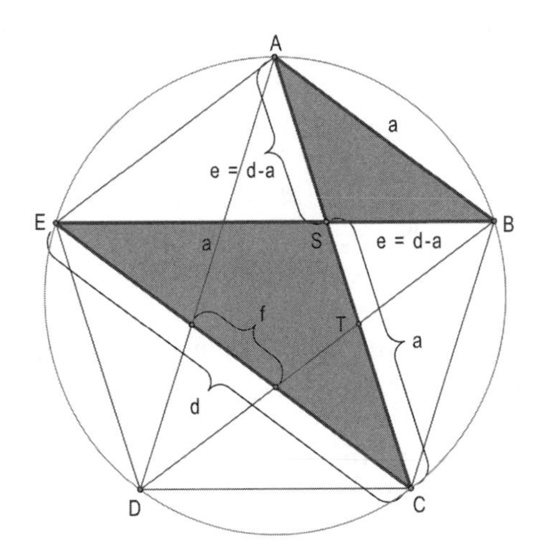


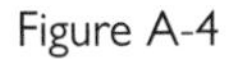
Figure A-4

This time we will use the following similar triangles: $\Delta ABS \sim \Delta CES$ to get $\frac{EC}{CS}=\frac{AB}{BS}=\frac{AB}{BE-ES}$. Using the length designations for figure A-4, we can write this as

$$\frac{d}{a}=\frac{a}{d-a}, \text{ or } d^2-ad-a^2=0.$$

Dividing by $a$, we get

$$\left(\frac{d}{a}\right)^2 - \frac{d}{a} - 1 = 0,$$

again the golden ratio equation, and in similar fashion, we get $d = a\phi$.

We can also see from figure A-4 that $a = e + f$, $d = a + e = 2e + f$. Therefore, $\frac{CS}{AS} = \frac{a}{d-a} = \phi$, which gives us: $a = \phi \cdot (d - a)$, or another way: $e = d - a = \frac{a}{\phi}$, or $a = e\phi$.

Since point $T$ partitions $AC$ into the golden section,

$$ST = f = a - e = AT - AS = a - \frac{a}{\phi} = a\left(1 - \frac{1}{\varphi}\right) = \frac{a}{\phi^2},$$

or, in another way, we can say that $a = f\phi^2$.

It then follows that $e$ and $f$ are also in the golden ratio.

$$\frac{e}{f} = \frac{AS}{ST} = \frac{\frac{a}{\phi}}{\frac{a}{\phi^2}} = \phi;$$

that is, $e = f\phi$.

Furthermore, we can also establish the sides of the two consecutive pentagons ($ABCDE$ and $PQRST$) (see fig. 4-64):

$$\frac{a}{f} = \frac{a}{\frac{a}{\phi^2}} = \phi^2,$$

which gives us $a = f\phi^2$, or $f = \frac{a}{\phi^2}$.

We represent the height, $b = AF$, of $\Delta ACD$ as follows:

$$b = d \cdot \cos\frac{\alpha}{2} = a\phi \cdot \cos\frac{36°}{2} = a\phi \cdot \sqrt{\frac{5+\sqrt{5}}{8}} = \frac{a}{2} \cdot \sqrt{5+2\sqrt{5}} = \frac{a}{2} \cdot \phi\sqrt{\phi^2+1}.$$

Again, we see the golden ratio everywhere in this configuration!

## Justification for Note on Page 150

To show that $\sqrt{5+2\sqrt{5}}=\sqrt{25+10\sqrt{5}}-\sqrt{10+2\sqrt{5}}$.

We begin by squaring both sides of the equation and seek to show the equality:

$$5+2\sqrt{5}=\left(\sqrt{25+10\sqrt{5}}-\sqrt{10+2\sqrt{5}}\right)^2.$$

$$=25+10\sqrt{5}\quad+10+2\sqrt{5}\quad-2\cdot\sqrt{25+10\sqrt{5}}\cdot\sqrt{10+2\sqrt{5}}$$

$$=25+10\sqrt{5}\quad+10+2\sqrt{5}\quad-2\cdot\sqrt{250+400+100\sqrt{5}+50\sqrt{5}}$$

$$=25+10\sqrt{5}\quad+10+2\sqrt{5}\quad-2\cdot\sqrt{350+150\sqrt{5}}$$

$$=25+10\sqrt{5}\quad+10+2\sqrt{5}\quad-2\cdot\sqrt{\left(15+5\sqrt{5}\right)^2}$$

$$=35+12\sqrt{5}-2\cdot\sqrt{\left(15+5\sqrt{5}\right)^2}$$

$$=35+12\sqrt{5}-2\left(15+5\sqrt{5}\right)$$

$$=5+2\sqrt{5}$$

## Pentagon's Rotation—Justification of Conclusions

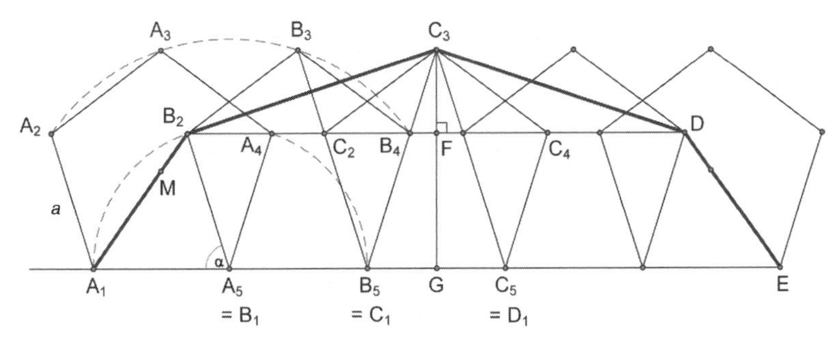


Figure A-5

Using figure A-5, we shall take a more detailed look at what is really happening here that allows us to draw the conclusions we drew in chapter 4.[6] By the first rotation of 72° at point $A_5$, we find that $A_1$ goes to $B_2$, $A_2$ goes to $B_3$, $A_3$ goes to $B_4$, $A_4$ goes to $B_5$ and $A_5$ stays at $B_1$. The angle of rotation, $\angle A_4A_5B_5 = 72°$, since each angle of the pentagon is 108°. The isosceles $\Delta A_1A_5B_2$, where $A_1A_5 = A_5B_2 = a$, has angles $\alpha = \angle A_1A_5B_2 = 72°$ and $\angle A_5A_1B_2 = \angle A_1B_2A_5 = 54°$. We can then generalize that continuing this process will give us $\angle B_2A_1E = \angle DEA_1$, or that in figure 4-71 we have $\angle BAE = \angle AED = 54°$. This, by the way, also establishes that the center point of the pentagon $M$ must lie on $A_1B_2$ (fig. A-5), or as we stated above for figure 4-71, $M$ must lie on $AB$, since it bisects $A_2A_1A_5$.

If we apply the law of cosines[7] to $\Delta A_1A_5B_2$, we get

$$\begin{aligned} A_1B_2^{\,2} &= A_1A_5^{\,2} + A_5B_2^{\,2} - 2A_1A_5 \cdot A_5B_2 \cdot \cos \angle A_1A_5B_2 \\ &= a^2 + a^2 - 2a^2 \cdot \cos\alpha \\ &= 2a^2(1-\cos\alpha) = 2a^2(1-\cos 72°) \\ &= 2a^2\left(1-\frac{\sqrt{5}-1}{4}\right) = 2a^2\left(1-\frac{1}{2\phi}\right). \end{aligned}$$

Therefore, $AB = A_1B_2 = \sqrt{\dfrac{5-\sqrt{5}}{2}} \cdot a = \sqrt{3-\phi} \cdot a$ (see fig. 4-71).

The diagonal $d = C_2C_4$ of pentagon $C_1C_2C_3C_4C_5$ is bisected by the point $F$.

We have $d = a \cdot \dfrac{\sqrt{5}+1}{2} = \phi \cdot a$; therefore, $C_2F = \dfrac{d}{2} = \dfrac{\phi \cdot a}{2}$.

When we apply the Pythagorean theorem to right $\Delta C_2C_3F$, we get

$$\begin{aligned} C_3F &= \sqrt{C_2C_3^{\,2} - C_2F^2} = \sqrt{a^2 - \left(\frac{d}{2}\right)^2} = \sqrt{a^2 - \frac{\phi^2 \cdot a^2}{4}} \\ &= \frac{a}{2} \cdot \sqrt{4-\phi^2} = \frac{a}{2}\sqrt{4-(\phi+1)} = \frac{a}{2} \cdot \sqrt{3-\phi}. \end{aligned}$$

We can also express $C_3F = \sqrt{\dfrac{5-\sqrt{5}}{8}} \cdot a = \dfrac{a}{2} \cdot \sqrt{\dfrac{5-\sqrt{5}}{2}}$.

Now we have shown that $C_3F = \frac{a}{2} \cdot \sqrt{3-\phi}$, and previously we showed that $A_1B_2 = a\sqrt{3-\phi}$, so we can now conclude that $C_3F = \frac{A_1B_2}{2}$.

Since $\angle A_5B_5B_3 = \angle A_5B_5C_2 = 72°$, we find $C_2$ is on the diagonal $B_3B_5$.

Recall that $C_2$ partitions the diagonal $B_2B_4$ into the golden ratio.

We know that triangle $B_2B_3C_2$ is a golden triangle, so it then follows that

$$B_2C_2 = \frac{B_2B_4}{\phi} = \frac{d}{\phi} = \frac{\phi \cdot a}{\phi} = a.$$

This then leads us to

$$B_2F = B_2C_2 + C_2F = a + \frac{d}{2} = a + \frac{\phi \cdot a}{2} = \left(1 + \frac{\phi}{2}\right) \cdot a = \frac{\sqrt{5}+5}{4} \cdot a.$$

We now apply the Pythagorean theorem to right $\Delta B_2C_3F$ to get

$$BC = B_2C_3 = \sqrt{B_2F^2 + C_3F^2} = \sqrt{\left(\frac{\sqrt{5}+5}{4}a\right)^2 + \left(\sqrt{\frac{5-\sqrt{5}}{8}}a\right)^2}$$

$$= \sqrt{\frac{5+\sqrt{5}}{2}} \cdot a = \sqrt{2+\phi} \cdot a.$$

It follows that

$$\frac{BC}{AB} = \frac{B_2C_3}{A_1B_2} = \frac{\sqrt{\frac{5+\sqrt{5}}{2}} \cdot a}{\sqrt{\frac{5-\sqrt{5}}{2}} \cdot a} = \frac{\sqrt{5+\sqrt{5}}}{\sqrt{5-\sqrt{5}}} = \phi.$$

Because of symmetry, we have $\frac{CD}{DE} = \phi$, which was what we set out to show.

As we seek to compare areas, let's consider the strange-looking pentagon $ABCDE$ (or in fig. A-5, $A_1B_2C_3DE$). The area of this pentagon is the sum of the trapezoid $A_1B_2DE$ and the isosceles $\Delta B_2C_3D$.

The area of the original pentagon is

$$A_{\text{Original pentagon}} = a^2 \cdot \frac{\sqrt{25+10\sqrt{5}}}{4} = \frac{a^2}{4} \cdot \sqrt{5(5+2\sqrt{5})} \quad \text{(see p. 154).}$$

For the height $h = C_3G$ of $\Delta B_5C_3G$, we have

$$h = d \cdot \cos\angle B_5C_3G = a\ \phi \cdot \cos\frac{36°}{2} = a\ \phi \cdot \sqrt{\frac{5+\sqrt{5}}{8}} = \frac{a}{2} \cdot \sqrt{5+2\sqrt{5}}.$$

Then the height of the trapezoid is

$$FG = C_3G - C_3F = \frac{a}{2} \cdot \sqrt{5+2\sqrt{5}} - \frac{a}{2} \cdot \sqrt{\frac{5-\sqrt{5}}{2}} = \frac{a}{2} \cdot \sqrt{\frac{5+\sqrt{5}}{2}}.$$

We have

$$B_2D = B_2F + FD = 2B_2F = 2 \cdot \frac{\sqrt{5}+5}{4} \cdot a = \frac{\sqrt{5}+5}{2} \cdot a.$$

The area of the trapezoid $A_1B_2DE$ is

$$\text{Area}_{A_1B_2DE} = \frac{1}{2}(A_1E + B_2D) \cdot FG$$

$$= \frac{1}{2}\left(5a + \frac{\sqrt{5}+5}{2} \cdot a\right) \cdot \frac{a}{2} \cdot \sqrt{\frac{5+\sqrt{5}}{2}} = a^2 \cdot \sqrt{\frac{95\sqrt{5}+325}{32}}$$

$$= \frac{a^2}{4} \cdot \sqrt{\frac{5\sqrt{5}(19+13\sqrt{5})}{2}}.$$

The area of $\Delta B_2C_3D$ is

$$\text{Area}_{\Delta B_2C_3D} = \frac{1}{2} B_2D \cdot C_3F = \frac{1}{2} \cdot \frac{\sqrt{5}+5}{2} \cdot a \cdot \frac{a}{2} \cdot \sqrt{\frac{5-\sqrt{5}}{2}}$$

$$= a^2 \cdot \sqrt{\frac{5\sqrt{5}+25}{32}} = \frac{a^2}{4} \cdot \sqrt{\frac{5(5+\sqrt{5})}{2}}.$$

The area of the strange-looking pentagon $A_1B_2C_3DE$ is

$$\text{Area}_{ABCDE} = \text{Area}_{A_1B_2DE} + \text{Area}_{\Delta B_2C_3D}$$

$$=\frac{a^2}{4}\cdot\sqrt{\frac{5\sqrt{5}(19+13\sqrt{5})}{2}}+\frac{a^2}{4}\cdot\sqrt{\frac{5(5+\sqrt{5})}{2}}$$

$$=\frac{a^2}{4}\cdot\sqrt{225+90\sqrt{5}}=\frac{a^2}{4}\cdot\sqrt{45\sqrt{5}(2+\sqrt{5})},$$

which is three times the original pentagon's area:

$$\frac{a^2}{4}\cdot\sqrt{5(5+2\sqrt{5})},$$

which we wished to demonstrate.

## Proof for the Height of the Rooflike Cap on a Cube

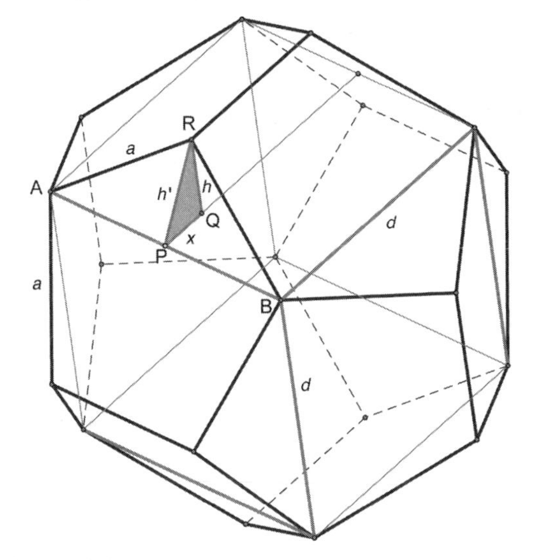


Figure A-6

Consider the right $\Delta APR$ with a leg length $AP=\frac{d}{2}$ and $PR=h'$ as well as hypotenuse $AR=a$. We also have the altitude from $R$ to the surface of the cube forming $\Delta PQR$ with sides $PQ=x$ and $QR=h$ as well as hypotenuse $PR=h'$. The edge of the cube ($d$) is the diagonal of the pentagon whose side has length $a$. Therefore, $d=\phi a$. With $d-2x=a$, we get

$$x=\frac{d-a}{2}=\frac{\phi a-a}{2}=\frac{a}{2}(\phi-1)=\frac{a}{2}\cdot\frac{1}{\phi}=\frac{a}{2\phi}.$$

By applying the Pythagorean theorem twice, we have $PR^2 = h'^2 = AR^2 - AP^2 = a^2 - \frac{d^2}{4} = a^2(1 - \frac{\phi^2}{4})$, which then gives the height

$$h' = \frac{a}{2} \cdot \frac{\sqrt{\phi^2 + 1}}{\phi}.$$

Then a second time:

$$QR^2 = h^2 = PR^2 - PQ^2 = h'^2 - x^2 = \frac{a^2}{4} \cdot \frac{\phi^2 + 1}{\phi^2} - \frac{a^2}{4\phi^2} = \frac{a^2}{4\phi^2}(\phi^2 + 1 - 1) = \frac{a^2}{4},$$

whereupon $h = \frac{a}{2}$.

If the inscribed cube has edge length equal to 1, or $d = 1$, then we immediately get $a = \frac{d}{\phi} = \frac{1}{\phi}$, and then $h = \frac{a}{2} = \frac{1}{2\phi}$.

## More Trigonometric Relationships in Terms of the Golden Section

(a) $\sin 18° = \frac{\sqrt{5} - 1}{4} = \frac{1}{2\phi} = \frac{1}{2}\sqrt{1 - \frac{1}{\phi}}.$

$$\sin 36° = \sqrt{\frac{5 - \sqrt{5}}{8}} = \frac{1}{2}\sqrt{\frac{5 - \sqrt{5}}{2}} = \frac{1}{2}\frac{\sqrt{\phi^2 + 1}}{\phi} = \frac{1}{2}\sqrt{2 - \frac{1}{\phi}}.$$

$$\sin 54° = \frac{\sqrt{5} + 1}{4} = \frac{\phi}{2} = \frac{1}{2}\sqrt{1 + \phi}.$$

$$\sin 72° = \sqrt{\frac{5 + \sqrt{5}}{8}} = \frac{1}{2}\sqrt{\frac{5 + \sqrt{5}}{2}} = \frac{1}{2}\sqrt{\phi^2 + 1} = \frac{1}{2}\sqrt{2 + \phi}.$$

(b) $\cos 18° = \sin(90° - 18°) = \sin 72° = \sqrt{\frac{5 + \sqrt{5}}{8}} = \frac{1}{2}\sqrt{\phi^2 + 1} = \frac{1}{2}\sqrt{2 + \phi}.$

$$\cos 36° = \sin(90° - 36°) = \sin 54° = \frac{\sqrt{5} + 1}{4} = \frac{\phi}{2} = \frac{1}{2}\sqrt{1 + \phi}.$$

$$\cos 54° = \sin(90° - 54°) = \sin 36° = \sqrt{\frac{5-\sqrt{5}}{8}} = \frac{1}{2}\frac{\sqrt{\phi^2+1}}{\phi} = \frac{1}{2}\sqrt{2-\frac{1}{\phi}}.$$

$$\cos 72° = \sin(90° - 72°) = \sin 18° = \frac{\sqrt{5}-1}{4} = \frac{1}{2\phi} = \frac{1}{2}\sqrt{1-\frac{1}{\phi}}.$$

(c) $\tan 18° = \dfrac{\sin 18°}{\cos 18°} = \sqrt{\dfrac{5-2\sqrt{5}}{5}} = \dfrac{1}{\phi\sqrt{\phi^2+1}}$, $\cot 18° = \dfrac{\cos 18°}{\sin 18°} = \sqrt{5+2\sqrt{5}} = \phi\sqrt{\phi^2+1}$.

$\tan 36° = \dfrac{\sin 36°}{\cos 36°} = \sqrt{5-2\sqrt{5}} = \dfrac{\sqrt{\phi^2+1}}{\phi^2}$, $\cot 36° = \dfrac{\cos 36°}{\sin 36°} = \sqrt{\dfrac{5+\sqrt{5}}{10}} = \dfrac{\phi}{\sqrt{\phi^2+1}}$.

$\tan 54° = \dfrac{\sin 54°}{\cos 54°} = \sqrt{\dfrac{5+2\sqrt{5}}{5}} = \dfrac{\phi^2}{\sqrt{\phi^2+1}}$, $\cot 54° = \dfrac{\cos 54°}{\sin 54°} = \sqrt{5-2\sqrt{5}} = \dfrac{\sqrt{\phi^2+1}}{\phi^2}$.

$\tan 72° = \dfrac{\sin 72°}{\cos 72°} = \sqrt{5+2\sqrt{5}} = \phi\sqrt{\phi^2+1}$, $\cot 72° = \dfrac{\cos 72°}{\sin 72°} = \sqrt{\dfrac{5-2\sqrt{5}}{5}} = \dfrac{1}{\phi\sqrt{\phi^2+1}}$.

(d) $\dfrac{\sin 18°}{\sin 36°} = \sqrt{\dfrac{5-\sqrt{5}}{10}} = \dfrac{1}{\sqrt{\phi^2+1}}$, $\dfrac{\sin 36°}{\sin 18°} = \sqrt{\dfrac{5+\sqrt{5}}{2}} = \sqrt{\phi^2+1}$.

$\dfrac{\sin 18°}{\sin 54°} = \dfrac{3-\sqrt{5}}{2} = \dfrac{1}{\phi^2}$, $\dfrac{\sin 54°}{\sin 18°} = \dfrac{\sqrt{5}+3}{2} = \phi^2$.

$\dfrac{\sin 18°}{\sin 72°} = \sqrt{\dfrac{5-2\sqrt{5}}{5}} = \dfrac{1}{\phi\sqrt{\phi^2+1}}$, $\dfrac{\sin 72°}{\sin 18°} = \sqrt{5+2\sqrt{5}} = \phi\sqrt{\phi^2+1}$.

$\dfrac{\sin 36°}{\sin 54°} = \sqrt{5-2\sqrt{5}} = \dfrac{\sqrt{\phi^2+1}}{\phi^2}$, $\dfrac{\sin 54°}{\sin 36°} = \sqrt{\dfrac{5+2\sqrt{5}}{5}} = \dfrac{\phi^2}{\sqrt{\phi^2+1}}$.

$\dfrac{\sin 36°}{\sin 72°} = \dfrac{\sqrt{5}-1}{2} = \dfrac{1}{\phi}$, $\dfrac{\sin 72°}{\sin 36°} = \dfrac{\sqrt{5}+1}{2} = \phi$.

$\dfrac{\sin 54°}{\sin 72°} = \sqrt{\dfrac{5+\sqrt{5}}{10}} = \dfrac{\phi}{\sqrt{\phi^2+1}}$, $\dfrac{\sin 72°}{\sin 54°} = \sqrt{\dfrac{5-\sqrt{5}}{2}} = \dfrac{\sqrt{\phi^2+1}}{\phi}$.

(e) $\frac{\sin 18^\circ}{\cos 36^\circ}=\frac{3-\sqrt{5}}{2}=\frac{1}{\phi^2}$, $\frac{\cos 36^\circ}{\sin 18^\circ}=\frac{\sqrt{5}+3}{2}=\phi^2$.

$\frac{\sin 18^\circ}{\cos 54^\circ}=\sqrt{\frac{5-\sqrt{5}}{10}}=\frac{1}{\sqrt{\phi^2+1}}$, $\frac{\cos 54^\circ}{\sin 18^\circ}=\sqrt{\frac{5+\sqrt{5}}{2}}=\sqrt{\phi^2+1}$.

$\frac{\sin 18^\circ}{\cos 72^\circ}=1$, $\frac{\cos 72^\circ}{\sin 18^\circ}=1$.

$\frac{\sin 36^\circ}{\cos 54^\circ}=1$, $\frac{\cos 54^\circ}{\sin 36^\circ}=1$.

$\frac{\sin 36^\circ}{\cos 72^\circ}=\sqrt{\frac{5+\sqrt{5}}{2}}=\sqrt{\phi^2+1}$, $\frac{\cos 72^\circ}{\sin 36^\circ}=\sqrt{\frac{5-\sqrt{5}}{10}}=\frac{1}{\sqrt{\phi^2+1}}$.

$\frac{\sin 54^\circ}{\cos 72^\circ}=\frac{\sqrt{5}+3}{2}=\phi^2$, $\frac{\cos 72^\circ}{\sin 54^\circ}=\frac{3-\sqrt{5}}{2}=\frac{1}{\phi^2}$.

## To Show That the Following Are True:

$$\pi=2\cdot\left(\arctan\frac{1}{\Phi^5}+\arctan\Phi^5\right)$$

**and**

$$\pi=6\arctan\frac{1}{\Phi}-2\arctan\frac{1}{\Phi^5}$$

*Proof*:[8]

We begin by using the known relationship:

$$\tan(\delta+\varepsilon)=\frac{\tan\delta+\tan\varepsilon}{1-\tan\delta\cdot\tan\varepsilon},$$

to get

$$\tan 2\alpha=\frac{2\tan\alpha}{1-\tan^2\alpha}\text{ (where }\tan^2\alpha\neq 1).$$

If we let $\delta = 2\alpha$ and $\varepsilon = \alpha$ in the above identity, we get

$$\tan 3\alpha = \tan(2\alpha + \alpha) = \frac{\tan 2\alpha + \tan\alpha}{1 - \tan 2\alpha \cdot \tan\alpha} = \frac{3\tan\alpha - \tan^3\alpha}{1 - 3\tan^2\alpha}.$$

If we now let $\alpha = \arctan\frac{1}{\phi}$, we get

$$\tan 3\alpha = \frac{3\tan\alpha - \tan^3\alpha}{1 - 3\tan^2\alpha} = \frac{3\tan\arctan\frac{1}{\phi} - \tan^3\arctan\frac{1}{\phi}}{1 - 3\tan^2\arctan\frac{1}{\phi}}$$

$$= \frac{3\cdot\frac{1}{\phi} - \frac{1}{\phi^3}}{1 - 3\cdot\frac{1}{\phi^2}} = \frac{3\cdot\phi^2 - 1}{\phi(\phi^2 - 3)} = \frac{3\phi + 2}{1 - \phi} = -\frac{5\sqrt{5} + 11}{2} = -\phi^5.$$

Since $\tan(180° - x) = \tan(\pi - x) = -\tan x$ we have

$\phi^5 = -\tan 3\alpha = \tan(\pi - 3\alpha)$, also $\pi - 3\alpha = \arctan\phi^5$,

that is, $3\alpha = \pi - \arctan\phi^5 = \pi - (\frac{\pi}{2} - \arctan\frac{1}{\phi^5}) = \frac{\pi}{2} + \arctan\frac{1}{\phi^5}$.

Therefore, we have $\pi - \arctan\phi^5 = \frac{\pi}{2} + \arctan\frac{1}{\phi^5}$, which when we multiply by 2 gives us

$$2\pi - 2\arctan\phi^5 = \pi + 2\arctan\frac{1}{\phi^5}.$$

By adding $(-\pi + 2\arctan\phi^5)$ to both sides of the equation, we get

$$\pi = 2\left(\arctan\frac{1}{\phi^5} + \arctan\phi^5\right) = 2\cdot\left(3\arctan\frac{1}{\phi} - \arctan\frac{1}{\phi^5}\right)$$

$$= 6\arctan\frac{1}{\phi} - 2\arctan\frac{1}{\phi^5}.$$

## FOR CHAPTER 5:

### Curiosity 1

In an equilateral triangle, $\Delta ABC$, each side of length $s$ is partitioned (with the same orientation) into the segments $a$ and $b$, which are in the golden ratio (fig. A-7). The result is that an inscribed equilateral triangle, $\Delta DEF$, is created with side length $c$.

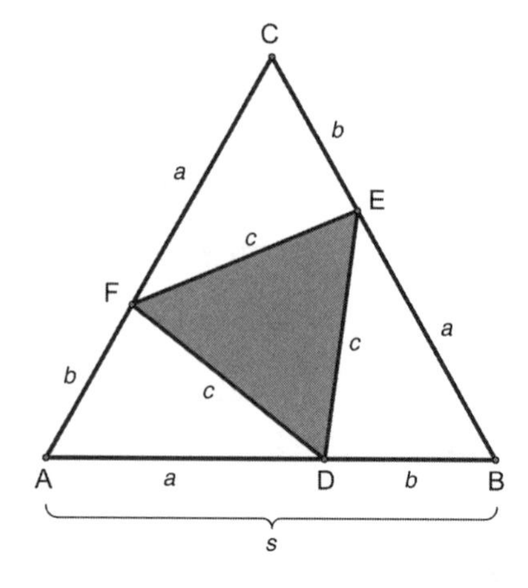


Figure A-7

Here are some of the appearances of $\phi$ in this figure:

1. $c = \frac{s}{\phi}\sqrt{1+\frac{1}{\phi^2}-\frac{1}{\phi}}$.

2. $\text{Area}_{\Delta DEF} = \frac{\sqrt{3}(\phi^2-\phi+1)}{4\phi^4}\cdot s^2$.

3. The ratio of the areas of the two equilateral triangles is

$$\frac{\text{Area}_{\Delta ABC}}{\text{Area}_{\Delta DEF}} = \frac{\phi^2}{1+\frac{1}{\phi^2}-\frac{1}{\phi}}.$$

4. The area of each of the three congruent triangles $\Delta ADF$, $\Delta BDE$, and $\Delta CEF$ is

$$\text{Area}_{\Delta ADF} = s^2 \frac{\sqrt{3}}{4\phi^3}.$$

5. The ratio of the areas of the original equilateral triangle to one of the three congruent triangles is

$$\frac{\text{Area}_{\Delta ABC}}{\text{Area}_{\Delta ADF}} = \frac{2}{\phi} + 3.$$

6. The ratio of the area of the smaller equilateral triangle to one of the three congruent triangles is

$$\frac{\text{Area}_{\Delta DEF}}{\text{Area}_{\Delta ADF}} = \frac{2}{\phi}.$$

## The Justifications:

1. We begin by getting the area of $\Delta ABC$ and $\Delta DEF$:

$$\text{Area}_{\Delta ABC} = \frac{1}{2} s^2 \cdot \sin 60° = \frac{1}{2} s^2 \cdot \frac{\sqrt{3}}{2} = \frac{s^2\sqrt{3}}{4}, \text{ and Area}_{\Delta DEF} = \frac{c^2\sqrt{3}}{4}.$$

The triangles $ADF$, $BDE$, and $CEF$ are congruent (by Side-Angle-Side, SAS).

Applying the law of cosines to triangle $ADF$:

$$DF^2 = c^2 = AD^2 + AF^2 - 2AD \cdot AF \cdot \cos 60°$$

$$= a^2 + b^2 - 2ab \cdot \tfrac{1}{2} = a^2 + b^2 - ab.$$

We partitioned the sides of the original equilateral triangle in the golden ratio. Therefore,

$$\frac{s}{a} = \frac{a}{b} = \phi.$$

It then follows that

$$c^2 = a^2 + \frac{a^2}{\phi^2} - \frac{a^2}{\phi} = a^2\left(1 + \frac{1}{\phi^2} - \frac{1}{\phi}\right),$$

which gives us

$$c = a\sqrt{1 + \frac{1}{\phi^2} - \frac{1}{\phi}} = \frac{a\left(\sqrt{10} - \sqrt{2}\right)}{2}$$

$$= \frac{s}{\phi}\sqrt{1 + \frac{1}{\phi^2} - \frac{1}{\phi}} = \frac{s\left(3\sqrt{2} - \sqrt{10}\right)}{2}.$$

2. $\text{Area}_{\Delta DEF} = \dfrac{c^2\sqrt{3}}{4} = \dfrac{s^2}{\phi^2}\dfrac{\sqrt{3}}{4}\left(1 + \dfrac{1}{\phi^2} - \dfrac{1}{\phi}\right)$

$$= \frac{s^2\sqrt{3}\left(\phi^2 - \phi + 1\right)}{4\phi^4} = \frac{s^2\left(7\sqrt{3} - 3\sqrt{15}\right)}{4}$$

$$\approx 0.1263514035 \cdot s^2.$$

3. $\dfrac{\text{Area}_{\Delta ABC}}{\text{Area}_{\Delta DEF}} = \dfrac{\frac{\sqrt{3}}{4} \cdot s^2}{\frac{\sqrt{3}}{4} \cdot c^2} = \left(\dfrac{s}{c}\right)^2 = \left(\dfrac{s}{\sqrt{1 + \frac{1}{\phi^2} - \frac{1}{\phi}} \cdot \frac{s}{\phi}}\right)^2 = \left(\dfrac{\phi}{\sqrt{1 + \frac{1}{\phi^2} - \frac{1}{\phi}}}\right)^2$

$$= \frac{\phi^2}{1 + \frac{1}{\phi^2} - \frac{1}{\phi}} = \frac{3\sqrt{5} + 7}{4} \approx 3.427050983.$$

4. $\text{Area}_{\Delta ADF} = \dfrac{1}{2} \cdot AD \cdot AF \cdot \sin 60° = \dfrac{1}{2}ab \cdot \dfrac{\sqrt{3}}{2} = \dfrac{\sqrt{3}}{4} \cdot \dfrac{a^2}{\phi} = \dfrac{\sqrt{3}}{4\phi} \cdot a^2$

$$= \frac{\sqrt{3}}{4\phi} \cdot \frac{s^2}{\phi^2} = \frac{s^2\sqrt{3}}{4\phi^3} = \frac{s^2\left(\sqrt{15} - 2\sqrt{3}\right)}{4}$$

$$\approx 0.1022204328 \cdot s^2.$$

5. $$\frac{\text{Area}_{\Delta ABC}}{\text{Area}_{\Delta ADF}}=\frac{\frac{\sqrt{3}}{4}}{\frac{\sqrt{15}-2\sqrt{3}}{4}}=\frac{\sqrt{3}}{\sqrt{15}-2\sqrt{3}}=\sqrt{5}+2=\frac{2}{\phi}+3\approx 4.236067977.$$

6. $$\frac{\text{Area}_{\Delta DEF}}{\text{Area}_{\Delta ADF}}=\frac{\frac{7\sqrt{3}-3\sqrt{15}}{4}}{\frac{\sqrt{15}-2\sqrt{3}}{4}}\cdot\frac{7\sqrt{3}-3\sqrt{15}}{\sqrt{15}-2\sqrt{3}}=\sqrt{5}-1=\frac{2}{\phi}\approx 1.236067977.$$

A check of the above can be made by taking the sum of the areas of the four triangles and showing that it is the area of the original equilateral triangle:

$$\text{Area}_{\Delta DEF}+3\cdot\text{Area}_{\Delta ADF}=\frac{7\sqrt{3}-3\sqrt{15}}{4}\cdot s^2+3\cdot\frac{\sqrt{15}-2\sqrt{3}}{4}\cdot s^2=\frac{\sqrt{3}}{4}\cdot s^2.$$

## Justification for Curiosity 19

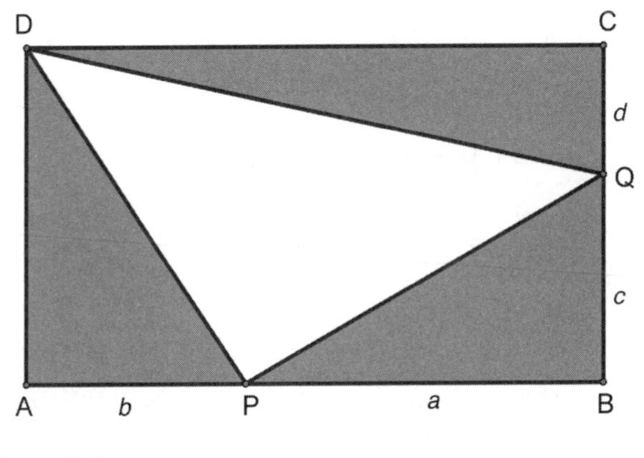


Figure A-8

We are given that the areas of the three shaded triangles (fig. A-8) are equal: $\text{Area}_{\Delta APD}=\text{Area}_{\Delta PBQ}=\text{Area}_{\Delta CDQ}$.

Therefore, $\frac{1}{2}\cdot b(c+d)=\frac{1}{2}\cdot ac=\frac{1}{2}\cdot(a+b)d$, and $b(c+d)=ac=(a+b)d$, which leads to $bc+bd=ac=ad+bd$.

We then get
$(a+b) : a=c : d$
$(c+d) : c=a : b.$

It follows that $(a+b) : (c+d)=b : d$
as well as $bc+bd=ac=ad+bd$; thus, $bc=ac$. That is, $a : b=c : d$.

For our purposes:

$$a=\frac{bc}{d} \text{ and } a=\frac{b(c+d)}{c}.$$

Therefore, $\frac{bc}{d}=\frac{b(c+d)}{c}$.

Multiplying both sides by $cd$ gives us $bc^2=bd(c+d)=bcd+bd^2$, or $bc^2=bcd+bd^2$, which, when divided by $b$, yields $c^2=cd+d^2$, or $c^2-d^2-cd=0$. Then dividing by $d^2$, we get $(\frac{c}{d})^2-\frac{c}{d}-1=0$. There appears our equation for the golden section. With $x=\frac{c}{d}$, we get $x^2-x-1=0$, giving us roots:

$$\frac{1}{2}\pm\sqrt{\frac{1}{4}+1}=\frac{1}{2}\pm\sqrt{\frac{5}{4}}=\frac{1\pm\sqrt{5}}{2}.$$

As we focus on the positive root, we have

$$x=\frac{c}{d}=\frac{\sqrt{5}+1}{2}=\phi.$$

Therefore, $c : d=a : b=(c+d) : c=(a+b) : a=\phi=\phi : 1$, which shows that $P$ and $Q$ partition $AB$ and $BC$, respectively, into the golden ratio.

## Construction for Curiosity 23

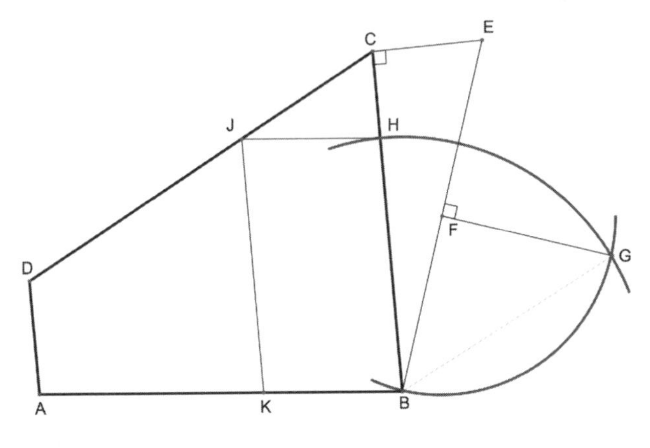


Figure A-9

The construction steps are as follows:

(1) Construct trapezoid $ABCD$ with $AD \parallel BC$ and $BC = 3AD$.

(2) Construct right $\Delta BCE$ with $CE = AD$ and $\angle BCE = 90°$.

(3) Construct the perpendicular bisector of $BE$ at its midpoint $F$ and then mark $G$ so that $FG = \frac{BE}{2}$.

(4) Construct a circle with center $B$ and radius $BG$ to intersect $BC$ at point $H$.

(5) Finally, construct parallelogram $BHJK$ with point $J$ on $CD$ and point $K$ on $AB$.

We then have $K$ partitioning the line segment $AB$ in the golden ratio. This can be easily justified, since with $AD = b$ and $BC = 3b$, we get

$BE = \sqrt{BC^2 + CE^2} = b\sqrt{10}$, and

$$BG = \sqrt{BF^2 + FG^2} = \sqrt{2BF^2} = \sqrt{2} \cdot \frac{BE}{2} = \sqrt{2}\frac{b\sqrt{10}}{2} = b\sqrt{5}.$$

Since

$$JK = \sqrt{\frac{9b^2 + b^2}{2}} = \sqrt{\frac{a^2 + b^2}{2}} = b\sqrt{5},$$

we have $JK = BH = BG$ as the root mean square between $a$ and $b$ when $a = 3b$. Thus, $AK : BK = \phi : 1$.

## Proof for Curiosity 24

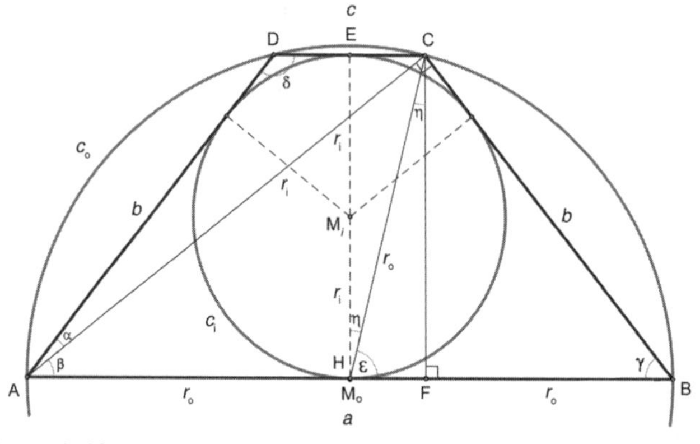

Figure A-10

We are given that $AB = a$, $BC = AD = b$, and $CD = c$. From the relationship of the tangents to the same (inscribed) circle, we have $a + c = b + b = 2b$, or $c = 2b - a$.

We let $\alpha = \angle CAD$, $\beta = \angle BAC$, $\gamma = \angle ABC$, $\delta = \angle ADC$, $\varepsilon = \angle BM_oC$, and $\eta = \angle CM_oE$; $E$ is the midpoint of $CD = c$. From point $C$, we construct a perpendicular $CF$ to $AB$. We have $CF = EM_o = 2r_i$. Also $r_i = M_iM_o$, the radius of the inscribed circle. Furthermore, $\angle BAD + \angle ADC = \angle ABC + \angle BCD = 180°$.

Since $\angle ACB$ is inscribed in a semicircle, it is a right angle and $\beta + \gamma = 90°$. We have isosceles $\Delta AM_oC$, with $AM_o = CM_o = r_o$, and $\angle M_oAC = \angle ACM_o = \beta$. The angles $\angle CM_oE$ and $\angle FCM_o$, as alternate interior angles of the parallel lines, are equal. Therefore, $\angle CM_oE = \angle FCM_o = \eta$.

For right $\Delta ACF$, therefore, $\angle FAC + \angle ACF = \angle FAC + \angle ACM_o + \angle M_oCF = 2\beta + \eta = 90°$, or $\eta = 90° - 2\beta$.

For $\Delta ACB$: $\quad \sin\angle BAC = \sin\beta = \dfrac{BC}{AB} = \dfrac{b}{2r_o}$, also $b = 2r_o \cdot \sin\beta$.

For $\Delta CEM_o$: $\quad \sin\angle CM_oE = \sin\eta = \dfrac{CE}{CM_o} = \dfrac{c}{2r_o}$, also $c = 2r_o \cdot \sin\eta$.

Let's consider $c=2b-a$, and then substitute. $c=2b-a=2\cdot 2r_o\cdot\sin\beta-2r_o=2r_o\cdot(2\sin\beta-1)$. On one side we have $c=2r_o\cdot\sin\eta$. On the other hand, we also have $2r_o\cdot(2\sin\beta-1)=2r_o\cdot\sin\eta=2r_o\cdot\sin(90°-2\beta)$ $\quad | \ 2r_o$.

$$
\begin{aligned}
2\sin\beta-1 &= \sin(90°-2\beta) && |\text{subtraction}^{9}\\
&= \sin 90°\cdot\cos 2\beta-\cos 90°\cdot\sin 2\beta\\
&= 1\cdot\cos 2\beta-0\cdot\sin 2\beta\\
&= \cos 2\beta && |\text{double angle formula for cosine}\\
&= \cos^2\beta-\sin^2\beta && |\text{the Pythagorean theorem}\\
&= 1-\sin^2\beta-\sin^2\beta\\
&= 1-2\sin^2\beta && |\text{add } 2\sin^2\beta-1\\
2\sin^2\beta+2\sin\beta-2 &= 0 && |\text{divide by 2}\\
\sin^2\beta+\sin\beta-1 &= 0.
\end{aligned}
$$

Substituting give us $x=\sin\beta$ and then appears the equation that we are by now quite familiar with, the equation for the golden section: $x^2+x-1=0$, where the only usable root is $x=\sin\beta=\frac{1}{\phi}$.

For $\Delta ACM_o$: $\varepsilon=\angle BM_oC$ and $\angle AM_oC=\angle AM_oE+\angle CM_oE=90°+\eta$.

Thus, $\varepsilon=90°+(90°-2\beta)=2\beta$ or $\beta=\frac{\varepsilon}{2}$.

To justify this, consider: $\sin\frac{\varepsilon}{2}=\sin\beta=\frac{1}{\phi}$ or $\beta=\frac{\varepsilon}{2}\approx 38.17°$.

$$\sin\varepsilon=\sin 2\beta=2\sin\beta\cdot\cos\beta=2\sin\beta\cdot\sqrt{1-\sin^2\beta}=\frac{2}{\phi}\cdot\sqrt{1-\frac{1}{\phi^2}}=\frac{2\sqrt{\phi}}{\phi^2};$$

thus, $\varepsilon\approx 76.35°$.
$\eta=90°-2\beta\approx 13.65°$.

From $\angle BAD=\angle ABC=\gamma$, we get $\beta+\gamma=90°$, or $\gamma=90°-\beta\approx 51.83°$.

Since $\alpha+\beta=\gamma$, we get $\angle CAD=\alpha=\gamma-\beta\approx 13.65°$.

We have two angles, $\angle BCD=\angle ADC=\delta$, and we get $\gamma+\delta=180°$, or $\delta=180°-\gamma\approx 128.17°$.

For the trapezoid's sides, $b=a\cdot\sin\beta$ and $c=M_0F=2r_0\cdot\sin\eta=a\cdot\sin\eta$, thus, $b=a\sin\beta=\frac{a}{\phi}$, or $c=a\sin\eta=a\sin(90°-2\beta)=a(1-2\sin^2\beta)=\frac{\phi-1}{\phi^2}a$.

For the radius of the circumscribed circle: $r_o = \frac{a}{2}$.

For the radius of the inscribed circle:

For $\Delta ACF$: $\sin\beta = \frac{CF}{AC} = \frac{EM_o}{AC} = \frac{2r_i}{AC}$; that is, $r_i = \frac{1}{2}\cdot AC\cdot \sin\beta = \frac{1}{2}\cdot AC\cdot\frac{1}{\phi}$.

For $\Delta ABC$: $AC^2 = AB^2 - BC^2 = a^2 - b^2 = a^2 - \left(\frac{1}{\phi}a\right)^2 = \frac{1}{\phi}a^2$.

Therefore, $AC = \frac{a}{\sqrt{\phi}} = \frac{a\sqrt{\phi}}{\phi}$.

This gives us $r_i = \frac{1}{2}\cdot AC\cdot\frac{1}{\phi} = \frac{a}{2}\frac{\sqrt{\phi}}{\phi}\cdot\frac{1}{\phi} = \frac{a\sqrt{\phi}}{2\phi^2}$.

## Explanation for Curiosity 25

We provide you with an overview of the solution and refer you to a more detailed version at the following website: http://www.mathekalender.de/info/loesungsheft_2009.pdf (see pp. 94–99: December 15, 2009, by Ingmar Lehmann and Elke Warmuth, Humboldt-Universität-Berlin—n.b. points $H$ and $F$ are switched in this reference).

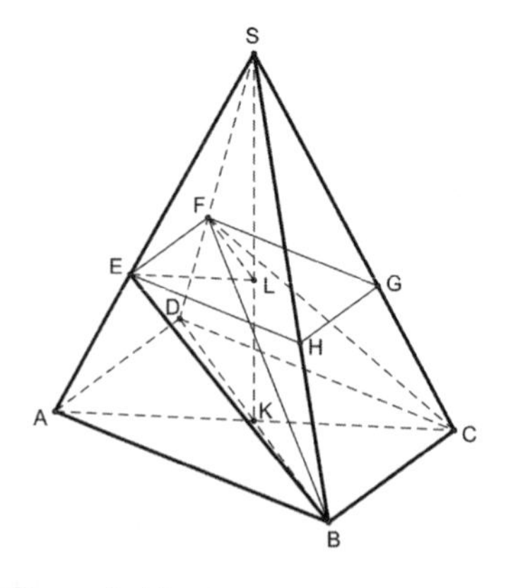


Figure A-11

We begin by using an auxiliary plane $EFGH$ parallel to the base $ABCD$. The height of the pyramid, $KS$, contains the point of intersection of the diagonals of rectangle $EFGH$. We shall let $AD = BC = a$, $AB$

= $CD = b$, and $AS = l$. Also, $KS = h$. There exists a value, $q$, where $0 < q < 1$, so that $ES = FS = ql$, $EF = GH = q \cdot a$, $LS = q \cdot h$, and $AE = DF = (1 - q)l$, $KL = (1 - q)h$. We then get $4h^2 + b^2 = 4l^2 - a^2$.

Pyramid $BCDF$: Volume = $\frac{1}{3} \cdot \frac{ab}{2} \cdot (1-q)h$

Pyramid $ADFEB$: Volume = $\frac{1}{3} \cdot \frac{1-q^2}{2} \cdot abh$

When we add these two volumes, we get the lower portion of the figure.

$$\frac{1}{3} \cdot \frac{ab}{2} \cdot (1-q)h + \frac{1}{3} \cdot \frac{1-q^2}{2} \cdot abh = \frac{1}{6} \cdot (2-q+q^2) \cdot abh.$$

Since we want this volume to be half the volume of the full figure, we get the following:

$$\frac{1}{6} \cdot (2-q-q^2) \cdot abh = \frac{1}{2} \cdot \frac{1}{3} \cdot abh, \text{ also } 2-q-q^2 = 1.$$

This is equivalent to $q^2 + q - 1 = 0$, where the positive root is the now-familiar $\frac{1}{\phi}$. This establishes that points $E$ and $F$ partition pyramid edges $AS$ and $DS$ into the golden section.

$$AS = l,\ AE = DF = (1 - q)l = (1 - \tfrac{1}{\phi})l,\ ES = FS = ql = \tfrac{1}{\phi}l,$$

Therefore, $\frac{AS}{ES} = \frac{l}{\frac{1}{\phi}l} = \phi$ and $\frac{ES}{AE} = \frac{\frac{1}{\phi}l}{\left(1-\frac{1}{\phi}\right)l} = \frac{1}{\phi-1} = \phi.$

## FOR CHAPTER 6:

### Connection between the Divergence Angle of the Real Number λ, and the Number of Visible Spirals (*Contact parastichy*)

The convergents $\frac{P_k}{Q_k}$ are the best rational approximations of $\lambda$, that is, all further fractions with a denominator smaller than $Q_{k+1}-1$ approximate $\lambda$ more poorly than $\frac{P_k}{Q_k}$ (Lagrange's theorem)[10]—see also Rosen.[11] In the case of the golden angle, Lagrange's theorem can be proven in an elementary fashion.[12]

Suppose we represent the fraction $\frac{y}{x}$ as the point $(x, y)$ of the fundamental lattice $\mathbf{Z}\times\mathbf{Z}$, we then obtain the following geometric interpretation of Felix Klein's (1849–1925)[13] development of the continuous fraction. The points with integer coordinates, which lie closer to a straight line (with the slope $\lambda$ in a restricted band $[0, x]\times\mathbf{R}$ of the fundamental lattice $\mathbf{Z}\times\mathbf{Z}$) than the previous points, are then essentially the points with the convergence coordinates $(Q_k, P_k)$ (fig. A-12).

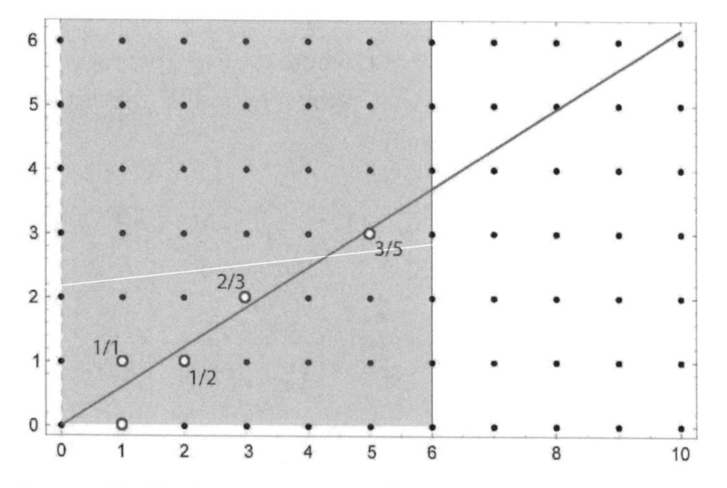


Figure A-12. The fundamental lattice $\mathbf{Z}\times\mathbf{Z}$ marked with the straight line $y=\lambda x$, where $\lambda=2-\phi\approx0.381966$. In the gray area lies the lattice point (5, 3), which is the point closest to the straight line.

Because the convergents $\frac{P_k}{Q_k}$ are also the best rational approximations of $\lambda$, then $\lambda \cdot Q_k \approx P_k$, and the value of the angle $\alpha_k := \alpha \cdot Q_k - 360° \cdot P_k$ is smaller than all $\alpha_n = \alpha \cdot n - 360° \cdot m$, with $n < Q_{k+1} - 1$. The additional growth $\Delta r_n = r_{n+1} - r_n$ of the radial component $r_n = \sqrt{n}$ of the points generated by the Vogel model decreases monotonically. For this reason, an area exists (see below) where the next neighbor of the point $X(Q_k)$ is the point $X(2Q_k)$, and the point $X(Q_k+1)$ is followed by the point $X(2Q_k+1)$, and so on. In this way, $Q_k$ spirals with the same rotational direction are generated as well as the arithmetic progression of the indexes of the points on the spiral.[14]

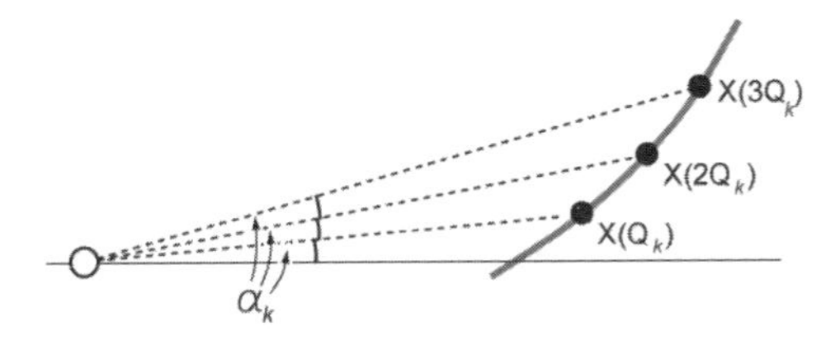


Figure A-13. Genesis of a parastichy.

# Notes

## Introduction

1. J. Kepler, *Gesammelte Werke*, vol. 1, *Mysterium cosmographicum. De stella nova*, ed. Max Caspar (Munich: C. H. Beck, 1938).

## Chapter 1: Defining and Constructing the Golden Ratio

1. In an article on aesthetics in the ninth edition of the *Encyclopedia Britannica*, 1875.

2. Euclid, *Elements*, book 2, proposition 11; book 6, definition 3, proposition 30; book 13, propositions 1–6.

3. There is reason to believe that the letter $\phi$ was used because it is the first letter of the name of the celebrated Greek sculptor Phidias (ca. 490–430 BCE) (in Greek: [Pheidias] ΦΕΙΔΙΑΣ or *Φειδίας*), who produced the famous statue of Zeus in the Temple of Olympia and supervised the construction of the Parthenon in Athens, Greece. His frequent use of the golden ratio in this glorious building (see chap. 2) is likely the reason for this attribution. It must be said that there is no direct evidence that Phidias consciously used this ratio.

The American mathematician Mark Barr was the first using the letter $\phi$ in about 1909 (see Theodore Andrea Cook, *The Curves of Life*

(New York: Dover Publications, 1979), p. 420. It should be noted that sometimes you will also find the lowercase $\phi$—or less commonly the Greek letter $\tau$ (tau), the initial letter of τομή (to-me or to-mĭ'), meaning *to cut.*

4. This construction is credited to Heron of Alexandria (ca. 40–ca. 120).

5. This theorem was originally demonstrated by Euclid in book 6 of his *Elements* § 3. The proof can also be found in A. S. Posamentier, J. H. Banks, and R. Bannister, *Geometry: Its Elements and Structure* (New York: McGraw-Hill, 1977).

6. It was observed by George Odom, a resident of the Hudson River Psychiatric Center, in the early 1980s (see J. van Craats, "A Golden Section Problem from the Monthly," *American Mathematical Monthly* 93, no.7 [1986]: 572 or C. Pritchard, ed., *The Changing Shape of Geometry* [Cambridge: Cambridge University Press, 2003], p. 294).

7. $AM = AG$, since they are two tangents to a circle from the same external point.

8. Florian Cajori, *A History of Mathematics*, 5th ed. (1894; repr., New York: Macmillan, 1999), p. 29.

9. Lee Dickey, "The 3-4-5 Right Triangle and the Golden Mean Discovered by Gabries Bosia," http://www.math.uwaterloo.ca/~lj dickey/geometry_corner/3-4-5.html (accessed October 12, 2011).

10. Hans Walser, "The Golden Section and Lattice Geometry," 2006, pp. 10–12, http://www.math.unibas.ch/~walser/Miniaturen/M48 _GS_and_lattice_geometry/GS_and_lattice_geometry.pdf.

11. Kurt Hofstetter, "A Simple Construction of the Golden Section," *Forum Geometricorum* 2 (2002): 65–66.

## Chapter 2: The Golden Ratio in History

1. The British egyptologist W. M. F. Petrie (1853–1942) established these measurements.

2. Herbert Westren Turnbull, *The Great Mathematicians* (London: Methuen, 1929).

3. Euclid's *Elements*, book 6, definition 3.

4. This "mean and extreme ratio" Euclid defines (book 6, definition 3) as follows: "A straight line [segment] is said to have been cut in *extreme and mean ratio* when, as the whole line [segment] is to the greater, so is the greater to the lesser."

5. Some mathematics books use the Greek letter $\tau$ (tau) representing the first letter of the Greek word meaning "to cut."

6. Leonardo da Vinci was a polymath: painter, sculptor, architect, musician, scientist, mathematician, engineer, inventor, anatomist, geologist, botanist, and writer.

7. *Wikipedia*, s.v. "List of works designed with the golden ratio," http://en.wikipedia.org/wiki/List_of_works_designed_with_the_golden_ratio (accessed October 12, 2011).

8. Dan Pedoe, *Geometry and the Visual Arts* (Harmondsworth, UK: Penguin, 1976; New York: Dover, 1983).

9. George Markowsky, "Misconceptions about the Golden Ratio," *College Mathematics Journal* 23, no. 1 (1992): 2–19.

10. Marguerite Neveux and H. E. Huntley, *Le nombre d'or: Radiographie d'un mythe suivi de La divine proportion* (Paris: Éditions du Seuil, 1995).

11. Roger Herz-Fischler, *A Mathematical History of the Golden Number* (Mineola, NY: Dover, 1998).

12. The pseudonym for Charles Édouard Jeanneret-Gris (1887–1965), a Swiss-French architect and artist.

13. Étienne Béothy (1897–1961), a French-Hungarian sculptor and artist.

14. Jo Niemeyer (1946–), a German graphic artist and architect.

15. See also the examples in A. S. Posamentier and I. Lehmann, *The Fabulous Fibonacci Numbers* (Amherst, NY: Prometheus Books, 2007), pp. 231–69; also see Ingmar Lehmann, "Fibonacci-Zahlen—Ausdruck von Schönheit und Harmonie in der Kunst," *Der Mathematikunterricht* 55, no. 2 (2009): 51–63.

## Chapter 3: The Numerical Value of the Golden Ratio and Its Properties

1. For the equation $ax^2+bx+c=0$, the formula is $x=\frac{-b\pm\sqrt{b^2-4ac}}{2a}$.

2. We offer the value of $\phi$ here only to 1,000-place accuracy.

3. We offer the value of $\frac{1}{\phi}$ here also to 1,000-place accuracy.

4. For more information on these ubiquitous numbers, see A. S. Posamentier and I. Lehmann, *The Fabulous Fibonacci Numbers* (Amherst, NY: Prometheus Books, 2007).

5. Paramanand Singh, "Ācārya Hemacandra and the (So-Called) Fibonacci Numbers," *Mathematics Education* 20, no. 1 (1986): 28–30.

6. Simon Jacob was one of the best-known German master calculators of his time. In 1557 he published a practice book on arithmetic calculation that gave insight to the art of calculation. (*Rechenbuch auf den Linien und mit Ziffern*, 1557; reprint Ein New und Wolgegründt Rechenbuch, 1612).

7. Peter Schreiber, "A Supplement to J. Shallit's Paper 'Origins of the Analysis of the Euclidean Algorithm,'" *Historia Mathematica* 22 (1995): 422–24.

8. The theory of continued fractions goes back to Leonhard Euler (1707–1783). A yet earlier appearance of continued fractions can be found in the seventeenth century when Christian Huygens (1629–1695) used them to construct gears to model the solar system and tried to approximate the relative speeds of the planets with as few numbers as possible.

9. A *unit fraction* is a fraction with the numerator 1.

10. This is done by considering the value of each portion of the continued fraction up to each plus sign, successively.

11. The values given in this chart are rounded to nine decimal places.

12. To prove this, we assume $L$ would be the limit of the series $\frac{F_{n+1}}{F_n}$.

$$L=\lim_{n\to\infty}\frac{F_{n+1}}{F_n}=\lim_{n\to\infty}\frac{F_{n-1}+F_n}{F_n}=\lim_{n\to\infty}\left(\frac{F_{n-1}}{F_n}+\frac{F_n}{F_n}\right)$$

$$= \lim_{n\to\infty}\left(\frac{F_{n-1}}{F_n}+1\right) = 1 + \lim_{n\to\infty}\frac{1}{\frac{F_n}{F_{n-1}}} = 1 + \frac{1}{L}.$$

With $L = 1 + \frac{1}{L}$, we get $L^2 - L - 1 = 0$. Then $L = \phi$ (because $\frac{F_{n+1}}{F_n} > 0$).

Therefore the series $\frac{F_{n+1}}{F_n}$ approaches the value of $\phi$, and the series $\frac{F_n}{F_{n+1}}$ approaches the value of $\frac{1}{\phi}$.

13. Jacques-Philippe Marie Binet, "Memoire sur l'integration des equations lineaires aux differences finies d'un ordre quelconque, a coefficients variables," *Comptes rendus de l'academie des sciences de Paris*, 17 (1843): 563.

14. For a derivation of this formula, see Posamentier and Lehmann, *Fabulous Fibonacci Numbers*, pp. 366–69.

15. The Bernoulli family is like a clan (of eight mathematicians in three generations)—famous and estranged!

16. Lucas is also well known for his invention of the Tower of Hanoi puzzle and other mathematical recreations. The Tower of Hanoi puzzle appeared in 1883 under the name of M. Claus. Notice that Claus is an anagram of Lucas. His four-volume work on recreational mathematics (1882–1894) has become a classic. Lucas died as the result of a freak accident at a banquet when a plate was dropped and a piece flew up and cut his cheek. He died of erysipelas (a superficial bacterial skin infection) a few days later.

17. To understand the precise sense in which the Fibonacci ratios are the best possible approximations to $\phi$, see Keith Ball, *Strange Curves, Counting Rabbits, and Other Mathematical Explorations* (Princeton, NJ: Princeton University Press, 2003), pp. 163–64.

18. If we assume $G$ would be the limit of the series $\frac{F_{n+2}}{F_n}$, then we have

$$G = \lim_{n\to\infty}\frac{F_{n+2}}{F_n} = \lim_{n\to\infty}\frac{F_n + F_{n+1}}{F_n} = \lim_{n\to\infty}\left(\frac{F_n}{F_n}+\frac{F_{n+1}}{F_n}\right) = \lim_{n\to\infty}\left(1+\frac{F_{n+1}}{F_n}\right) = 1 + \phi = \phi^2,$$

$$\text{or } G = \lim_{n\to\infty} \frac{F_{n+2}}{F_n} = \lim_{n\to\infty}\left(\frac{F_{n+2}}{F_{n+1}} \cdot \frac{F_{n+1}}{F_n}\right) = \lim_{n\to\infty}\frac{F_{n+2}}{F_{n+1}} \cdot \lim_{n\to\infty}\frac{F_{n+1}}{F_n} = \phi \cdot \phi = \phi^2.$$

$$\text{Therefore, } \lim_{n\to\infty}\sqrt{\frac{F_{n+2}}{F_n}} = \phi.$$

19. A proof of this can be found in I. J. Good, "A Reciprocal Series of Fibonacci Numbers," *Fibonacci Quarterly* 12, no. 4 (1974): 346.

20. *Complex numbers* are in the form of $a+bi$, where $a$ and $b$ are real numbers and $i=\sqrt{-1}$.

## Chapter 4: Golden Geometric Figures

1. For example, Gustav Fechner (1801–1887), a German experimental psychologist, began a serious inquiry to see if the golden rectangle had a special psychological aesthetic appeal. Fechner made thousands of measurements of commonly seen rectangles, such as those of playing cards, writing pads, books, windows, and so on. He found that most had a ratio of length to width that was close to $\phi$ (*Zur experimentalen Ästhetik* [On Experimental Aesthetics] [Leipzig, Germany: Breitkopf & Haertel, 1876]). For more information on this subject, see A. S. Posamentier and I. Lehmann, *The Fabulous Fibonacci Numbers* (Amherst, NY: Prometheus Books, 2007), pp. 115–17.

2. Claudi Alsina and Roger B. Nelsen, *Math Made Visual: Creating Images for Understanding Mathematics* (Washington, DC: Mathematical Association of America, 2006), pp. 77, 156.

3. This can also be written as: $\phi^2 = \sum_{i=1}^{\infty} \frac{i}{\phi^{i+1}}$.

4. Marjorie Bicknell-Johnson and Duane DeTemple, "Vizualizing Golden Ratio Sums with Tiling Patterns," *Fibonacci Quarterly* 33, no. 4 (1995): 298–303; James Metz, "The Golden Staircase and the Golden Line," *Fibonacci Quarterly* 35, no. 3(1997): 194–97.

5. On the Cartesian plane, the equation of the spiral would be: $x(\theta)=r(\theta)\cdot\cos\theta=a\cdot e^{k\theta}\cdot\cos\theta$ and $y(\theta)=r(\theta)\cdot\sin\theta=a\cdot e^{k\theta}\cdot\cos\theta$.

6. Descartes was a French mathematician, who is responsible for the field of analytic geometry, which is done on a "Cartesian plane," named for its founder. Descartes mentioned this spiral in 1638 correspondences with another French mathematician, Marin Mersenne (1588–1648), who is famous for his work with prime numbers.

7. The equation for the golden spiral is $r(\theta)=e^{-\frac{\ln\phi}{\pi/2}\cdot\theta}$ $(\approx e^{-0.0053467798\theta})$.

8. A tessellation is a tiling of a plane with polygons where there are no points uncovered and no overlap between polygons.

9. Notice that $F_0=0$ and $F_{-1}=1$.

10. See the spiral with quarter circles in a golden rectangle, which approximates the golden spiral.

11. Herta T. Freitag and Sahib Singh, "Golden Radii," *Fibonacci Quarterly* 32, no. 4 (1994): 376.

12. The Togo flag was designed by the artist Paul Ahyi (1930–2010), who claims to have attempted to have the flag constructed in the shape of a golden rectangle.

13. Isosceles $\Delta CDM$, where $CM=DM=r$. Therefore, $\angle CDM+\angle DCM=2\cdot\angle CDM=2\psi=180°-\phi=108°$, and $\angle CAD=\angle CED=\angle CBD=\alpha=36°$.

14. From past discussions, we have $\sin 36°=\sqrt{\frac{5-\sqrt{5}}{8}}=\frac{1}{2}\cdot\frac{\sqrt{\phi^2+1}}{\phi}$, and then

$$\frac{1}{\sin 36°}=\sqrt{\frac{10+2\sqrt{5}}{5}}=\frac{2\phi}{\sqrt{\phi^2+1}}.$$

15. Note that this form is often seen: $a\cdot\sqrt{\frac{5+\sqrt{5}}{10}}=\frac{a}{10}\cdot\sqrt{50+10\sqrt{5}}$.

16. We will use this form of the value of $a$ in chapter 5:

$$a=\frac{r}{2}\cdot\sqrt{10-2\sqrt{5}}.$$

17. For more on the golden ellipse, see H. E. Huntley, "The Golden Ellipse," *Fibonacci Quarterly* 12, no. 1 (1974): 38–40, and M. G. Monzingo, "A Note on the Golden Ellipse," *Fibonacci Quarterly* 14, no. 5 (1976): 388.

18. A proof of this formula can be found in H. S. M. Coxeter, *Introduction to Geometry*, 2nd ed. (New York: Wiley, 1989).

19. Hans Walser, *The Golden Section* (Washington, DC: Mathematical Association of America, 2001).

20. The artist who made this painting was Jacopo de' Barbari (ca. 1440–1516), a friend of Luca Pacioli.

21. A *polyhedral angle* is an angle formed by three or more planes meeting at a point.

22. The volume of a regular pyramid is one-third the product of the area of the base and the height.

23. *Wikipedia*, s.v. "Kepler-Poinsot polyhedron," http://www.en.wikipedia.org/wiki/Kepler-Poinsot_polyhedron (accessed October 12, 2011).

24. Monte Zerger, "The Golden State—Illinois," *Journal of Recreational Mathematics* 24, no. 1 (1992): 24–26.

## Chapter 5: Unexpected Appearances of the Golden Ratio

1. The *triangle inequality* states that the sum of any two sides of a triangle must be greater than the third side.

2. This figure was produced by Jo Niemeyer, a contemporary artist who has focused much of his work on the golden section. See http://www.partanen.de.

3. For further exploration of the arbelos, see A. S. Posamentier and I. Lehmann, *π: A Biography of the World's Most Mysterious Number* (Amherst, NY: Prometheus Books, 2004), pp. 211–13.

4. You can find a proof in A. S. Posamentier, *Advanced Euclidean Geometry* (Hoboken, NJ: John Wiley, 2002), pp. 128–30.

5. A. S. Posamentier, *Math Charmers: Tantalizing Tidbits for the Mind* (Amherst, NY: Prometheus Books, 2003).

6. See Stuart Dodgson Collingwood, ed., *Diversions and Digressions of Lewis Carroll* (New York: Dover, 1961), pp. 316–17.

7. Alternate interior angles of parallel lines are congruent.

8. See Alfred S. Posamentier and Ingmar Lehmann, *The Fabulous Fibonacci Numbers* (Amherst, New York: Prometheus Books, 2007), p. 45, item 11.

9. This line *DMK* also bisects angle *CDE*.

10. For a randomly selected triangle, where *A* is the area, *p* is the perimeter, and *r* is the radius of the inscribed circle,

$r=\frac{2A}{p}=\frac{2A}{a+b+c}=\frac{A}{s}$, where *s* is the semiperimenter. In the right

triangle *CDF*, is $A=\frac{ab}{2}$, and by the Pythagorean theorem,

$c=\sqrt{a^2+b^2}$.

Then $r=$

$$\frac{2A}{p}=\frac{ab}{a+b+\sqrt{a^2+b^2}}\cdot\frac{a+b-\sqrt{a^2+b^2}}{a+b-\sqrt{a^2+b^2}}=\frac{a+b-\sqrt{a^2+b^2}}{2}=\frac{a+b-c}{2}.$$

11. J. A. H. Hunter, “Triangle Inscribed in a Rectangle,” *Fibonacci Quarterly* 1 (1963): 66.

12. Remember, $\phi^2=\phi+1$.

13. $\Delta ABC \sim \Delta CBD$, $\frac{CB}{AB}=\frac{DB}{CB}$, or $\frac{\phi}{\phi+1}=\frac{DB}{\phi}$.

Then $DB=\frac{\phi^2}{\phi+1}=\frac{\phi+1}{\phi+1}=1$.

14. $\angle BAC=\arcsin\frac{1}{\phi}(\approx 38.17°)$.

15. See A. S. Posamentier, *The Pythagorean Theorem: The Story of Its Power and Glory* (Amherst, NY: Prometheus Books, 2010).

16. A right pyramid is a pyramid whose axis is perpendicular to its base. The axis is the line joining the vertex of the pyramid to the midpoint of the base.

17. In general, the perpendicular $C_{n-1}C_n$ at $C_n$ intersects line through *B* and $C_{n-1}$ at point $C_{n+1}$.

## Chapter 6: The Golden Ratio in the Plant Kingdom

1. This chapter was contributed by Heino Hellwig, who is a research assistant at the DFG-Research Center MATHEON at the Humboldt University in Berlin, Germany. Chapter and all images/figures reprinted courtesy of Heino Hellwig.

2. K. F. Schimper, "Beschreibung des Symphytum Zeyheri und seiner zwei deutschen Verwandten der *S. bulborum* Schimper und *S. tuberosum*" *Jacqu. Geiger's Magazin für Pharmacie* 29 (1830): 1–92.

3. H. S. M. Coxeter, *Introduction to Geometry*, 2nd ed. (New York: Wiley, 1989).

4. H. Vogel, "A Better Way to Construct the Sunflower Head," *Math Bioscience* 44 (1979): 179–89.

5. This can be elegantly described with polar coordinates as follows:

$$\begin{pmatrix} \alpha_n \\ r_n \end{pmatrix} = \begin{pmatrix} n \cdot \alpha \\ \sqrt{n} \end{pmatrix}.$$

6. It is particularly elegant to observe on the graph the characterization of complex numbers: $z = \lambda + i \cdot h$.

7. G. van Iterson, *Mathematische und mikroskopisch-anatomische Studien über Blattstellungen* (Jena: Gustav-Fischer-Verlag, 1907).

8. S. King, F. Beck, and U. Lüttge, "On the Mystery of the Golden Angle in Phyllotaxis" *Plant, Cell and Environment* 27 (2004): 685–95.

9. E. J. H. Corner, *Das Leben der Pflanzen* (Lausanne: Editions Rencontre, 1971), p. 91.

10. Ibid., p. 90.

11. J. N. Ridley, "Computer Simulation of Contact Pressure in Capitula," *Journal of Theoretical Biology* 95 (1982): 1–11.

12. I. Adler, "A Model of Contact Pressure in Phyllotaxis," *Journal of Theoretical Biology* 45 (1974): 1–79.

13. Simon Schwendener, *Mechanische Druckkräfte bewegen die*

*Primordia in Positionen mit maximalen Abstand voneinander* [Mechanical pressure forces move the primordial into the position with maximal separation distance] (1878).

14. J. Kepler, *Von sechseckigen Schnee*, ed. L. Dunsch (Dresden: Hellerau-Verlag, 2005).

15. Hubert Airy, "On Leaf-Arrangement," *Proceedings of the Royal Society of London* 21 (1873): 176–79.

16. R. Snow, "Problems of Phyllotaxis and Leaf Determination," *Endeavour* 14 (1955): 190–99.

17. W. Hofmeister, "Allgemeine Morphologie der Gewächse," *Handbuch der Physiologischen Botanik* 1 (1868): 405–664.

18. H. Hellwig, R. Engelmann, and O. Deussen, "Contact Pressure Models for Spiral Phyllotaxis and Their Computer Simulation," *Journal of Theoretical Biology* 240, no. 3 (2006): 489–500.

19. J. C. Schoute, "Beiträge zur Blattstellungslehre I," *Die Theorie. Rec. Trav. Bot. Neerl.* 10 (1913): 153–339.

20. S. Douady and Y. Couder, "Phyllotaxis as a Dynamical Self-Organizing Process (Part I, II, III)," *Journal of Theoretical Biology* 178 (1996): 255–312.

21. P. Atela, C. Golé, and S. Hotton, "A Dynamical System for Plant Pattern Formation," *Journal Nonlinear Science* 12, no. 6 (2002): 641–76.

## Chapter 7: The Golden Ratio and Fractals

1. This chapter was written by Dr. Ana Dias, associate professor, Department of Mathematics, Central Michigan University. Chapter and all images/figures reprinted courtesy of Dr. Ana Dias.

2. The square fractal is described by Hans Walser in his book *The Golden Section* (first American edition published by the Mathematical Association of America, 2001).

## Appendix: Proofs and Justifications of Selected Relationships

1. This is done by considering the value of each portion of the continued fraction up to each plus sign, successively.

2. The number $e$ is the base of the system of natural logarithms. It is the limit of

$$\left(1+\frac{1}{n}\right)^n$$

as $n$ increases without limit. The symbol $e$ was introduced by the Swiss mathematician Leonhard Euler (1707–1783) in 1737. In 1761, the German mathematician Johann Heinrich Lambert (1728–1777) showed that $e$ is irrational, and in 1873, the French mathematician Charles Hermite (1822–1901) proved $e$ is a transcendental number. A transcendental number is a number that is not the root of *any* integer polynomial equation, meaning that it is not an algebraic number of any degree. This definition guarantees that every transcendental number must also be irrational.

3. For more on the various representations of $\pi$, see A. S. Posamentier and I. Lehmann, *π: A Biography of the World's Most Mysterious Number* (Amherst, NY: Prometheus Books, 2004).

4. Dietrich Reuter, "'Goldene Terme,' nicht nur am regulären Fünf- und Zehneck," *Praxis der Mathematik* 26 (1984): 298–302. Reuter chooses another method to solve the problem.

5. $AE = a$ (pentagon side),
$AD = d$ (diagonal of the pentagon, or side of the pentagram),
$AM = r$ (radius of the circumscribed circle),
$FM = \rho$ (radius of the inscribed circle),
$AF = b$ (pentagon height; also bisects the opposite side, and bisects the opposite angle),
$AG = c$ (height, perpendicular bisector and angle bisector for $\Delta ABE$, and $\Delta ARS$),
$AR = e$ (exterior portion of pentagram side), and
$RS = f$ (side of smaller pentagon). See figure 4-57.

6. Duane W. DeTemple, "A Pentagonal Arch," *Fibonacci Quarterly* 12, no. 3 (1974): S235–36. However, our development differs from the one presented by DeTemple.

7. The law of cosines is a relationship among the angles and sides of triangle, that is: $c^2=a^2+b^2-2ab\cos C$. When the angle $C$ is 90°, we get the Pythagorean theorem, since the $\cos 90°=0$.

8. Paul S. Bruckman, "A Piece of Pi," *Fibonacci Quarterly* 39, no. 1 (2001): 92–93.

9. These trigonometric formulas are to be recalled for this item.

$\sin(x-y)=\sin x\cdot\cos y-\cos x\cdot\sin y$

$\cos 2x=\cos^2 x-\sin^2 x$

$\sin^2 x+\cos^2 x=1$

10. Joseph-Louis de Lagrange (1736–1813).

11. K. H. Rosen, *Elementary Number Theory and Its Applications* (Menlo Park, CA: Addison-Wesley, 1988).

12. K. Ball, *Strange Curves, Counting Rabbits and Other Mathematical Explorations* (Princeton, NJ: Princeton, University Press, 2006).

13. Felix Klein, *Ausgewählte Kapitel der Zahlentheorie*, vol. 1 (Leipzig: Teubner, 1907).

14. K. Azukawa and T. Yuzawa, "A Remark of the Continued Fraction Expansion of Conjugates of the Golden Section," *Mathematics Journal of Toyama University* 13 (1990): 165–76.

# Index